T0343372

Maximizing Machinery Uptime

Maximizing Machinery Uptime

Fred K. Geitner and
Heinz P. Bloch

ELSEVIER

AMSTERDAM • BOSTON • HEIDELBERG • LONDON
NEW YORK • OXFORD • PARIS • SAN DIEGO
SAN FRANCISCO • SINGAPORE • SYDNEY • TOKYO
Gulf Publishing Press is an imprint of Elsevier

Gulf Professional Publishing is an imprint of Elsevier
30 Corporate Drive, Suite 400, Burlington, MA 01803, USA
Linacre House, Jordan Hill, Oxford OX2 8DP, UK

Library of Congress Cataloging-in-Publication Data
Geitner, Fred K.
  Maximizing machinery uptime / Fred K. Geitner and Heinz P. Bloch.
    p. cm.
  Includes index.
  ISBN-13: 978-0-7506-7725-7 (case bound : alk. paper)
  ISBN-10: 0-7506-7725-2 (case bound : alk. paper) 1. Industrial
management. 2. Machinery in the workplace. 3. Organizational
effectiveness. I. Bloch, Heinz P., 1933– II. Title.
  HD31.G393 2006
  658.5'14—dc22

                                           2006000796

British Library Cataloguing-in-Publication Data
A catalogue record for this book is available from the British Library.

ISBN 13: 978-0-7506-7725-7
ISBN 10: 0-7506-7725-2

For information on all Gulf Professional Publishing
publications visit our Web site at www.books.elsevier.com

Transferred to Digital Printing 2009

Contents

Acknowledgments

We are indebted to several individuals and companies for granting permission to use material they had previously published. Our thanks go to Jim Corley for his relevant case studies involving Weibull analysis; the Logistics Technology Support Group of the Carderock Division and the Naval Surface Warfare Center in Bethesda, Maryland, for their permission to use excerpts from their *Handbook of Reliability Prediction Procedures for Mechanical Equipment*; John Sohre, whose experience-based numerical classification of factors influencing machinery reliability have helped us in the past ("Predicting Reliability of Turbomachinery"); Maurice Jackson and Barry Erickson for their pertinent observations and recommendations on how to evaluate the merits of certain features on centrifugal pumps; Stan Jakuba for his solid explanations of failure mode and effect analysis; General Electric for a publication explaining the concept of reliability index numbers; Karl Ost of Degussa, Hüls, Germany, for his contribution to life-cycle cost analysis of process pumps; Abdulrahman Al-Khowaiter, Aramco Oil Company, Saudi Arabia, for authorship of a section on the application of mechanical engineering principles to turbomachinery, reciprocating process compressor, and coupling guard design; Uri Sela, Sequoia Engineering and Design Associates, for his thoughts on quality machinery design installation and effective machinery monitoring; Messrs. Hasselfeld and Korkowski for permission to use their treatise on pump base plate grouting; Paul Barringer for his section on reliability policies and, in the section on continuous improvement, CROW/AMSAA reliability plotting; John S. Mitchell for his contribution to asset management philosophy; Robert J. Motylenski for the section on proven turnaround practices; Ben Stevens of OMDEC for defining the role of computerized maintenance management systems in achieving machinery uptime; Hussain Al-Mohssen of Aramco for his detailed description of a continuous improvement effort involving gas turbine flange bolting; Abdulaziz Al-Saeed, Aramco, for a contribution on efforts pertaining to turbomachinery train coupling guard design improvement;

and L. C. Peng for explaining the misunderstandings and pitfalls of some intuitive fixes to equipment-connected piping.

Our special thanks go to Bill Moustakakis who agreed to compile both theory and case histories dealing with machinery piping. We know from years of experience that this subject merits far more of the reliability professional's time and attention if true long-term machinery uptime is to be achieved.

Preface

The profitability of modern industrial and process plants is significantly influenced by uptime of the machines applied in their numerous manufacturing processes and support services. These machines may move, package, mold, cast, cut, modify, mix, assemble, compress, squeeze, dry, moisten, sift, condition, or otherwise manipulate the gases, liquids, and solids which move through the plant or factory at any given time. To describe all imaginable processing steps or machine types would, in itself, be an encyclopedic undertaking and any attempt to define how the reliability of each of these machine types can be assessed is not within the scope of this text.

However, large multinational petrochemical companies have for a number of years subjected such process equipment as compressors, extruders, pumps, and prime movers, including gas and steam turbines, to a review process which has proven cost-effective and valuable. Specifically, many machines proposed to petrochemical plants during competitive bidding were closely scrutinized and compared in an attempt to assess their respective strengths and vulnerabilities and to forecast life-cycle performance; the goal was to quantify the merits and risks of their respective differences, and finally to combine subjective and objective findings in a definitive recommendation. This recommendation could take the form of an unqualified approval, or perhaps a disqualification of the proposed equipment. In many cases, the assessment led to the request that the manufacturer *upgrade* his machine to make it meet the purchaser's objectives, standards, or perceptions.

This text wants to build on the philosophy of its predecessor, *An Introduction to Machinery Reliability Assessment* (ISBN 0-88415-172-7) by the authors. It outlines the approach that should be taken by engineers wishing to make reliability and uptime assessments for any given machine. It is by no means intended to be an all-encompassing "cook book" but aims, instead, at highlighting the principles that over the years have worked well for the authors. In other cases, it gives typical examples of what to look for, what to investigate, and when to go back to the equipment manufacturers with questions or an outright challenge.

We begin by directing our readers' attention to practical assessment techniques such as machinery component uptime prediction and life-cycle cost analysis. Then, in order to emphasize that *the promise of machinery uptime begins at the drawing board,* we would like to take our readers through the various life cycles of process machinery starting at specification and selection, then moving into the operational and maintenance environment and finally dwelling on continuous improvement efforts as one of the premier processes for uptime assurance.

We wish to acknowledge the constructive suggestions received from John W. Dufour and Dr. Helmut G. Naumann, who reviewed the manuscript for the first edition of *An Introduction to Machinery Reliability Assessment* (1990). Their comments certainly helped to improve the original as well as this current text.

Chapter 1
Introduction

Ask any plant manager in the world if he is interested in plant safety and he will answer in the affirmative. Ask him about his desire to produce reliably and he will probably give you the same answer. But interests and desires are not always aligned with a thoughtful and consistent implementation strategy and some of our readers will have to examine to what extent they are – or are not – in tune with Best-of-Class (BOC) practices.

Over the years, we have come to appreciate that reliability improvement and machinery uptime are virtual synonyms. To achieve uptime optimization, the machinery specification and actual design must be right. The machine must be operated within its design envelope. It must also be maintained correctly.

This harmonizes with the various editions of our text *Machinery Failure Analysis and Troubleshooting* (ISBN 0–88415–662–1) where we emphasize that, to capture high reliability, plant equipment has to be free of

- design defects
- fabrication deficiencies
- material defects
- assembly or installation flaws
- maintenance errors
- unintended operation
- operator error.

Indeed, and as we shall see, these seven failure categories are implicitly recognized whenever a facility is being planned and put into service. They are also recognized when performing failure analysis, because all failures of all machines will fit into one or more of these seven failure categories. It should be noted that the three major frames or boxes of Figure 1-1 contain these categories as well.

1

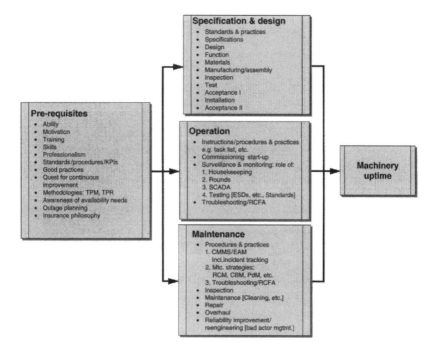

Figure 1-1. Elements contributing to machinery uptime.

But that is not the full story. Certainly a plant organization uses and manages the functional endeavors described as Specification & Design, Operation, and Maintenance. It is easy to visualize that various subcategories exist and that these, too, must somehow be managed. But they are properly managed only by a few, and we call them the BOC performers. These leading plants are reliability-focused, whereas the "business as usual" plants are stuck in an outdated cycle of repeat failures. We chose to label the latter as repair-focused.

In essence, it is our purpose to highlight the various issues that need to be addressed by plants that wish to achieve, optimize, and sustain machinery uptime. To that end, this text describes what BOC companies are doing. Likewise, a bit of introspection may point out where the reader has an opportunity to improve.

Prerequisites for Capturing Future Uptime

There are important prerequisites for achieving machinery uptime. Much reliability-related work must be done – and is being done – by BOC companies *before* a plant is built. Reliability audits and reviews are part of this effort and must be adequately staffed. The cost of these endeavors is part of a reliability-focused project. Moreover, the cost estimates and appropriation requests for such projects are never based on the initial cost of least expensive machinery. Instead, they are always based on data obtained from bidders that build reliability into their equipment. Competent machinery engineers assist in the bid evaluation process and assign value to maintenance cost avoidance and reliability improvement features to Bidder A over Bidder B [1].

Yet, not always are owners going for the lowest first cost. When it is evident that an existing plant is in trouble or in obvious need of improvement, equipment owners very often switch tactics and go for "high tech." They then procure the latest fad hardware and software. They belatedly attempt to institute crafts training and look to older retirees for instant improvement. To teach maintenance procedures or whatever other topic, they often engage teacher-trainers that have once worked for companies with name recognition, preferably ones that advertise their products or prowess on TV. But while some of these teacher-trainers have sufficient familiarity with process machinery to know why the client-owner experiences repeat failures, others do not. As an example, just ask some of these teacher-trainers to explain why authoritative texts consider oil slinger rings an inferior lube application method for many pumps used in process plants. Then, sit back and listen to their answers. The short-term solution entails working only with competent, field-experienced, and yet analytically trained, reliability consultants. The long-term solution is to groom one's own talent and skills.

Grooming Talent and Skills

Many managers fail to see the need to groom talent, to hire and hold on to people with the ability, motivation, and desire to learn all there is to be learned about a technical subject. They often delude themselves into believing that they can always hire a contractor to do the work, but do not realize that few contractors are better informed or better qualified than their own, albeit often ill-prepared employees. Managers often fail to recognize that machinery uptime optimization is ultimately achieved by talent that is deliberately groomed. This "groomed talent" includes people who are keenly interested in reading technical journals and the proceedings of technical symposia and conferences. This "groomed talent" relentlessly pursues self-training as well as outside training opportunities.

In essence, then, good managers nurture good people. Good managers challenge their technical employees to become subject-matter experts. They encourage these employees to map out their own training plans and then facilitate implementing these plans. Good managers will see to it that these employees, from young maintenance technicians to wizened senior engineers, become valuable and appreciated contributors. They also see to it that good technical employees are respected and rewarded accordingly.

A good workforce must have rock-solid basic skills. It would be of no benefit to buy better bearings and then allow unacceptable work practices to persist. Work practices must conform to certain standards and these standards must be put in writing. Then, these standards must be transformed into checklists or similar documents that are used at the workbench or in the field location where such work is being performed. Management's role includes allocation of resources to produce the requisite standards and verifying that they are being consistently applied. The standards and checklists must become part of a culture that builds basic skills. Moreover, the standards must be adhered to with determination and consistency. They should not be compromised as an expedient to reach the limited short-term goal of "just get it running again quickly." Neither should compliance with standards be allowed to become just one more of the many temporary banner exhortations that fizzle out like so many "flavors of the month."

By far the most important organizational agent in accomplishing the long-term reliability objectives of an industrial enterprise is totally focused on employee training. While this requirement may be understood to cover all employees regardless of job function, we are here confining our discussion to a plant's reliability workforce. A good organization will map out a training plan that is the equivalent of a binding contract between employer and employee. There has to be accountability in terms of proficiency achieved through this targeted training.

But before we delve into this training-related subject, we must explore current trends and recent inclinations that largely focus on procedural issues. We must also examine sound organizational setups as they relate to achieving optimized machinery uptime.

Sound Organizational Setup Explained

Smart organizations use a dual ladder of advancement, as discussed a little later in this chapter. However, regardless of whether or not a dual career path approach is used, two short but straightforward definitions are in order:

1. The function of a maintenance department is to routinely maintain equipment in operable condition. It is thus implied that this department is tasked with restoring equipment to as-designed or as-bought condition.

2. Reliability groups are involved in structured evaluations of upgrade opportunities. They perform life-cycle cost studies and develop implementation strategies whenever component upgrading makes economic sense.

For a reliability improvement group to function most effectively, its members have to be shielded from the day-to-day preventive and routine equipment repair and restoration involvement. Best Practices Plants often issue guidelines or predefine a trigger mechanism that prompts involvement by the reliability group. Examples might include equipment that fails for the third time in a given 12-month period, equipment distress that has or could have caused injury to personnel, failures that caused an aggregate loss in excess of $20,000, and so forth.

There must be a true quest for real improvement, not the quest for reciting and invoking improvement methodologies. While the quest for continuous, real, and lasting improvement is to be commended, the quest for merely invoking continuous improvement methodologies often turns into a chase after the elusive "magic bullet." All employees and all job functions must embrace the pursuit of real and lasting improvement. This collaborative effort is no different from the desire to have a reliable automobile. In addition to the fundamental design being right, the driver-operator and maintenance technician must do their part if acceptable "automobile uptime" is to be achieved. However, while every job function in a reliability-focused plant must participate in this quest, the process must be defined and supervised by enlightened managers.

Regarding the quest for continuous improvement methodologies, we have seen a veritable alphabet soup of acronyms come and go since the early references to Predictive Maintenance (PdM) in the mid-1950s. An "ME" campaign (meaning Manufacturing Excellence) was among them; few people at the affected location remember it. In 1975, a campaign aimed at making "every man a manager" was instituted in some plants known to the authors; it, too, failed miserably. While striving toward self-directed workforces is a laudable goal, it requires a core of competent and well-informed people.

As of 2005, PdM has survived and TPM, TPR, and ODR/OSS are foremost among the early twenty-first-century reliability methodologies and initiatives. But the point is that while it is OK to have one's method-ologies or even advanced technology-related procedures right, it is *not* OK to neglect the basics, the fundamentals of machinery reliability and optimized uptime. There will never be high reliability and optimized uptime where mechanics and technicians either lack the understanding or are not practicing the basics.

Finally, we should always recall that it is *not* OK to understand or per-haps blindly follow methodologies while, at the same time, disregarding common sense. The authors disagree with the notion expressed by some that in modern industry there is no longer a place for preventive maintenance (PM). Yet, we know only too well that modern industry cannot confine its practices to PM alone. Other approaches must supplement PM and even the question "who's doing PM" must be examined.

PdM, TPM, TPR, and ODR/DSS Explained

Routine preventive maintenance has served industry since the Industrial Revolution in the late eighteenth century. And PM still has its place in the many thousands of instances where avoiding failure by prevention of defect development, i.e. PM, makes more economic sense than allowing flaws to develop to the point where they become detectable, but also irreversible. An excellent example is changing oil in an automobile. This kind of PM is surely more cost-effective than keeping the same oil in the crankcase for many years, but analyzing it periodically for metal chips. While such periodic analyses would constitute PdM, that type of PdM makes no economic sense. Yet, properly used in an overall program of uptime optimization, PdM is indeed relevant, important, and representative of best practices.

By the mid-1950s, PdM, with its instrumentation routines aimed at spotting developing defects, came into being. PdM encompasses vibration monitoring and analysis, thermographic and ultrasonic examinations and inspections, and a host of other methods. All of these are intended to predict failure progression to the point where planned equipment shutdowns would prevent major damage and excessive downtime.

However, in order to maintain the equipment in optimal condition, new and progressive maintenance techniques needed to be established and a measure of "fine tuning" looked attractive. Such "fine tuning" involves the cooperation of equipment and process support personnel, equipment operators, and equipment suppliers. As was shown in the automobile uptime analogy, these three must again work together to eliminate equipment breakdowns, reduce scheduled downtime, and maximize asset utilization for optimum achievement of throughput and product quality.

Assuming it is being properly implemented, Total Productive Maintenance (TPM) can provide the methods and work processes to measure and eliminate much of the non-productive time. Once TPM has been successfully implemented, a facility is considered ready to progress to Total Process Reliability (TPR). Total Process Reliability views every maintenance event as an opportunity to upgrade manufacturing processes, hardware, software, work and operating procedures, and even management and supervisory methods. On the equipment level, TPR practitioners would always (!) be in a position to answer the two all-important questions: (i) is upgrading possible and (ii) is upgrading justified by prevailing economics.

Total Productive Maintenance often involves the use of an information management system, planned maintenance activities, emphasis on preventive maintenance, assessing equipment utilization to eliminate non-essential assets (reducing numbers of equipment), operator and mechanic training, to some extent decentralizing asset responsibility, and operator-ownership of equipment through basic care – a concept that leads into Operator-Driven Reliability (ODR). In turn, ODR might lead to Decision Support Systems (DSS).

Reliability-Focused Plants and Operator Involvement

We believe that process plants worldwide can be divided into those that are repair-focused and those that are decidedly reliability-focused. The former will have trouble surviving, whereas the latter will stay afloat with considerably less difficulty. Repair-focused facilities emphasize parts replacement and have neither the time nor the inclination to make systematic improvements. Rarely do they identify why the parts failed, and rarer yet do they implement the type of remedial action that makes repeat failures a thing of the past. Reliability-focused plants, on the other hand, view every repair event as an opportunity to upgrade. Whenever cost-justified, this upgrading is being done by adhering more closely to smarter work processes, by following better procedures, by selecting superior components, implementing better quality controls, using more suitable tools, etc. That, then, gets at the heart of maximizing machinery uptime.

Upgrade measures are employed with considerable forethought by reliability-focused companies. These companies will first identify the feasibility of such measures and will then determine their cost-effectiveness and quantitative justification. To do this effectively and over the long haul, they will employ trained engineers. The term "trained engineers" implies that they are informed researchers and readers that use analytical methods to make sound, experience-based decisions. Companies hold on to trained, highly motivated engineers by creating and nurturing a work environment that is conducive to high employee morale. Intelligent, highly productive operators are part of this work environment.

Since even the best-trained engineers cannot go it alone, they are given competent help. With that in mind, reliability-focused companies recognize the critically important role of the equipment or process operator. Best-in-Class companies are, therefore, poised to pursue ODR initiatives. Operator contributions are necessary because operators are the first to notice deviations from normal operation. They, the operators, are best equipped to understand the interactions between process and equipment behavior.

Operators need training. Their responsibilities and accountabilities must be defined and "institutionalized." Institutionalizing means that their job functions and actions, their responses and the implementation steps they follow must become mandatory routines as opposed to optional routines. More than two decades ago, plants in California and Texas experimented with this concept; they called it the multi-skill approach and assigned operators certain ODR tasks.

Operator-Driven Reliability is nearly always part of a generally applied maintenance plan: A distinct group of activities that *makes* things happen, rather than simply suggesting what *should* happen. In the *Handbook of Industrial Engineering*, author Ralph Peters outlines a number of common-sense steps. He strongly recommends starting with an over-all strategic maintenance plan like TPM or RCM (Reliability-Centered Maintenance) and asks that the interested entity include defined goals and objectives for ODR within this plan. A top-notch reliability-focused facility would understand that ODR is a deliberate process for gaining commitment from operators to:

- Keeping equipment clean and properly lubricated
- Keeping fasteners tightened
- Detecting and reporting symptoms of deterioration
- Providing early warnings of catastrophic failures
- Making minor repairs and being trained to do them
- Assisting maintenance in making selected repairs
- Start with necessary communication between maintenance, operators, and the rest of the total operation to gain commitment and internal cooperation
- Develop list of major repairs in the future
- Utilize leadership-driven, self-managed teams, e.g. "reliability improvement teams"
- Develop written and specific team charter
- Have teams evaluate/determine the best methods for operator cleaning, lubrication, inspection, minor repairs, and level of support during repairs
- Develop written procedures for operators and include them in quality and maintenance guides
- Evaluate the current predictive and preventive maintenance procedures and include those that the operator can do as part of ODR
- Document startup, operating, and shutdown procedures along with commissioning and changeover practices

- Consider quality control and health, safety, and environment requirements
- Document operator training requirements and what maintenance groups must do to support these requirements
- Develop operator training certification to validate operator-performed tasks.

Modern process plants train their operating technicians to have a general understanding of the manufacturing processes, process safety, basic asset preservation, and even interpersonal skills.

Operator involvement in reliability efforts ensures the preservation of a plant's assets. Operator activities thus include the electronic collection of vibration and temperature data and spotting deviations from the norm. Operator activities do not, however, encompass data analysis; data analysis is the reliability technician's task. Additional activities include routine mechanical tasks such as the replacement of gauges and sight glasses, and assisting craftspeople engaged in the verification of critical shutdown features and instruments. Also, operating personnel participate in electric motor testing and electric motor connecting/disconnecting routines.

The creation of functional departments tasked with both data capture and data analysis should be closely examined. Such departments may not be efficient; they risk involving expert analyst personnel in mundane data collection routines. It should be noted that operators are the first line of defense, the first ones to spot deviations from normal operation. For optimum effectiveness, they should be used in that capacity, i.e. data collection should be assigned to operators.

Supporting the Operator

ODR must be given tangible support by virtually every one of the other job functions represented at a specific facility. This recognition should logically lead to the development of well thought-out and appropriately configured DSS.

Decision Support Systems might be described as an advanced, multi-faceted asset management system which aims at automating an industrial reliability maintenance decision-making process. This process integrates monitoring and diagnostic approaches that include Distributed Control Systems (DCS), Computerized Maintenance Management Systems (CMMS), internal and external websites, and the many other sources of the company's own internal knowledge. Once successfully implemented, a sound, well-developed DSS will be a powerful source of information allowing rapid and exact equipment and process diagnosis, failure analysis, corrective action mapping, and so forth. It will turn the operator into a knowledge worker who will be supported by true expert systems.

Awareness of Availability Needs and Outage and Turnaround Planning

Another prerequisite for maximizing machinery uptime is being aware of the availability needs of one's plant. If production is seasonal or not sensitive to shutdown frequency or duration (within reason, of course), it makes little economic sense to demand the maximum in machinery availability. There cannot be any one simple rule covering the many possibilities and options, and management personnel must seek input from knowledgeable reliability professionals.

As an example, a plastics extruder that must stay on line for very long periods of time without shutdown may have to be equipped with a non-lubricated coupling connecting it to its driver. Conversely, a plastics extruder employed in a process requiring its helical screw rotor to be exchanged for a different one during monthly changes to substantially different plastic products could be equipped with a less expensive gear coupling that might have to be re-greased every month.

Being aware of one's equipment availability needs is also important for intelligent planning of downtime events for inspection and repair. Outage planning (sometimes called turnaround, also called "IRD" for inspection and repair downtime) is closely related to awareness of availability. It boggles the mind how often management neglects this issue. It defies common sense to buy the cheapest equipment and then expect long, trouble-free operation without shutdowns. A plant that bought bare-bones machinery must expect more outages than a plant that thoroughly investigated the life-cycle cost of better machinery and then carefully specified this equipment before placing purchase orders.

There are certain ethylene plants that, in 2004, operated with 8-year outage intervals while others barely made it to 5 years. The reader will intuitively know which of the two had, at the design and inquiry stages, pre-invested in detailed machinery reliability assessment efforts. Attempts by the 5-year plant to move into the 8-year category are costly and slow. To again use an automobile analogy, buying a certain model with a six-cylinder engine will cost less than buying it with eight cylinders, but the incremental cost of later converting from six to eight cylinders will be far greater.

Modern outage planning uses in-plant reliability data acquired over time. Without data, such planning will involve considerable guessing. Using data from one's own operations and from similar plants and equipment elsewhere, the scope and mandate of these activities is to impart reliability, availability, and maintainability in methodical and even mathematical fashion. Needless to say, this will not be done by default; instead, it requires management involvement and stewardship.

Insurance and Spare Parts Philosophies

In the early 2000s, a very competent reliability professional explained that his company continues to have issues with its spare parts philosophy and overall parts management. He described a situation that is very common today:

> Unfortunately, what we have done to ourselves over the last 20 years is a piecemeal approach that is too frequently found wanting. The plant inevitably stays down for two days when it should only have been down for 18–24 hrs after an unplanned shutdown. I am now being further challenged by being asked to set up the spares for our new world-scale methanol plant. Surely the spares that we stock for a syngas turbine should be somewhat generic. The fact that we have three different turbine manufacturers simply means getting the relevant part numbers/serial numbers to the warehousing people to complete an administrative exercise as all the other factors, i.e. risk, production loss etc., are similar.

Each plant differs from the next one in certain respects. Although two refineries or fertilizer plants may represent identical designs, they are not likely to have identically trained or motivated staff. One plant takes perhaps greater risks in areas where operating prudence should be practiced. Some plants allow adequate time for turbine warm-up while others use the incredibly risky "full speed ahead on lukewarm" approach. Or, although professing to perform failure analysis, many plants will replace failed parts before even understanding why the part failed in the first place. In doing so, they set themselves up for repeat failures.

Some facilities employ structured and well-supervised maintenance supervision, work execution, and follow-up inspection, while others are quite remiss in allocating time and resources to these pursuits. Also, one plant may be located in a geographic area blessed with competent repair shops while the other is not. Smart plants do a considerable amount of pooling of major turbomachinery spares, i.e. several plants have access to a common spare. Moreover, some plants have found it prudent to specify and procure certain turbo equipment diaphragms made from readily repairable steel rather than difficult-to-repair cast iron. Some will only purchase steam turbine blading that represents prior art, while others will buy prototype blade contours that promise perhaps a fraction of a per cent higher energy efficiency. Certain blades falling into this category are then subjected to high operational stresses and are prone to fail prematurely. Even well-designed turbine blades are at risk if the steam supply system is unreliable or deficient in some ways.

Needless to say, the list could go on. Any reasonable determination of recommended spare parts must include not only consideration of the above but also an analysis of prior parts consumption trends and an assessment of storage practices, to name but a few key items. It is no secret that most users are reluctant to share their field experience and related pertinent information by publishing it. Broadcasting past mistakes, existing shortcomings, and underperformance threatens the job security of plant management. Conversely, educating others as to the details that had ensured past successful operations is frowned upon as "sharing a competitive advantage with the enemy." The answer? Experience shows that competent consultants with lots of practical field experience should be engaged to periodically audit HP and major chemical plants. That is the only logical answer to the question of spare parts stocking in a highly competitive environment. To the best of our knowledge, there is no magic computer program that can manipulate the almost endless number of variables that must be weighed and taken into consideration to determine how many spares are needed in petrochemical plants.

Reliability-Focus versus Repair-Focus

To be profitable, an industrial facility must abandon its repair focus and move toward becoming almost exclusively reliability-focused. There are many ways to reach this goal and the best path to success may depend on a facility's present state of affairs, so to speak. Here, then, is just one more reminder. Assuming you want to move toward best practices (BP) and are – pardon the suggestion – a "Room-for-Improvement" (RFI) plant, you may wish to compare your present organizational lineup and its effectiveness against BP pursued and implemented at process plant locations elsewhere.

A comparison of repair-focused plants with reliability-focused facilities is in order. It should be realized that conscientiously maintaining reliability focus is synonymous with implementing the desire to optimize machinery uptime.

- The reliability function at repair-focused facilities is not generally separated from the plant maintenance function. At repair-focused plants, traditional maintenance priorities and "fix it the way we've always done it" mentality win out more often than warranted. In contrast, reliability-focused facilities know precisely when upgrading is warranted and cost-justified. Again, they view every maintenance event as an opportunity to upgrade and are organized to respond quickly to proven opportunities.
- The reward system at repair-focused plants is often largely production-oriented and is not geared toward consistently optimizing the bottom-line life-cycle-cost (LCC) impact. At repair-focused

facilities the LCC concept is not applied to upgrade options. This differs from reliability-focused facilities that are driven by the consistent pursuit of longer-term LCC considerations. Here, life-cycle costing is applied on both new and existing (worthy of being considered for upgrading) equipment.

- At repair-focused companies, reliability professionals have insufficient awareness of the details of successful reliability implementations elsewhere. The situation is different at reliability-focused facilities that provide easy access to mentors and utilize effective modes of self-teaching via mandatory(!) exposure to trade journals and related publications. Management at these BOC facilities arranges for frequent and periodic "shirt-sleeve seminars." These informal in-plant seminars are actually briefing sessions that give visibility to the reliability technicians' work effort. They disseminate technical information in single-sheet laminated format and serve to upgrade the entire workforce by slowly changing the prevailing culture.

- Lack of continuity of leadership is found at many repair-focused plants. These organizations do not seem to retain their attention span long enough to effect a needed change from the present repair focus to the urgently needed reliability focus. The influence of both mechanical and I&E equipment reliability on justifiably coveted process reliability does not always seem to be appreciated at repair-focused plants. On the other hand, we know of no BP organization (top quartile company) that is repair-focused. Experts generally agree that successful players must be reliability-focused to survive in the coming decades.

- Some of the most successful BP organizations have seen huge advantages in randomly requiring maintenance superintendents and operations superintendents switching jobs back and forth. There is no better way to impart appropriate knowledge and "sensitivity" to both functions.

- At repair-focused facilities, failure analysis and effective data logging are often insufficient and generally lagging behind industry practices. Compared to that, BP organizations interested in machinery uptime extension involve operations, maintenance, and project/reliability personnel in joint failure analysis and logging of failure cause activities. A structured and repeatable approach is being used and accountabilities are understood.

- At the typical RFI facility, the plant where there is "room for improvement," there are gaps in planning functions and process-mechanical coordinator (PMC) assignments. There is also an apparent emphasis on cost and schedule that allows non-optimized equipment and process configurations to be installed and, sometimes, replicated. At RFI plants, reliability-focused installation standards are rarely invoked and responsible owner follow-up on contractor or vendor work is practiced infrequently.

- Best Practices organizations actively involve their maintenance and reliability functions in contractor follow-up. Life-cycle cost considerations are given strong weight. Also, leading BP organizations have contingency budgets that can be tapped in the event that cost-justified debugging is required. They do not tolerate the notion that operations departments must learn to live with a constraint.
- A reliability-focused BP organization will be diligent in providing feedback to its professional workforce. The typical repair-focused company does not use this information route.

Mentoring, Resources, and Networking

Occasionally, even a repair-focused organization has both Business Improvement and Reliability Improvement teams in place. As it plans to move toward BOC status, the repair-focused plant must make an honest appraisal of the effectiveness of these teams. Their value obviously hinges on the technical strength and breadth of experience of the various team members.

At the typical repair-focused location, maintenance-technical personnel are often unfamiliar with helpful written material that could easily point them in the right direction. As an example, repair-focused companies often use only one mechanical seal supplier. Moreover, access to the manufacturer is sometimes funneled entirely through a distributor.

In contrast, BP or BOC organizations have full access to the design offices of several major mechanical seal manufacturers. They have acquired, and actively maintain, a full awareness of competing products. They will find sound and equitable means to select whichever seal configuration, material choice, etc. necessary to meet specified profitability objectives. This is reflected in their contract with a seal alliance partner.

At repair-focused companies, a single asset may require costly maintenance work effort every year, while another, seemingly identical asset, lasts several years between shutdowns. This paradox is tackled and solved at BP organizations. They provide access to mentors whose assistance will lead to true root cause failure analysis (RCFA). The result is authoritative and immensely cost-effective definition of what is in the best interest of the company. Based on experience and analysis, this could be repeat repair, upgrading, or total replacement.

Repair-focused plants seem to "re-invent the wheel," or use ineffective and often risky trial-and-error approaches. Reliability-focused multi-plant or international organizations make extensive use of networking.

Relatively informal, very low cost Network Newsletters use input from grass-roots contributors who gain "visibility" and "name recognition" by being eager to communicate their successes to other affiliates. A Network Chairperson is being used to communicate with plant counterparts. This job function is assigned to an in-plant specialist on a rotational basis.

Many well-intentioned companies endeavor to identify and implement the best, or most appropriate, reliability organization. Some opt to divide their staff along traditional lines into Technical Services, Operations, and Maintenance divisions, departments, or just plain work functions. They often place their reliability personnel under the Maintenance Management umbrella, but then have second thoughts when reliability professionals end up immersed in fighting the "crisis of the day," as it were.

While it has been our experience that organizational alignments are considerably less important than the technical expertise, resourcefulness, motivation, and drive of individual employees, there are obvious advantages to an intelligent lineup. What, then, is an "intelligent lineup," or sound organizational setup?

Dual Career Paths at Top Companies

Top performing companies have created two career paths for their personnel. Where two career paths exist, upward mobility and rewards or recognition by promotion are possible in either the administrative or technical ladders of advancement. This dual ladder represents perhaps the only sound and proven way to keep key technical personnel in such industries as hydrocarbon processing. Some engineers would not want to become managers, and there are not enough management openings to promote all competent engineers to such positions.

Where there are two career paths, there is income and recognition parity between such administrative and technical job functions as

Administrative side	*Technical side*
Group Leader	Project Engineer
Section Supervisor	Staff Engineer
Senior Section Supervisor	Senior Staff Engineer
Department Head	Engineering Associate
Division Manager	Senior Engineering Associate
Plant Manager	Scientific Advisor
Vice President	Senior Scientific Advisor

Recognition and reward approaches have much to do with management style. There are many gradations and cultural differences that make one approach preferred over the next one. It is not possible to either know or judge them all. Suffice it to say that a thoughtless reward and recognition system is a serious impediment to employee satisfaction.

More Keys to a Productive Reliability Workforce

An under-appreciated workforce is an unmotivated, unhappy, and inefficient workforce. Such a workforce will rarely, if ever, perform well in areas of safety and reliability. How, then, will the interdependent safety, reliability, and profitability goals be achieved?

Forty years ago, world-renowned efficiency expert Dr. W. Edwards Deming provided the answer. He stipulated 14 "Points of Quality" that fully met the objectives of both employer and employee and are as true and relevant today as they have ever been. Deming had aimed his experience-based recommendations at the manufacturing industries and we transcribed his 14 points into wording that might find listening ears in the process plant reliability environment [2]. Here is our expanded recap:

1. As was brought out earlier in this text, view every maintenance event as an opportunity to upgrade. Investigate its feasibility beforehand; be proactive.
2. Ask some serious questions when there are costly repeat failures. There needs to be a measure of accountability. Recognize, though, that people benefit from coaching, not intimidation.
3. Ask the responsible worker to certify that his or her work product meets the quality and accuracy standards stipulated in your work procedures and checklists. That presupposes that procedures and checklists exist.
4. Understand and redefine the function of your purchasing department. Support this department with component specifications for critical parts, then insist on specification compliance. "Substitutes" or non-compliant offers require review and approval by the specifying reliability professional.
5. Define and insist on daily interaction between process (operations), mechanical (maintenance), and reliability (technical and project) workforces.
6. Teach and apply RCFA from the lowest to the highest organizational levels.
7. Define, practice, teach, and encourage employee resourcefulness. Maximize input from knowledgeable vendors and be prepared to pay them for their effort and assistance. Do not "re-invent the wheel."
8. Show personal ethics and evenhandedness that are valued and respected by your workforce.
9. Never tolerate the type of competition among staff groups that causes them to withhold critical information from each other or from affiliates.
10. Eliminate "flavor of the month" routines and meaningless slogans.

11. Reward productivity and relevant contributions; let it be known that time spent at the office is in itself not a meaningful indicator of employee effectiveness.
12. Encourage pride in workmanship, timeliness, dependability, and the providing of good service. Employer and employee honor their commitments.
13. Map out a program of personal and company-sponsored mandatory training.
14. Exercise leadership and provide direction and feedback.

"CARE" – Deming's Method Streamlined and Adapted to Our Time

In early 2000, a Canadian consulting company [3] developed a training course that brings Deming's method into new focus. They concluded that companies can be energized with empathy and, using the acronym CARE, conveyed the observation that companies excel when management gives consistent evidence of

- Clear direction and support
- Adequate and appropriate training
- Recognition and reward
- Empathy.

Although mentioned last, *empathy* is the cornerstone of the approach. But, let us first consider the other letters.

The Letter "C": Clear Direction Via Role Statement

Regarding the first letter of the acronym, "C," we believe that clear direction involves role statements and training plans. A lack of role statements for reliability professional can lead to inefficiency and encourages being trapped in a cycle of "fire-fighting." Not having written role statements deprives the entire organization of a uniform understanding of roles and expectations for reliability professionals. Not having a role statement may turn the reliability professional into a maintenance technician, a person who is more involved in maintaining the status quo than a person engaged in true failure avoidance through engineered component and systems upgrades. Clearly then, BP organizations use role statements as a roadmap to achieving mutually agreed-upon goals. Among other things, this allows meaningful performance appraisals.

The four CARE items represent rather fundamental principles of management. Still, while empathy forms the foundation, it alone will not deliver full results for any organization. The drive toward certain success starts with clear direction and support. Clear direction must be put down in writing. For example, and as was alluded to earlier in this chapter, reliability professionals must receive this clear direction in the form of a role statement [3]. Their role might include, but not be limited to, those mentioned below.

1. Assistance role
   - Establishment of equipment failure records and stewardship of accurate data logging by others. Know where we are in comparison with BOC performers.
   - Review of preventive maintenance procedures that will have been compiled by maintenance personnel.
   - Review of maintenance intervals. Understand when, where, and why we deviate from BP.
2. Evaluation of new materials and recommendation of changes, as warranted by LCC studies.
3. Investigation of special, or recurring equipment problems. Example:
   - Ownership of failures that occur for the third time in any 12-month period.
   - Coaching others in RCFA.
   - Definition of upgrade and failure avoidance options.
4. Serving as contact person for original equipment manufacturers.
   - Understanding how existing equipment differs from models that are being manufactured today.
   - Being able and prepared to explain if upgrading existing equipment to state-of-art status is feasible and/or cost-justified.
5. Serving as contact person for other plant groups.
   - Communicate with counterparts in operations and maintenance departments.
   - Participate in Service Factor Committee meetings.
6. Develop priority lists and keep them current.
   - Understand basic economics of downtime. Request extension of outage duration where end-results would yield rapid payback.
   - Activate resources in case of unexpected outage opportunities.
7. Identify critical spare parts.
   - Arrange for incoming inspection of critical spare parts prior to placement in storage locations.
   - Arrange for inspection of large parts at vendor's/manufacturer's facilities prior to authorizing shipment to plant site.
   - Define conditions allowing procurement from non-OEMs.

8. Review maintenance costs and service factors.
   - Compare against Best-in-Class performance.
   - Recommend organizational adjustments.
   - Compare cost of replacing versus repairing; recommend best value.

9. Periodically communicate important findings to local and affiliate management.
   - Fulfill a networking and information-sharing function.
   - Arrange for key contributors to make brief oral presentations to mid-level managers (share the credit, give visibility to others).

10. Develop training plans for self and other reliability team contributors.

The above listing represents a role statement for equipment reliability engineers. While it represents a summary that can be expanded or modified to address specific needs, it is representative of the written "clear direction" that is being taught in the CARE program.

The "support" element is re-enforced in items 9 and 10, above. In one highly successful and profitable company, an astute plant manager organized a mid-level management "steering committee" which every week invited a different lower-level employee to make a ten-minute presentation on how they performed their work. The vibration technician explained how early detection of flaws saved the company time and money, an instrument technician demonstrated the key ingredients of an on-line instrument testing program, etc. Each reliability issue or program had a management sponsor or "champion," who saw to it that a program stayed on track, and that organizational and other obstacles were removed.

The Letter "A"

Next, there is a melding of "Clear Direction and Support" with "Adequate and Appropriate Training." How so? Well, training plans were initiated by the employee, which means he or she had to give considerable thought to long-term professional growth. The initial proposal by the employee was reviewed, supplemented, modified, often amplified, but always given top priority by management.

In addition to structured self-training, a reliability professional at BOC plants prepares "shirt-sleeve seminars" – training sessions lasting perhaps ten minutes. He rolls up his sleeves and, at the end of an assembly of personnel for safety talks, presents a reliability and uptime optimization topic to those present. At shirt-sleeve seminars, key learnings are being discussed and disseminated. These key learnings include reminders that reliability principles must be consistently employed by everyone. Site management must verify continuity of this dissemination effort and endorse the application of reliability principles such as consistent use of checklists.

But training, of course, must go beyond "shirt-sleeve" seminars. Best Practices organizations encourage salaried professionals to submit their projected training plans, both long term and short term, in writing. These plans are then critically reviewed and employer requirements reconciled with an employee's developmental needs. Input from competent consultants is often enlisted. Best Practices organizations make active and consistent use of what they have learned.

Note that our earlier statements on "clear direction and support" introduced the training issue. Let us face it, we are losing the ability to apply basic mathematics and physics to equipment issues in our workplace situations. As an example, hundreds of millions of dollars are lost each year due to erroneous lubrication techniques alone. The subject is not dealt with in a pragmatic sense in the engineering colleges of industrialized nations. The connection between Bernoulli's law taught in high-school physics classes and the proper operation of constant level lubricators is lost on a new generation of computer-literate engineers. Managers chase after the "magic bullet" – salvation must be in "high tech," they think.

That is an incredibly costly misconception. We have truly neglected to understand the importance of the non-glamorous basics. We are no longer interested in time-consuming details. We have encouraged our senior contributors to retire early. All too often, no thought is given to the consequences. Assumptions are made that one could hire contractors to do the thinking for us and not many decision-makers see the fallacy in this reasoning. It should be obvious that at times contract personnel are even less qualified, or have less incentive, to determine the LCC of different alternatives and address the root causes of repeat failures of machinery. We have become "big picture" men, from the maintenance technician all the way up to the company CEO. We cannot be bothered by details, have no time for details, and are not rewarded for dealing with details.

But, as some outstanding performers have clearly shown, attention to detail is perhaps the most important step they took to get to the top. They have developed and continue to insist on adequate and appropriate training. This training deals with not only concepts and principles, but hundreds of details as well.

Employees of Best-in-Class companies develop their own short- and long-range training plans. Time and money are budgeted and the training plan signed off by the employee and his or her manager. A training plan has the status of a contract. It can only be altered by mutual consent.

The training plan for a machinery-technical employee was published in *Improving Machinery Reliability* [1] and typically consists of four columns, as replicated below.

Career Years	"Knowledge of"	"Work Capability in"	"Leading Expertise in"
1	Company organization	Interpretation of flow sheets, piping & instrument diagrams	
	Rotating equipment types	Elementary technical support tasks, e.g. alignment, vibration monitoring	
	Company's communication routines	Essential computer calculations	
	Relevant R&D studies, vendor capabilities, in-house technical files		
2	Pump and compressor design	Design specification consulting & support	
	Machinery reliability appraisal techniques	Machinery performance testing	
	Gear design	Start-up assistance, all-fluid machines	
	Major refining processes		
3	Machinery design audits	Company standards updates	
	Machinery piping	General technical service tasks elementary troubleshooting	
	Major chemical processes	Machine–electronic interfaces	
4	Materials handling equipment	General troubleshooting "shirt-sleeve seminars" (conduct informal training)	
	Hyper compressors	Machinery quality assessment and verification	
5	Thin-film evaporators	Start-up advisory tasks	Machinery optimization
		Appraisal documentation update tasks	Machinery maintenance
	Plastics extruders	Hyper compressor specifics	
6	Fiber processing equipment	Machinery design audits	Machinery selection
7	Patent and publication matters	Technical publications	Machinery failure analysis

A career development training plan was developed along the same lines [3]. Here is the format we have seen for imparting knowledge to new, intermediate, and advanced machinery engineers.

I. NEW ENGINEER (Plant mechanical engineer hiree)
   Years 1 and 2, possibly years 1 through 5.

   A. *On-the-job training*
      Rotational assignments within the plant in various groups to be exposed to different job functions for familiarization. Areas to be covered should include machinery, mechanical, inspection, electrical, instrumentation, operations, maintenance, etc.

   B. *In-house training* (Applicable to headquarters/central engineering locations)
      Plant and/or corporate standards development/revisions and updates

      • Courses in the above
      • Courses dealing with industry standards (API, NEMA, NPRA, etc.)
      • Machinery (compressors, pumps, steam and gas turbines, gears, turboexpanders, etc.)
      • Failure analysis and troubleshooting (Seven Root Cause method, "FRETT")
      • Practical lubrication technology for machinery
      • Machinery vibration monitoring and optimized analysis
      • Predictive monitoring (lube oil analysis, valve temperature monitoring, etc.).

   C. *Outside training pursuits* (Suggested minimum once/year, preferred frequency twice/year)

      1. General vendor-type information courses. Examples:

         • A major manufacturer's gas turbine maintenance seminar
         • Major mechanical seal manufacturers' training courses
         • A major manufacturer's compressor technology, selection, application, and maintenance seminar
         • Compressor Control (Anti-Surge) and Turbomachinery Governor Control courses
         • A major turbomachinery manufacturer's lube and seal oil systems maintenance course
         • Coupling manufacturer's training course, etc.

      2. Texas A&M University Turbomachinery Symposium

3. Texas A&M University International Pump Users Symposium
4. Professional Advancement courses in

- Machinery Failure Analysis and Prevention
- Machinery Maintenance Cost Saving Opportunities
- Compressor and Steam Turbine Technology
- Machinery for Process Plants
- Reciprocating Compressor Operation and Maintenance
- Piping Technology
- Practical Mechanical Engineering Calculation Methods.

D. *Personal training* (Mandatory review of tables of contents of applicable trade journals, books, conference proceedings, etc. Mandatory collection and cataloging of copies of articles that are of potential future value). Here are some examples of trade journals that often prove useful to equipment reliability professionals:

- *Hydrocarbon Processing*
- *Maintenance Technology*
- *Oil and Gas Journal*
- *Chemical Engineering*
- *Control Design*
- *Gas Turbine World*
- *Chemical Processing*
- *Hydraulics and Pneumatics*
- *Power Engineering*
- *Pumps and Systems*
- *Evolution (SKF Bearing Publication)*
- *Reliability*

- *Mechanical Engineering*
- *Diesel Progress*
- *Diesel & Gas Turbine Worldwide*
- *Distributed Power*
- *Sound and Vibration*
- *Lubes and Greases*
- *Sulzer Technical Review*
- *Plant Services*
- *World Pumps*
- *Compressor Tech Two*
- *Practicing Oil Analysis*
- *NASA Tech Briefs*

Books to be reviewed should include texts on machinery reliability assessment (which include checklists and procedures and popular texts on pumps), Weibull analysis, reciprocating and metering pumps, electric motor texts, books dealing with gear technology, etc. We refer the reader to the Bibliography at the end of this chapter.

II. INTERMEDIATE ENGINEER (Plant Mechanical/Machinery Engineer), years 3 through 5, possibly 3 through 8.

A. *Rotational assignment.* Two-year assignment at affiliate location, possibly at Central Engineering or Company Headquarters.

- Involvement in field troubleshooting and upgrading issues
- Familiarization with equipment, work procedures, data logging practices, etc.
- Spare parts procurement practices (probability studies)
- Life-cycle costing involvement
- Maintainability and surveillability input
- Structured networking involvement (provide feedback to other groups).

B. *Outside training pursuits.*

- Extension of earlier exposure
- Attendance at relevant trade shows and exhibitions (provide feedback to others)
- Attendance at ASME, NPRA, STLE, and related conferences (provide feedback)
- Speaker at local ASME/STLE/Vibration Institute meetings.

C. *Personal training and continuing education.*

- Develop short articles for trade journals and/or similar publications
- Develop short courses (initial aim: in-plant presentations, intra-affiliate presentations)
- Advanced self-study of material on probability, statistics, automation, management of change
- Studies in applicable economics.

III. ADVANCED ENGINEER (Corporate Specialist, Core Engineering Specialist), years 9 and more, depending on exposure and achievements under II – A/B/C, above.

- International conferences (speaker/participant)
- Peer group interfaces (e.g., on discussion panels, industry standards committees, etc.)
- Develop and present technical papers at national/international engineering conferences
- Pursue book publishing opportunities (case histories, teaching tools, work procedures)
- Regular contributions to trade journals
- Development of consultant skills.

The Letter "R": Recognition and Reward

One of the most important and seemingly little known facts is that most professional employees seek different employment for reasons other than better pay. This situation is analogous to divorces. Few marriages break up because of the intense desire to find a new partner whose income exceeds that of the previous one. Most marriages break up because of lack of respect, untruthfulness, immoral or insensitive conduct, or just plain incompatibility. Most employer–employee relationships are wrecked for the same reasons.

Recognition and reward often come in the form of sincere expressions of appreciation for whatever good qualities or commendable performance are displayed by the employee. A few well-chosen words given privately are usually better than public praise. All too often, public praise generates envy in others and may make life more difficult to the recipient of praise. Rewards in the form of Certificates of Recognition to be hung on the office wall come perilously close to being meaningless and employers would be wise to consider how these pieces of paper are perceived. If you want to do something positive for the employee, give him or her a certificate for $300 worth of technical books, or a $200 gift certificate for dinner at an upscale restaurant, or a new floor covering or whatever reaffirms that the employee's contributions are valued.

Several major petrochemical companies frequently reward top technical performers with a bonus of $5000 for exceptional resourcefulness, or the implementation of cost-saving measures, being "doers" instead of "talkers." There is nothing a company likes more than having its professional employees go on record with a firm, well-documented recommendation for specific action, rather than compiling lists of open-ended options for managers to consider. Top technical performers do just that: They make solidly researched recommendations, showing their effect on risk reduction and downtime avoidance, or demonstrating their production and quality improvement impact.

Empathy: The Overlooked Contributor to Asset Preservation

The last item, empathy, is by far the most important and also the most neglected. Yet, it represents the foundation of the CARE concept. Without empathy, without the ability to put oneself into the shoes of the people one manages, a manager will never know them, certainly will not understand them, and will never bring them to their full potential as employees and people.

Empathy is an understanding so intimate that the feelings, thoughts, and motives of a fellow human being are readily comprehended by another. You may think that this "intimate understanding" has no place at the office or on the factory floor. Think again.

Say an employee is late coming to work and the manager rebukes her before, or instead of, tactfully inquiring as to the reason for the tardiness. Assume that this employee has a sick child at home. Does the rebuke make her a more efficient or happier worker? We all know the answer to that question.

Let us say the manager would understand how empathy works, or would remember how he would like to be treated if it were his child that is sick. Let us say the manager would, therefore, offer the employee such options as doing the work at or from her home. The most likely result of his showing empathy and compassion would be that instead of getting 80% efficiency out of the unhappy worker at the office, he gets 120% efficiency from the appreciative worker at home. All parties would benefit from empathy and compassion in the workplace.

We are fully aware of the standing objection to empathy: "The workers will take advantage of me. I would look like a pushover, and not like the firm leader that I want to project." Let us just end the discussion by stating unequivocally that the vast majority of professional employees respond better to kindness than to harshness. Using such traits as compassion, cooperation, communication, and consideration will result in a more productive, satisfied, motivated, and loyal workforce than many managers could ever imagine.

Yes, empathy is doing more to retain this most valuable asset, your professional employees, than money, slogans, exhortations, and threats. Empathy, indeed, is the foundation of the ingredients of CARE [4], and is the hallmark of a long-term BOC company. And so, as we move into the more purely technical topics and chapters of this text, let us never lose sight of the importance of the "people aspects" in capturing and optimizing machinery uptime.

References

1. Bloch, H. P., *Improving Machinery Reliability*, 3rd edn. Woburn, Massachusetts: Butterworth-Heinemann Publishing Company, 1998.
2. *Heinz Bloch Shortcuts to Machinery Reliability Improvement*, Version 2, October 2000. CD-ROM, 0–7506–7424–5, Butterworth-Heinemann Publishing Company, www.bh.com/engineering.
3. Systems Approach Strategies (SAS), www.systemsapproach.com.
4. Larkin, M. and Shea-Larkin, D., *CARE – Energize Your Company With Empathy*. Toronto, Ontario, Canada: Stoddard Publishing Company Ltd., 2000.

Bibliography

Bloch, H. P., *Practical Guide to Compressor Technology*. New York: McGraw-Hill, 1996.

Bloch, H. P., *Practical Guide to Steam Turbine Technology*. New York: McGraw-Hill, 1996.

Bloch, H. P., *Practical Lubrication for Industrial Facilities*. Lilburn, Georgia: Fairmont Press, 1999.

Bloch, H. P. and Geitner, F., *Machinery Failure Analysis and Troubleshooting*, 3rd Edition. Woburn, Massachusetts: Butterworth-Heinemann Publishing Company, 1996.

Bloch, H. P. and Hoefner, J. J., *Reciprocating Compressor Operation and Maintenance*. Woburn, Massachusetts: Butterworth-Heinemann Publishing Company, 1996.

Bloch, H. P. and Soares, C., *Turboexpander Technology and Applications*. Woburn, Massachusetts: Butterworth-Heinemann Publishing Company, 2001.

Dufour, J. W. and Ed Nelson, W., *Centrifugal Pump Sourcebook*. New York: McGraw-Hill, 1992.

Eisenmann, R. C. Sr and Eisenmann, R. C. Jr, *Machinery Malfunction Diagnosis and Correction*. Upper Saddle River, New Jersey: Prentice-Hall, Inc., 1998.

Sofronas, A., *Analytical Troubleshooting of Process Machinery and Pressure Vessels*. Hoboken, NJ: Wiley, 2005.

Peters, R. W., Maintenance management and control. *Handbook of Industrial Engineering*. New York: John Wiley & Sons, 2001, pp. 1585–1623.

Chapter 2
The meaning of reliability

In order to set the stage for our readers we would like to define some basic terms used in this text.

Machines are man-made concrete *systems* consisting of a totality of orderly arranged and functionally connected elements. A system is characterized by having a boundary to its environment. The system's connection to its environment is maintained by *input* and *output* parameters. Each system can usually be subdivided into two or more subsystems. Generally, these subdivisions may be made with a varying degree of detail depending on our overall purpose. Consider, for example, the "clutch coupling" system shown in Figure 2-1. We would usually find this "system" as an assembly within a machine. However, if we wanted to investigate the system from a functional point of view, we could dissect it into the subsystems "elastic coupling" and "clutch." These subsystems, in turn, could be broken down into system components or individual parts.

For the purpose of reliability assessments we have found the following definitions useful.

System and Mission

A system is any composite of hardware or software items that work together to perform a mission or a set of related missions. A mission is the external "goal" of a system. A function in turn is the internal "purpose" of a system or system components needed to accomplish the mission. A complex system may be made up of two or more groupings of hardware or software items, each of which has a distinct role in performing the mission of the system. The definition of function and mission in a given

28

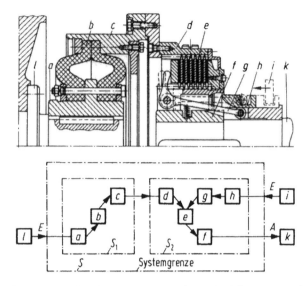

Figure 2-1. System "coupling." *a–h* are system elements, *i–l* are connecting elements, S is the total system, S_1 the subsystem "elastomeric coupling," S_2 the subsystem "clutch," E inputs, A outputs [1].

case is frequently subject to personal interpretation but should be as thorough as possible. Consider, for example, the oil system of an oil-injected rotary screw compressor (Fig. 2-2). Cursory examination may lead to the definition of the system mission or function as "Supplying oil for lubrication and cooling to the compressor." A better idea would be to subdivide this "function" into at least four related but distinct subfunctions and subsystems:

Subfunctions	*Subsystems*
1. Oil admission when compressor is running	1. Oil system – oil stop valve, item 28
2. Oil cooling and temperature control	2. Oil system/water system – coolers
3. Oil filtering	3. Oil system – filters
4. Air/oil separation	4. Oil/air system – separators

This more thorough breakdown will lead to a better understanding of the system mission as well as its function. The example also reveals that there are several functions that are performed simultaneously by one system, subsystem, or their components. It stands to reason that one would want to determine primary and secondary functions in these cases and rank them according to their criticality values.

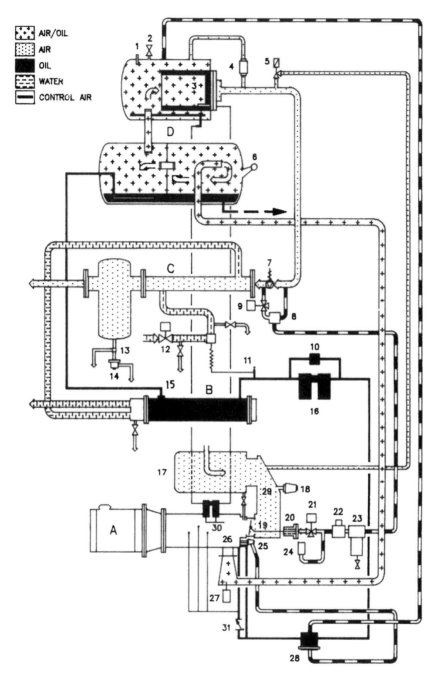

Figure 2-2. System diagram of a two-stage oil-flooded screw compressor (Demag).

Assembly and Part

An assembly is any functional component that can be disassembled into two or more subordinate components without disrupting permanent physical bonds [2]. A simple example of a mechanical seal assembly drawing is shown in Figure 2-3. The components of an assembly may be any combination of subassemblies or parts. A part in turn is defined as any hardware item that cannot be disassembled into subordinate components without severing permanent physical bonds. We have already seen how an assembly can be investigated regarding its functional characteristics. It is important in machinery reliability assessment to consider the geometric aspects of machinery parts. We introduce the term "element" [3] to define four internal functions used in machinery assemblies. There are four types of elements:

1. Transmitting elements, such as gear-tooth surfaces.
2. Constraining, confining, and containing elements, such as bearings or seals.
3. Fixing elements, such as threaded fasteners.
4. Elements that have no direct functions but which are inevitably needed to support the above functions (e.g., gear wheels or bearing supports).

The term "component" is used almost interchangeably with "assembly." However, "component" will have a somewhat more independent or stand-alone character. Machinery components, for example, are clutches, couplings, drive belts, gear boxes, or pneumatic and hydraulic systems.

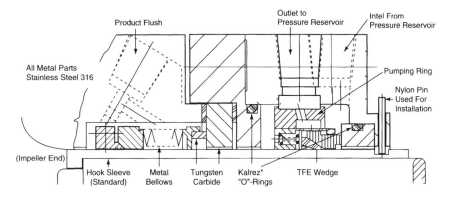

Figure 2-3. Mechanical shaft seal assembly (EG&G Sealol).

Assembly Hierarchy

From the foregoing it can easily be understood that machinery systems have a hierarchical structure (see Table 2-1). Assembly hierarchy describes the organization of system hardware elements into assembly levels. Assembly levels descend from the top – or system level – on the basis of functional and sometimes static relationships (see also Fig. 2-4). Thorough reliability assessments are carried out in reverse hierarchic sequence: first, we take a look at the lowest-level components; then the components of the next-highest level are assessed, and so on until the top level (the system level) has been reached.

Table 2-1
Assembly hierarchy

System level	Example
System	Screw compressor package
Subsystem	Compressor or driver
Assembly	Gear assembly
Part	See parts list
Element	Gear tooth, bearing, bolt

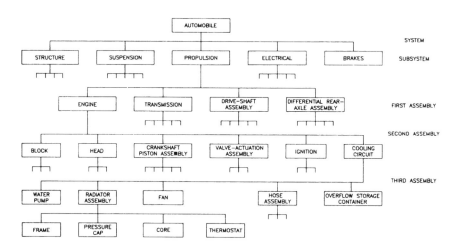

Figure 2-4. Assembly hierarchy for an automotive engine cooling circuit [4].

Failure

Machinery systems are subject to failure. In its simplest form, failure can be defined as any change in a machinery part or component which causes it to be unable to perform its intended function or mission satisfactorily. A popular yardstick for measuring *failure experience* of machinery parts, assemblies, components, or systems is to determine a failure rate. Failure rate is obtained by dividing the number of failures experienced on a number of homogeneous items, also called "population," within a time period, by the population. For example, if we had 10 injection pumps, and 3 failed during a period of 12 months, our failure rate (λ) would be:

$$\lambda = \frac{3 \text{ failures}}{10 \text{ machine-year}} = 0.3 \frac{\text{failures}}{\text{machine-year}}$$

or

$$\lambda = \frac{0.3}{365 \times 24} = 0.000034 \quad \text{or} \quad (34 \times 10^{-6}) \text{ failures per machine-hour}$$

For reliability assessments, failures are frequently classified as either chargeable or non-chargeable. A chargeable failure, for example, would be a failure that can be attributed to a defect in design or manufacture. A non-chargeable failure would be a failure caused by exposure of the part to operational, environmental, or structural stresses beyond the limits specified for the design. Other non-chargeable failures are those attributable to operator error or improper handling or maintenance.

Other terms used in the context of machinery failure experience are "malfunction" and "fault" that should, when used, be clearly defined.

Failure Mode

A failure mode is the appearance, manner, or form in which a machinery component failure manifests itself [5]. It should not be confused with the failure *cause*, as the former is the effect and the latter the cause of the failure event.

Failure modes can be defined for all levels of the system and the assembly hierarchy. For example, deterioration of the oil stop valve

(Fig. 2-2, item 28) of the oil-injected compressor system could have one of the following failure modes:

1. Fail open. *Consequence*: The compressor is flooded and cannot be started.
2. Fail close. *Consequence*: The compressor will shutdown due to high discharge temperature.
3. Fail not fully open or fully closed. *Consequence*: Gradual deterioration of system performance.

The causes of these failure modes could either be common, such as dirt or foreign objects in the valve, or specific to each failure mode – a broken return spring would keep the valve open, insufficient discharge pressure would keep it closed, and so forth.

Service Life

Service life designates the time-span during which a product can be expected to operate safely and meet specified performance standards, when maintained in accordance with the manufacturer's instructions and not subjected to environmental or operational stresses beyond specified limits [6]. The service life for a given machinery part represents a prediction that no less than a certain proportion of the machinery system or its components will operate successfully for the stated time period, number of cycles, or distance traveled. Service life is clearly a probabilistic term subject to a confidence limit. A good example is anti-friction bearings. Since a bearing failure generally results in the failure of the machine in which it is installed, bearing manufacturers have made a considerable effort to identify the factors that are responsible for bearing failures. A typical equation for determining ball bearing service life shows the rated life to be inversely proportional to the rotational speed of the inner ring and the third power of the applied radial load. Rated life in this case is the so-called L_{10} life, which is the number of bearing revolutions, or the number of working hours at a certain rotational speed and load, which will be reached or exceeded by 90% of all bearings.

Reliability

Reliability, finally, in general terms, is the ability of a system or components thereof to perform a required function under stated conditions for a stated period of time. It is also apparent that "reliability" is frequently used as a characteristic denoting a probability of success or success ratio [7]. This means that it may be stated that:

1. A component or piece of machinery should operate successfully for X hours on $Y\%$ of occasions on which it is required to operate; or
2. A machine should not fail more frequently than X times in Y running hours; or
3. The mean life of a population of similar components or machinery should be equal to or greater than Y hours with a standard deviation of S hours.

Maintainability

Many machinery components are designed to receive some form of attention during their life. The goal is to compensate for the effects of wear or to allow for the replacement of consumable or sacrificial elements. The ease with which this kind of work can be done is termed "maintainability." The operational and organizational function of this work is called "maintenance." Maintenance possibilities are illustrated in Fig. 2-5. It has been shown that, if maintenance on process machinery has to be performed at all, predictive maintenance is the most cost-effective mode [8–10].

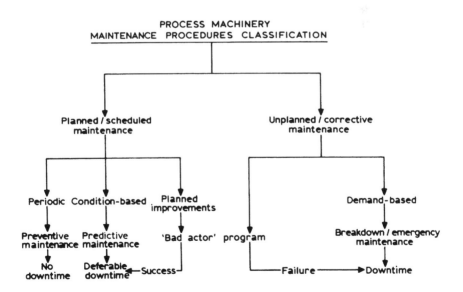

Figure 2-5. Process machinery maintenance procedures classification.

Maintainability then is the ability of an item, under stated conditions of use, to be retained in, or restored to, a state in which it can perform its required functions, when maintenance is performed under stated conditions and using prescribed procedures and resources [11].

Maintainability has a direct influence on the reliability of machinery systems. We will see that maintainability parameters must be considered an integral part of the machinery reliability assessment effort.

Surveillability

Surveillability is closely related to maintainability and will receive the same attention within the overall reliability assessment activity. We have already stated that process machinery maintenance can be optimized by practicing condition-based or predictive maintenance. Surveillability is the key. It is defined as a quantitative parameter that includes:

- accessibility for surveillance;
- operability if required;
- ability to monitor machinery component deterioration;
- provision of indicating and annunciation devices.

Availability

Maintainability together with reliability determine the availability of a machinery system. Availability is influenced by the time demand made by preventive and corrective maintenance measures. Maintenance activities which are performed during planned downtimes or on-line without affecting operation do not have an impact on availability. Availability (A) is measured by:

$$A = \frac{\text{MTBF}}{\text{MTBF} + \text{MTTR}} \tag{2.1}$$

where MTBF = mean time between failures,
   MTTR = mean time to repair or mean repair time.

Figure 2-6 shows the relationship of the concepts just discussed.

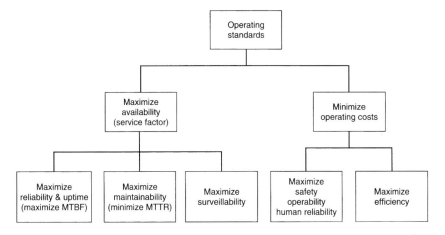

Figure 2-6. Reliability and uptime relationships.

References

1. Dubbel, *Taschenbuch für den Maschinenbau*, K. H. Grote, and J. Feldhusen, (Publishers), Berlin, 2000, ISBN: 3-540-22142-5.
2. Moss, M. A., *Designing for Minimal Maintenance Expense*. New York: Marcel Dekker, 1985, p. 13.
3. Yoshikawa, H. and Taniguchi, N., Fundamentals of mechanical reliability and its application to computer aided design. *Annals of the CIRP*, **24** (1), 1975, p. 298.
4. Moss, op. cit., p. 64.
5. Bloch, H. P. and Geitner, F. K., *Machinery Failure Analysis and Troubleshooting*. Houston: Gulf Publishing, 1983, pp. 2–3.
6. Moss, op. cit., p. 16.
7. British Standards Institution, op. cit., p. 14.
8. Finley, H. F., Maintenance management for today's high technology plants. *Hydrocarbon Processing*, January 1978, pp. 101–105.
9. Bloch and Geitner, op. cit., p. 610.
10. Matteson, T. D., Overhauling our ideas about maintenance. *Mechanical Engineering*, May 1986, pp. 86–88.
11. British Standards Institution, op. cit., p. 19.

Chapter 3
Uptime as probability of success

Probabilistic thinking is based on very old ideas which go back to De Mere [1], La Place [2], and Bayes [3]. What is probability and how does it relate to frequency, statistics and, finally, machinery reliability? The word probability has several meanings. At least three will be considered here.

One definition of probability has to do with the concept of equal likelihood. If a situation has N equally likely and mutually exclusive outcomes, and if n of these outcomes are event E, then the probability $P(E)$ of event E is:

$$P(E) = \frac{n}{N} \tag{3.1}$$

This probability can be calculated *a priori* and without doing experiments.

The example usually given is the throw of an unbiased die, which has six equally likely outcomes – the probability of throwing a one is 1:6. Another example is the withdrawal of a ball from a bag containing four white balls and two red ones – the probability of picking a red one is 1:3. The concept of equal likelihood applies to the second example also, because, even though the likelihoods of picking a red ball and a white one are unequal, the likelihoods of withdrawing any individual ball are equal.

This definition of probability is often of limited usefulness in engineering because of the difficulty of defining situations with equally likely and mutually exclusive outcomes.

A second definition of probability is based on the concept of relative frequency. If an experiment is performed N times, and if event E occurs on n of these occasions, then the probability of $P(E)$ of event E is:

$$P(E) = \lim_{n \to \infty} \frac{n}{N} \tag{3.2}$$

$P(E)$ can only be determined by experiment. This definition is frequently used in engineering. In particular it is this definition which is implied when we estimate the probability of failure from field failure data [4].

Thus, when we talk about the measurable results of probability experiments – such as rolling dies or counting the number of failures of a machinery component – we use the word "frequency." The discipline that deals with such measurements and their interpretation is called statistics. When we discuss a state of knowledge, a degree of confidence, which we derive from statistical experiments, we use the term "probability." The science of such states of confidence, and how they in turn change with new information, is what is meant by "probability theory."

The best definition of probability in our opinion was given by E. T. Jaynes of the University of California in 1960:

> Probability theory is an extension of logic, which describes the inductive reasoning of an idealized being who represents degrees of plausibility by real numbers. The numerical value of any probability $(A|B)$ will in general depend not only on A and B, but also on the entire background of other propositions that this being is taking into account. A probability assignment is 'subjective' in the sense that it describes a state of knowledge rather than any property of the 'real' world. But it is completely 'objective', in the sense that it is independent of the personality of the user: two beings faced with the same total background and knowledge must assign the same probabilities.

Later, Warren Weaver [5] defined the difference between probability theory and statistics:

> Probability theory computes the probability that 'future' (and hence presently unknown) samples out of a 'known' population turn out to have stated characteristics.
>
> Statistics looks at a 'present' and hence 'known' sample taken out of an 'unknown' population, makes estimates of what the population may be, compares the likelihood of various populations, and tells how confident you have a right to be about these estimates.
>
> Stated still more compactly, probability argues from populations to samples, and statistics argues from samples to populations.

Whenever there is an event E which may have outcomes E_1, $E_2, \ldots, E_n$, and whose probabilities of occurrence are $P_1, P_2, \ldots, P_n$, we can speak of the set of probability numbers as the "probability distribution" associated with the various ways in which the event may occur. This is a very natural and sensible terminology, for it refers to the way in which the available supply of probability (namely unity) is "distributed" over the various things that may happen.

Consider the example of the tossing of six coins. If we want to know "how many heads there are," then the probability distribution can be shown as follows:*

No. of heads	0	1	2	3	4	5	6
Probability	1/64	6/64	15/64	20/64	15/64	6/64	1/64

These same facts could be depicted graphically (Fig. 3-1). Accordingly, we arrive at a probability curve versus "frequency" as a way of expressing our state of knowledge.

As another application of the probability-of-frequency concept, consider the reliability of a specific machine or machinery system. In order to quantify reliability, frequency type numbers are usually introduced. These numbers are mean times between two failures or MTBF, for instance, which are based on failures per trial or per operating period. Usually they are referred to in months.

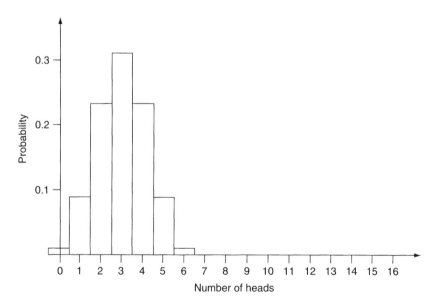

Figure 3-1. Distribution of probabilities measured by the vertical height as well as by areas of the rectangles (the six-coin case).

---

* See Appendix A for calculation of probability values.

Most of the time we are uncertain about what the MTBF is. "All machines and their components are not created equal," their load-cycles and operating conditions are unknown, and maintenance attention can vary from neglect to too frequent intervention. Consider for instance the MTBF of a sleeve bearing of a crane trolley wheel. At a MTBF of 18 months (30% utilization factor), early failures can be experienced after 6 months. The longest life experience may be five times the shortest life (see Fig. 3-2). Even though the data were derived from actual field experience, we cannot expect exact duplication of the failure experience in the future. Therefore, Figure 3-2 is our probabilistic model for the future of a similarly designed, operated, and maintained crane wheel.

It is important to distinguish the above idea from the concept of "frequency of frequency." Let R denote the historical reliability of an individual designated machine, selected at random from a population of similar machines. The historical reliability of a machine is defined as:

$$R = 1 - \frac{H_1}{H_1 + H} \tag{3.3}$$

where $H_1 =$ total time on forced outage (h),
$\quad\quad H =$ total service time (h).

We can build a frequency distribution using historical reliability for each machine showing what fraction of the population belongs to each reliability increment. If the population is large enough we can express this distribution as a continuous curve – a "frequency density" distribution, $\Phi(R)$. The units of $\Phi(R)$ are consequently frequency per unit R, or fraction of population per unit reliability.

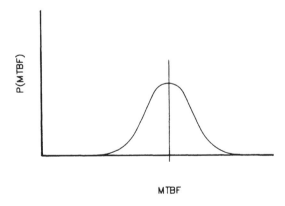

Figure 3-2. Probability-of-frequency curve for a machinery component.

This curve is an experimental quantity. It portrays the variability of the population, which is a measurable quantity. The value of *R* varies with the individual selected. It is a truly fluctuating or random variable.

Contrast this with the relationship shown in Figure 3-2, where we selected a specific machinery component and asked what its future reliability would be. That future reliability is the result of an experiment to be done. It is not a random variable: it is a definite number not known at this time. This goes to show that we must distinguish between a frequency distribution expressing the variability of a random variable and a probability distribution representing our state of knowledge about a fixed variable.

A third definition of probability is degree of belief. It is the numerical measure of the belief which a person has that an event will occur.

Often this corresponds to the relative frequency of the event. This need not always be so for several reasons. One is that the relative frequency data available to the individual may be limited or non-existent. Another is that although somebody has such data, he or she may have other information which causes doubt that the whole truth is available. There are many possible reasons for this.

Several branches of probability theory attempt to accommodate personal probability. These include ranking techniques, which give the numerical encoding of judgments on the probability ranking of items. Bayesian methods allow probabilities to be modified in the light of additional information [6].

The key idea of the latter branch of probability theory is based on Bayes' Theorem, which is further defined below.

In basic probability theory, $P(A)$ is used to represent the probability of the occurrence of event *A*; similarly, $P(B)$ represents the probability of event *B*. To represent the joint probability of *A* and *B*, we use $P(A \wedge B)$, the probability of the occurrence of both event *A* and event *B*. Finally, the conditional probability, $P(A|B)$, is defined as the probability of event *A*, given that *B* has already occurred.

From a basic axiom of probability theory, the probability of the two simultaneous events *A* and *B* can be expressed by two products:

$$P(A \wedge B) = P(A) \times P(B|A) \tag{3.4}$$

$$P(A \wedge B) = P(B) \times P(A|B) \tag{3.5}$$

Equating the right sides of the two equations and dividing by $P(B)$, we have what is known as Bayes' Theorem:

$$P(A|B) = P(A) \times [P(B|A)/P(B)] \tag{3.6}$$

In other words, it says that $P(A|B)$, the probability of A with information B already given, is the product of two factors: the probability of A prior to having information B, and the correction factor given in the brackets. Stated in general terms:

Posterior probability $\propto$ Prior probability $\times$ Likelihood

where the symbol $\propto$ means "proportional to" [7]. This relationship has been formulated as follows:

1. The A_i's are a set of mutually exclusive events for $i = 1 \ldots n$.
2. $P(A_i)$ is the prior probability of A_i before testing.
3. B is the observation event.
4. $P(B|A_i)$ is the probability of the observation, given that A_i is true.

Then

$$P(A_i|B) = \frac{P(A_i)P(B|A_i)}{\sum\limits_{i}^{n} P(A_i)P(B|A_i)} \tag{3.7}$$

where $P(A_i|B)$ is the posterior probability or the probability of A_i now that B is known. Note that the denominator of equation 3.7 is a normalizing factor for $P(A_i|B)$ which ensures that the sum of $P(A_i|B) = 1$.

As powerful as it is simple, this theorem shows us how our probability – that is, our state of confidence with respect to A_i – rationally changes upon getting a new piece of information. It is the theorem we would use, for example, to evaluate the significance of a body of experience in the operation of a specific machine.

To illustrate the application of Bayes' Theorem let us consider some examples. If C represents the event that a certain pump is in hot oil service and G is the event that the pump has had its seals replaced during the last year, then $P(G|C)$ is the probability that the pump will have had its seals replaced some time during the last 3 years *given that it is actually in hot oil service*. Similarly, $P(C|G)$ is the probability that a pump did have its seals replaced within the last 3 years *given that the pump is in hot oil service*. Clearly there is a big difference between the events to which these two conditional probabilities refer. One could use equation 3.6 to relate such pairs of conditional probabilities.

Although there is no question as to the validity of the equation, there is some question as to its applicability. This is due to the fact that it involves a "backward" sort of reasoning – namely, reasoning from effect to cause.

Example: In a large plant, records show that 70% of the bearing vibration checks are performed by the operators and the rest by central inspection. Furthermore, the records show that the operators detect a problem 3% of the time while the entire force (operators and central inspectors) detect a problem 2.7% of the time. What is the probability that a problem bearing, checked by the entire force, was inspected by an operator? If we let A denote the event that a problem bearing is detected and B denote the event that the inspection was made by an operator, the above information can be expressed by writing $P(B) = 0.70$, $P(A|B) = 0.03$, and $P(A) = 0.027$, so that substitution into Bayes' formula yields:

$$P(B|A) = \frac{P(B) * P(A|B)}{P(A)} \tag{3.8}$$

Numerically, this is:

$$P(B|A) = \frac{(0.70)(0.03)}{(0.027)} = \frac{0.021}{0.027} = \frac{7}{9} = 78\%$$

This is the probability that the inspection was made by an operator given that a problem bearing was found [8].

According to the foregoing, our understanding of reliability here is a probability rather than merely a historical value. It is statistical rather than individual.

References

1. Weaver, W., *Lady Luck. The Theory of Probability*. Garden City: Doubleday, 1963, pp. 45–52.
2. Weaver, op. cit., p. 251.
3. Weaver, op. cit., pp. 308–309.
4. Kaplan, S. J. and Garrick, B. J., Try probabilistic thinking to improve powerplant reliability. *Power*, March 1980, pp. 56–61.
5. Weaver, op. cit., pp. 306–307.
6. Lees, F. P., *Loss Prevention in the Process Industries*. Boston: Butterworth, 1980, Vol. 1, p. 81.
7. Henley, E. J. and Kumamoto, H., *Reliability Engineering and Risk Assessment*. Englewood Cliffs: Prentice-Hall, 1981, p. 256.
8. Freund, J. E. and Williams, F. J., *Elementary Business Statistics*. Englewood Cliffs: Prentice-Hall, 2nd edn, 1972.

Chapter 4
Estimating machinery uptime

The reliability of a machinery system may be mathematically described by defining distribution functions using discrete and random variables. An example of a discrete variable is the number of failures in a given time interval. Examples of continuous random variables are the time from part installation to failure or the time between successive equipment failures.

This approach has been particularly useful in the field of electronic engineering where it has been applied to the design and evaluation of electronic devices. Using reliability theory one can estimate the reliability of complex electronic systems. Calculation methods, specific to electronic systems, make use of failure probability data compiled for this purpose.

To evaluate electronic component reliability, the concept of constant failure rate is used, that is failure rates of electronic components remain constant during the useful life of the component. However, this is frequently not the case when evaluating mechanical component reliability. There are several reasons for this. It is, for example, an established fact that in many cases machinery components follow an increasing failure rate pattern. Another reason is the fact that machinery components are not well standardized. Finally, there seem to be many more failure modes experienced by machinery parts than by electronic parts. Consequently, reliability data for mechanical components and assemblies is scarce, and, when available, caution is advised. From this it follows that there is no accurate method available for absolute reliability prediction that takes the specific nature of machinery systems into account. As we will see later, it seems that only *relative* reliability predictions can be made for machinery. What is the specific nature of machinery? Figure 4-1 illustrates a machinery system by comparing it with an electric system. Consider, for example, the reliability of a tribo-mechanical system* in which wear

---

* A system with parts in rubbing contact.

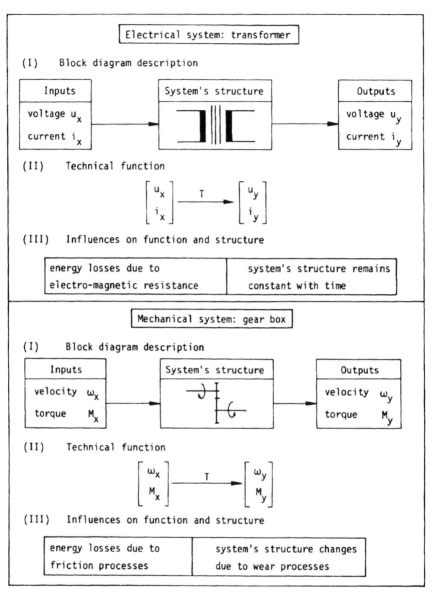

Figure 4-1. Comparison of the characteristics of an electrical and a mechanical system [1]. (Reprinted from Czichos, H., *Tribology—A Systems Approach to the Science and Technology of Friction, Lubrication, and Wear*, 1978, p. 26, Fig. 3-1, by courtesy of Elsevier Science Publishers, Physical Sciences & Engineering Div.)

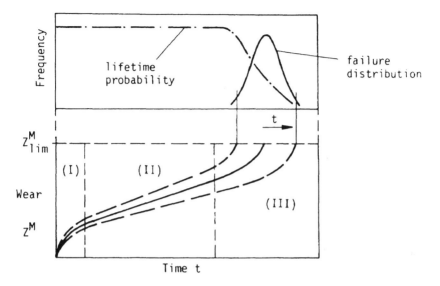

Figure 4-2. Wear curves and failure distribution [1].

behavior is a function of time. Three main characteristics may be determined for the loss–output wear rates of such a system [1]:

1. Self-accommodation ("running-in")
2. Steady-state
3. Self-acceleration ("catastrophic damage")

These three phase changes in the system behavior may follow each other in time (Fig. 4-2). Here, Z_{lim}^M denotes a maximum allowable level of wear loss. At this level the system structure has changed in such a way that the functional input–output relationship of the system has been severely disturbed. Repeated measurements show random data variations as indicated by the dashed lines in Fig. 4-2. A distribution of the "life" of the system or a failure distribution can be derived from sample functions of the wear process.

Earlier, we familiarized ourselves with the concept of relative frequency. The reader is referred to Figure 3-2, which for convenience, is reproduced in Figure 4-3. If we wish to determine the probability of failure occurring between the times t_b and t_c, we multiply the y-axis value by the interval $(t_c - t_b)$. Figure 4-3 is also called a probability density function where the equation of the curve is denoted by $f(t)$. As an example, if $f(t) = 0.6 \exp(-0.6t)$, we obtain the curve shown in Figure 4-4, a negative exponential distribution which will be dealt with later.

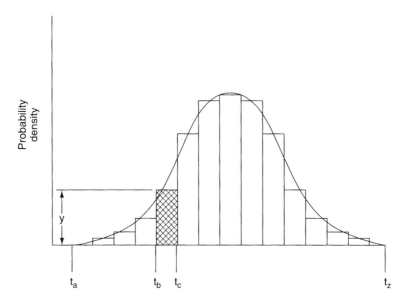

Figure 4-3. Probability density function.

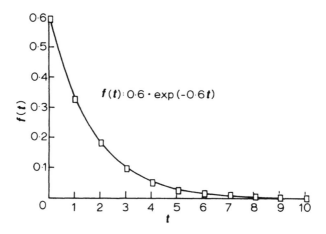

Figure 4-4. Negative exponential distribution.

Returning to Figure 4-3, the probability of a failure occurring between t_b and t_c is the area of the hatched portion of the distribution. This area is the integral between t_b and t_c of $f(t)$ or:

$$\int_{t_b}^{t_c} f(t)dt \tag{4.1}$$

Consequently, the probability of a failure occurring between times t_a and t_z is:

$$\int_{t_a}^{t_z} f(t)dt = 1 \qquad (4.2)$$

We stated earlier that the failure distributions of different types of machinery systems are not the same. Even the failure distributions of identical machines may not be the same if they are subjected to different levels of Force, Reactive Environment, Temperature, and Time (FRETT). There are a number of well-known probability density functions which have been found in practice to describe the failure characteristics of machinery (see Fig. 4-5) [2].

The cumulative distribution function. In reliability estimations we want to determine the probability of a failure occurring before some specified time *t*. This probability can be calculated by using the appropriate density function as follows:

$$\text{Probability of failure before time} = \int_{-\infty}^{t} f(t)dt \qquad (4.3)$$

The integral $\int_{-\infty}^{t} f(t)dt$ is termed $F(t)$ and is called the cumulative distribution function. One can state that as *t* approaches infinity, $F(t)$ approaches unity.

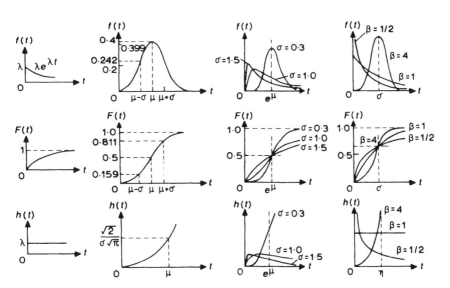

Figure 4-5. Density, cumulative distribution, and hazard functions of the exponential, normal, log-normal and Weibull distributions.

The reliability function. The function complementary to the cumulative distribution function is the reliability function, also called survival function. This function can be used to determine the probability that equipment will survive to a specified time t. The reliability function is denoted as $R(t)$ and is defined by:

$$R(t) = \int_t^\infty f(t)dt \qquad (4.4)$$

and, obviously:

$$R(t) = 1 - F(t) \qquad (4.5)$$

The failure rate or hazard function. The last type of function derived from the other functions is the hazard function. It has other names in the literature, such as intensity function, force of mortality, and also failure rate in a certain context. It is denoted as $h(t)$ and defined as:

$$h(t) = \frac{f(t)}{R(t)} = \frac{f(t)}{1 - F(t)} \qquad (4.6)$$

The hazard function is a conditional probability that a system will fail during the time t and dt under the condition that the system is safe until time t. Someone once had a simple explanation of the hazard function. It was made by analogy. Suppose someone takes an automobile trip of 200 mil and completes the trip in 4 hr. The average travel rate was 50 mph, although the person drove faster at some times and slower at others. The rate at any given instant could have been determined by reading the speed indicated on the speedometer at that instant. The 50 mph is analogous to the failure rate and the speed of any point is analogous to the hazard rate.

The foregoing definitions rely on some rather involved mathematics. The reader is referred to Green and Bourne [3] and Henley and Kumamoto [4] for more detailed explanations. However, we believe that there is no need to burden oneself with the mathematics of failure distributions. As we will see later, there has been considerable progress in the application of computerized models and appropriate software.

Specific distribution functions. A number of distributions have been proposed for machinery failure probabilities. Their definitions in terms of density function, cumulative distribution function, and hazard rate are depicted in Figure 4-5.

The exponential distribution is the most important function due to its wide acceptance in the reliability analysis work of electronic systems. As shown in Figure 4-5, this function is defined as:

$$f(t) = \lambda \cdot \exp(-\lambda t) \quad \text{for} \quad t \geqslant 0 \qquad (4.7)$$

where

$$\lambda = \frac{1}{\mu} \tag{4.8}$$

The exponential distribution is an appropriate model where failure of an item is due not to deterioration as a result of wear, but rather to random events. This feature of the exponential distribution also implies a constant hazard rate. The exponential distribution has been successfully applied as a time-to-failure model for complex systems consisting of a large number of components in series, none of which individually contributes significantly to the total failure density [5]. This distribution is often used because of its universal applicability to systems that are repairable. Many kinds of electronic components follow an exponential distribution. Machinery parts behave in this mode when they succumb to brittle failure. For example, Figure 4-6 shows that Diesel engine control unit failures followed an exponential distribution.

The normal distribution. Although the normal distribution has only limited applicability to life data, it is used where failures are due to wear processes. The hazard or failure rate of this distribution cannot be expressed in a simple form.

The lognormal distribution is defined by:

$$f(t) = \frac{1}{t\sigma\sqrt{2\pi}} \exp \frac{-[\log(t/t_{50})]^2}{2\sigma^2} \tag{4.9}$$

where t_{50} = median = $\exp(u)$,
$\quad u$ = the mean of the logarithms of the times to failure,
$\quad \sigma$ = standard deviation.

The limited applicability of normal distribution to life data has been mentioned [7]. This is not the case for the *lognormal distribution* which enjoys wide acceptance in reliability work. It has been applied in machinery maintainability consideration and where failure is due to crack propagation or corrosion. Nelson and Hayashi [8] give an exhaustive account of stress–temperature related furnace tube failure phenomena modeled by the lognormal distribution.

The Weibull distribution is defined by two parameters – η, the nominal or characteristic life,* and a constant β, a non-dimensional shape parameter. A typical Weibull distribution fit for life of a ball bearing is shown in Figure 4-7.

---

* Also called scale factor.

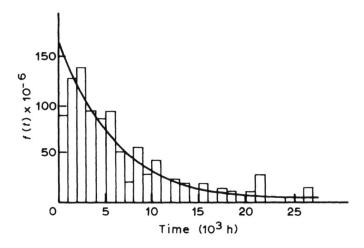

Figure 4-6. Density function $f(t)$ of the failure of diesel engine control units [6].

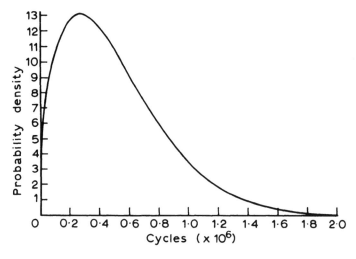

Figure 4-7. Weibull function for a ball bearing [9]. (Reprinted from Sidall, J.N., *Probabilistic Engineering Design*, 1983, p. 361, Fig. 11-3, by courtesy of Marcel Dekker, Inc.)

The ability of the Weibull function to model failure distributions makes it one of the most useful distributions for analyzing failure data. If the shape parameter $\beta > 1$, an increasing $h(t)$ is indicated which is symptomatic of wear-out failures. Where $\beta = 1$, we find an exponential function, which obviously is a special case of the Weibull distribution. With $\beta = 1$, a constant hazard or failure rate is indicated.

Where $\beta = 2$, this means that $h(t)$ is linearly increasing with t. The resulting distribution is a special case of the Weibull function known as the Rayleigh distribution [5].

If $\beta < 1$, a decreasing failure rate $h(t)$ is indicated. This would be typical for machinery components where run-in or initial self-accommodation takes place. Mechanical shaft seals would be a typical example.

The mean and standard deviation of the Weibull distribution involves complex calculations. For most engineering problems where the shape factor is greater than 0.5, they can be found from:

$$\mu = \eta(0.9 + 0.1/\beta^3) \qquad (4.10)$$

$$\sigma = \mu/\beta \qquad (4.11)$$

In cases where the shape factor is greater than 1, the mean is nearly equal to characteristic life (η). The error involved in this assumption will generally be small compared to other errors stemming from the quality of data.

One difficulty in attempting to fit theoretical distribution to failure or "life" data arises when a part or an assembly is subject to different failure modes. Table 4-1 lists some of the basic machinery component failure

Table 4-1
Selected basic machinery component failure modes and their statistical distributions

	Probability distribution		
Basic failure mode	Exponential	Normal	Weibull
1.0 Force/stress			
1.1 Deformation			●
1.2 Fracture	●		
1.3 Yielding	●		
2.0 Reactive environment			
2.1 Corrosion		●	●
2.2 Rusting			●
2.3 Staining	●		
3.0 Temperature/thermal			
3.1 Creep		●	
4.0 Time effects			
4.1 Fatigue			●
4.2 Erosion			●
4.3 Wear		●	●

modes and shows the distributions they tend to follow. There are three different possibilities in which these failure modes appear:

1. Simultaneously with some time differences. Fit corrosion, for instance, in an anti-friction (ball) bearing would appear as wear first and then as corrosion. Sidall [9] shows how to evaluate simultaneous failure mode occurrences in the context of failure distributions.
2. Failure modes occur singularly and exclusive of others. This is a somewhat theoretical assumption that we will not deal with any further.
3. A more realistic model can be created by assuming that failure modes occur consecutively in time. A commonly accepted concept is shown in Figure 4-8. In this curve, called the bathtub curve, three conditions can be distinguished: (1) early or infant mortality failures, (2) random failures, and (3) wear-out failures.

Condition 1 describes the early time period of a machinery system or part by showing a decreasing failure rate over time. It is usually assumed that this period of "infant mortality" or "burn-in" is caused by the existence of material and manufacturing flaws together with assembly errors. Parts or systems that would exclusively exhibit this behavior would fit a Weibull distribution with $\beta < 1$. Condition 2, the area of constant failure rate, is the region of normal performance. This period is termed "useful life," during which time only random failures will occur. Parts or systems that would exclusively exhibit this failure behavior would fit a Weibull distribution with $\beta = 1$ or for that matter an exponential distribution. Condition 3 ($\beta > 1$) is characterized by an increase of failure

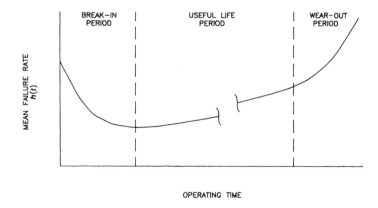

Figure 4-8. Mean failure rate curve as a function of time.

rate with time. As mentioned before, failures may be due to aging and wear out.

It has been said that the bathtub curve concept is purely theoretical and only serves the purpose of promoting a better understanding of failure events. However, real-world examples can be cited in connection with non-repairable parts. For example, if a large number of light bulbs or anti-friction bearings operate continuously, some fail due to defects. During the useful life, occasional random failures occur, but most survive to old age, when the failure rate rises.

The bathtub curve pertaining to repairable components and their systems is rarely discussed in the reliability literature [10]. However, most machinery components follow this curve. Its time axis is the cumulative operating time of the equipment, not the time interval between failures. The major difference between this curve and the non-repairable curve is that it continues indefinitely or until the equipment is removed from service because it is uneconomical to repair.

Estimation of Failure Distributions for Machinery Components

The data required to determine failure distributions are the individual times to failure of the equipment.

The procedure is to convert the data to become representative of the cumulative failure distribution $F(t)$. This is done by plotting times to failure against $F(t)$ on a scale which corresponds to the distribution to be fitted. For the exponential distribution this would be:

$$F(t) = 1 - \exp(-\lambda t) \tag{4.12}$$

Consequently:

$$t = \frac{1}{\lambda} \ln \frac{1}{1 - F(t)} \tag{4.13}$$

A plot of $1/[1 - F(t)]$ on a log scale against time on a linear scale produces a straight line. For the Weibull distribution:

$$\ln(t) = \frac{1}{\beta} \ln \ln \frac{1}{1 - F(t)} + \ln \eta \tag{4.14}$$

For most distributions, special graph papers are available which allow direct plotting of $F(t)$ versus t (Fig. 4-9 illustrates a Weibull graph). Nelson [11] describes distributions and the fitting of life or failure data. We encourage our readers to investigate the possible use of computer

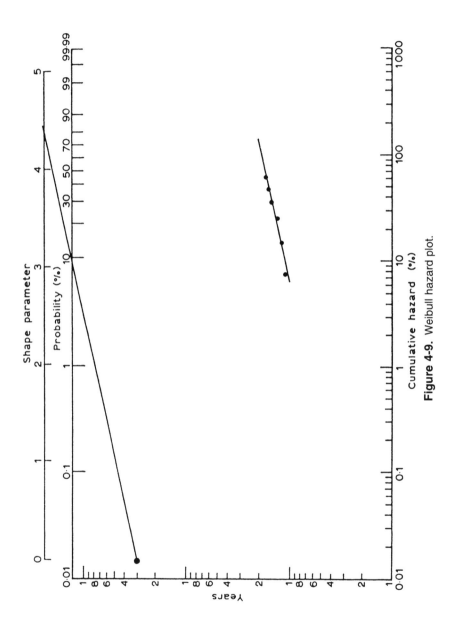

Figure 4-9. Weibull hazard plot.

software packages developed for the statistical analysis of data relating to the failures and successful performance of machinery or components. Their analysis capabilities range from simple calculations such as mean life, to the fitting of Weibull and other distribution models.*

Application of Failure Distributions

The application of failure distributions for reliability predictions has been described in numerous references. With the emergence of improved data bases there is a new interest in these applications. Exhaustive information covering the application of distribution functions to equipment maintenance, replacement, and reliability decisions can be obtained, for example, from Jardine [14].

Our first *example* will cover a replacement decision in connection with large (>1500 hp) electric motors in a petrochemical process plant. The motors considered for replacement had served this particular plant well for 18 years, but failure experience with similar motors at the same time had raised doubt in the owner's mind as to whether or not an 18-year-old motor could still be called reliable. All motors were 4000 kVA, 3 phase, 60 cycle, pipe-ventilated squirrel cage induction motors.

The failure experience of similar motors is listed in Table 4-2. Motors shown as having failed are denoted by a superscript (^{a}). These motors had stopped suddenly on-line through winding failures. Mean forced outage penalties were in the neighborhood of 1600 k$ considering the availability or unavailability of motor rewind shops and materials. The cost of an emergency rewind amounted to 125 k$, whereas the cost of a preventive rewind was 100 k$ with no penalty cost for loss of production. The problem was simply to balance the cost of preventive rewinds against their benefits. In order to do this one needs to determine the optimal preventive replacement age of the motor windings to minimize the total expected cost of replacement per unit time. Obviously, one requires a probabilistic model of the motor winding life in order to make a reliability assessment.

Obtaining the Weibull Function

The Weibull function was obtained by plotting the data contained in Table 4-2 on appropriate Weibull paper (Fig. 4-9).

The plotting method used has been proposed by Nelson [11] for "multiply censored" life data consisting of times to failure on failed units, and

---

* "RECLODE" program, University of Windsor, Ontario.

Table 4-2
Large motor winding failures: Failure data and hazard calculation

Motor	Rank	Years	Hazard	Cumulative hazard
C-70	20	8		
C-71A	19	8		
C-71B	18	8		
P-70A	17	8		
P-70B	16	8		
P-71	15	8		
C-25	14	10		
C-11	13	11[a]	7.69	7.69
C-52	12	12[a]	8.33	16.03
C-13	11	13[a]	9.09	25.12
C-31	10	13		
C-53	9	15[a]	11.11	36.23
C-41	8	16[a]	12.58	48.73
C-91	7	17[a]	14.29	63.01
C-32A	6	17		
C-32B	5	17		
C-01	4	18[b]		
C-30	3	18		
C-50	2	18		
C-51	1	18		

[a] Winding failure.
[b] Preventive winding replacement.

running times – called censoring times – on unfailed units. The method is known as hazard plotting. It has been used effectively to analyze field and life test data on products consisting of electronic and mechanical parts ranging from small electric appliances to heavy industrial equipment. The hazard plotting method originally appeared in Nelson [12], which also contains more details.

Steps

1. The n times, or years in our case, are placed in order from the smallest to the largest as shown in Table 4-2. The times are labeled with reverse ranks, that is the first time is labeled n, the second labeled $n-1, \ldots$, and the nth is labeled 1. The failure times are each marked by a superscript ([a]) to distinguish them from the censoring times.
2. Calculate a hazard value for each failure as $100/k$, where k is its reverse rank. The hazard values for the large motor winding failures are shown in Table 4-2. For example, for the winding failure after

13 years, the reverse rank is 11 and the corresponding hazard value is $100/11 = 9.1\%$.

3. Proceed to calculate the cumulative hazard value for each failure as the sum of its hazard value and the cumulative hazard value of the preceding failure. For instance, for the motor failure after 13 years of operation, the cumulative hazard value of 25.12 is calculated by adding the hazard value of 9.1 to the cumulative hazard value of 16.03 of the preceding failure.

4. For plotting purposes, the hazard paper of a theoretical distribution of time to failure was chosen. The Weibull distribution seemed appropriate. On the vertical axis of the Weibull hazard paper, make a time-scale that includes the sample range of failure times (i.e., years).

5. Plot each failure time vertically against its corresponding cumulative value on the horizontal axis. The plot of the large motor winding failures is shown in Figure 4-9. If the plot of the sample times to failure is reasonably straight on a hazard paper, one can conclude that the underlying distribution fits the data adequately. By eye, fit a straight line through the data points (Fig. 4-9). Practical advice and more tips on making hazard plots are given by Nelson [12] and King [13].

A hazard plot provides information on:

- the percentage of items failing by a given age;
- percentiles of the distribution;
- the behavior of the failure rate of the units as a function of their age;
- distribution parameters.

In our context we are mainly interested in the distribution parameters. We already know that the Weibull distribution has an increasing or decreasing failure rate depending on whether its shape parameter has a value greater than, equal to, or less than 1. To obtain the shape parameter, β, draw a straight line parallel to the fitted line so it passes through the dot in the upper left-hand corner of the paper and through the shape parameter scale. Nautical chart parallel rulers are ideally suited for this task. Figure 4-9 shows the result. The value on the shape parameter scale is the estimate and is $\hat{\beta} = 4.3$.* A β-estimate of 4.3 suggests that the winding failure rate increases with age – that is, in a wear-out mode. It also suggests that the machines should be rewound at some age when they are too prone to failure.

---

* The circumflex or "hat" symbol (ˆ) means "estimated" value.

In order to estimate the other parameter of the Weibull function η, we enter the hazard plot on the cumulative hazard scale at 100 or 63% on the probability scale. If we move up the fitting line on Figure 4-9 and then sideways to the time-scale, we find the scale parameter, that is 18.5 years.

We now proceed to define the Weibull distribution function that describes the large motor winding population. We write:

$$f(t) = \frac{\beta}{\eta - l} \left(\frac{t - l}{\eta - l} \right)^{\beta - 1} \exp - \left(\frac{t - l}{\eta - l} \right)^{\beta}, t \geqslant l \qquad (4.15)$$

then:

$$R(t) = \exp - \left(\frac{t - l}{\eta - l} \right)^{\beta} \qquad (4.16)$$

where t is the age of the motors
    η is the characteristic life
    β the shape parameter
    l is the location parameter

The location parameter, l, takes into account that our motors did not begin to fail before age 9–10. On the other hand, l would be equal to zero when it is expected that failures appear as soon as an item is placed into service and:

$$h(t) = \frac{f(t)}{R(t)} \qquad (4.6)$$

Applying the estimated parameters $\hat{\beta}$ and $\hat{\eta}$, Figure 4-10 was produced by using a simple computer program. After having made this reliability assessment one can now proceed to work the economic decision of how to optimize motor replacement.

Construction of the Replacement Model

The construction of the replacement model is credited to A. K. S. Jardine [14] and A. D. S. Carter [15].

1. C_p is the cost of preventive replacement.
2. C_f is the cost of forced outage replacement.

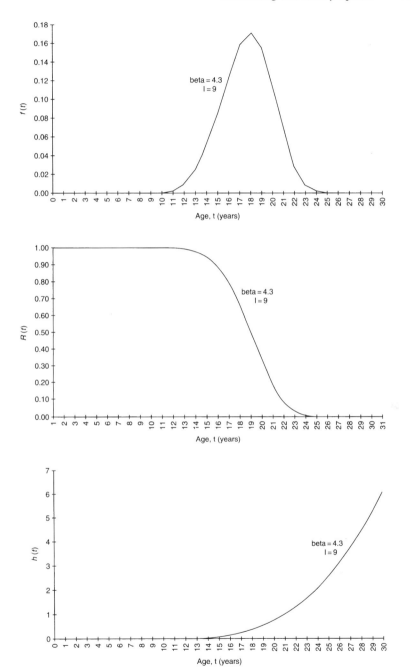

Figure 4-10. (a) Probability density curve for large motor windings; (b) reliability curve for large motor windings; (c) hazard curve for large motor windings.

3. $f(t)$ is the probability density function of the failure times of the motor windings.
4. The replacement strategy is to preventively replace the motors or their windings once they have reached a specified age t_p. Also, there will be replacement upon failures as necessary. This strategy is shown in Figure 4-11.
5. The goal is to determine the optimal replacement age of the motor windings to minimize the total expected replacement cost per unit time.

The equation describing the model of relating replacement age t_p to total expected replacement cost per unit time is:

$$C(t_p) = \frac{C_p \times R(t_p) + C_f \times [1 - R(t_p)]}{t_p \times R(t_p) + \int_{-\infty}^{t_p} tf(t)dt} \tag{4.17}$$

For the motor winding replacement case:

$$C_p = 100\,k\$$$
$$C_f = 1600\,k\$\ 1125\,k\$ = 1725\,k\$$$

The numerical solution to the problem is presented in Table 4-3. The various columns of Table 4-3 show the values of the variables in equation 4.17 as a function of t_p. Finally, Figure 4-12 illustrates $C(t_p)$ and shows that the optimal decision would have been to preventively rewind the company's large motors after 11–12 years.

The petrochemical company obviously missed out on optimizing its large motor rewind strategy, given the validity of the Weibull function based model. The question arose whether or not it would now be economical, into the 19th year of their large motor operations, to plan for a

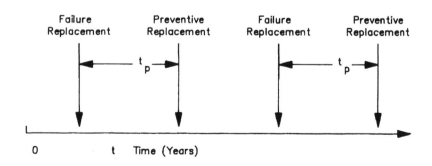

Figure 4-11. Large motor replacement strategy.

Table 4-3
Calculation results for large motor replacement case

t_p	$R(t_p)$	$1 - R(t_p)$	$\int t f(t_p)dt$	$C(t_p)$	$h(t_p)$	Incentive k$
0						
1	1.00	0.00	0.00	100.00	0.00	0.00
2	1.00	0.00	0.00	50.00	0.00	0.00
3	1.00	0.00	0.00	33.33	0.00	0.00
4	1.00	0.00	0.00	25.00	0.00	0.00
5	1.00	0.00	0.00	20.00	0.00	0.00
6	1.00	0.00	0.00	16.67	0.00	0.00
7	1.00	0.00	0.00	14.29	0.00	0.00
8	1.00	0.00	0.00	12.50	0.00	0.00
9	1.00	0.00	0.00	11.11	0.00	0.00
10	1.00	0.00	0.00	10.01	0.00	0.46
11	1.00	0.00	0.02	9.27	0.00	4.57
12	0.99	0.00	0.09	9.28	0.01	17.40
13	0.98	0.02	0.32	10.68	0.03	44.96
14	0.94	0.06	0.86	14.26	0.05	93.90
15	0.87	0.13	1.89	20.76	0.10	171.37
16	0.76	0.24	3.59	30.55	0.17	285.02
17	0.62	0.38	6.02	43.29	0.26	442.84
18	0.45	0.55	9.00	57.71	0.38	653.20
19	0.29	0.71	12.09	71.68	0.54	924.80
20	0.15	0.85	14.75	82.92	0.73	1266.61
21	0.07	0.93	16.60	90.11	0.98	1687.89
22	0.02	0.98	17.60	93.58	1.27	2198.16
23	0.00	1.00	18.00	94.76	1.63	2807.17
24	0.00	1.00	18.12	95.02	2.04	3524.90
25	0.00	1.00	18.14	95.06	2.53	
26	0.00	1.00	18.15	95.06	3.09	
27	0.00	1.00	18.15	95.06	3.73	
28	0.00	1.00	18.15	95.06	4.46	
29	0.00	1.00	18.15	95.06	5.28	
30	0.00	1.00	18.15	95.06	6.20	

preventive rewind of their three oldest motors during an upcoming shut-down. Using the relationship: Incentives or cost of insurance $= C_f \times h(t_p)$ annual penalties for the next five years* were determined as shown in column 7 of Table 4-3. These amounts were in turn claimed as credits in a discounted cash flow (DCF) analysis. They felt it was a sound decision to preventively rewind their three old motors during the shutdown.

---

* The average time between process unit shutdowns.

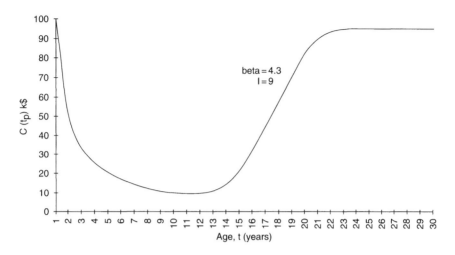

Figure 4-12. Expected replacement cost as a function of time.

Our second example involves the analysis of predominant failure regimes of process plant machinery.[†] Here our goal was to arrive at appropriate intervals for rebuilding pumps and motors in a major petrochemical plant.

The maintenance philosophy in this plant required that pumps and motors be rebuilt on a periodic basis of either the running hours or time in service. The rebuild criterion was water pumps to be rebuilt after 8000 hr, crude oil pumps after 16,000 hr, and motors after 20,000 hr. A unit was also rebuilt after five years if it had not reached its run-time limit. The primary assumption behind a criterion such as this is that machines deteriorate or wear out with time or during operation and should be removed from service prior to failure.

In order to evaluate the rebuild criteria just mentioned, the bathtub curve must be determined that characterizes the life of pumps and motors by establishing the relationship of failure rate with time.

Here, Hazard Analysis is used to analyze the run-time data to establish the predominant failure distribution and the median life of the units. The time to failure for each incident is plotted against the summation of the hazard function. This is shown in Figure 4-13. When the data are plotted on log-log format, the results are somewhat similar to that of the bathtub curve.

If proper Weibull hazard plotting paper, as shown in Figure 4-9, is not available, β can be determined as the reciprocal of the slope of the

---

[†]Courtesy James E. Corley, the MITRE Corp., New Orleans, Louisiana.

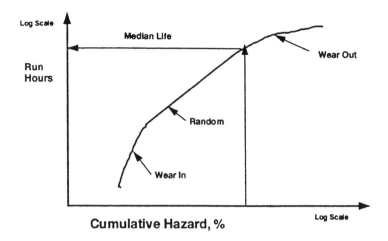

Figure 4-13. Hazard analysis plot.

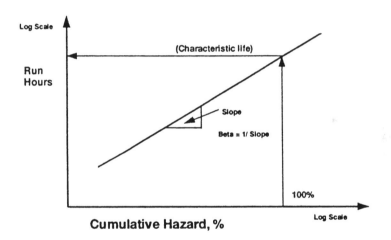

Figure 4-14. Hazard analysis plot showing the Weibull parameters, h and b.

plotted data as demonstrated in Figure 4-14. Consequently, for a hazard that decreases with time, a "wear-in" failure rate, the data will plot on a curve that has a slope greater than unity ($\beta < 1$). For failures that follow a constant failure rate distribution, the data points will fall on a curve that has a slope of unity ($\beta = 1$), and for "wear-out" distributions, the slope is less than unity ($\beta > 1$).

For large populations, all three distributions may be present and the data will resemble a bathtub that is inverted and rotated upward by 45°.

However, when the data plot as a straight line on log-log coordinates, a Weibull distribution can be fitted to the data.

Because the failures of the pumps and motors in this study generally have several different failure modes, the data may not fall on a straight line that can be fitted to a single Weibull distribution. For those cases, the life of the population will be taken as the median life (μ), or that time when 50% of the units have failed. On a hazard plot, the median life is determined where the fitted curve crosses the "Cumulative Hazard" value of 100%, which is equivalent to a 63.2% failure probability.

As in our previous example, we have to analyze data that are multiply censored, that is, failure data that are incomplete. Our group of machines being analyzed may have many machines that have not yet failed. They either may still be in operation or may have been removed from service for reasons other than failure. The machines that were removed for rebuild are in this category. This group of machinery has accumulated significant running hours, and although these machines have not failed, their running time must be factored into any life estimate.

Data Sources

Several sources of data were used to determine the life of pumps and motors. The most useful information is contained in the "Run-time Report," a report that is generated from a computer data base that contains the current running hours of the equipment that was installed previously in a given location. From this information a hazard analysis can be made. However, the use of the Run-time Report by itself as a source of failure data has a significant weakness in that it does not record whether a machine was removed because it failed or because it was taken out to be rebuilt or for some other reason.

To determine if a machine was removed from service because of failure or because it had met one of the rebuild criteria, a report was available in which the rebuilds were scheduled for work based upon their running or installed hours. It is general practice to schedule a rebuild when a machine reaches 80% of the run-time criteria or when it would exceed the 5-year criteria in the next year. This report is issued yearly and lists those machines that are scheduled for rebuild in the coming year. By cross-referencing information in both the Run-time Report and the Rebuild Schedule, it was generally possible to determine whether a machine was removed because it met a rebuild criterion or whether it failed. If a unit could not be found on the Rebuild Schedule and conversations with field personnel could not rule out a failure, the machine was considered to have failed in service. Because of the state of the data, a degree of judgment was sometimes required in making this assessment. As long as the number of data points is large, a few errors in the data should not significantly influence the overall conclusions drawn from the analysis. This would not be true if a small number of machines or data points was considered.

The Rebuild History Report summarizes the run-time of a given unit every time it is rebuilt. The data from this report were used to determine the effect that rebuilding had on machine life. Although the report included information for both pumps and motors, only motors were considered in this analysis. Pumps were excluded from this analysis because the data included a mix of carbon steel and stainless steel (SS) pumps. A material change for some pumps occurred during the time period of the data. It was felt that the improvement in pump life due to the introduction of SS pumps precluded a meaningful comparison of the units based on the number of rebuilds.

For the purpose of determining the failure rate and life of the equipment, the Rebuild History Report suffers from the same problems as the Run-time Report. The data do not indicate if the motor had failed prior to removal from service. The run-time data from the report were cross-referenced with those of the Rebuild Schedule to arrive at a judgment as to whether or not a machine had failed. Because of the large number of data points, some errors in discriminating between failures and non-failures should not affect the overall conclusions. However, for machines with over five rebuilds, there are too few data points for valid conclusions to be drawn.

The documentation of rebuild information in the Rebuild History Report did not include rebuild information prior to the date that the report was issued. Much of the equipment had been reworked several times before that date; thus, any analysis concerning the cumulative hours and rebuilds on the motors would be incomplete. The analysis cannot therefore determine the life of the original motors and is thus restricted to effect of rebuilds on machinery life after several unknown number of rebuilds. This limitation is of a particular concern when the existing motor life is compared with industry experience.

Analysis of Run-Time Data

To assist in manipulating the data for a Hazard Analysis, several Relational Data Bases (RDB) were constructed that contained machinery information to correlate pump and motor reliability. For both pumps and motors, a RDB was constructed that included such factors as location, machine type, service, speed, and performance. A separate "Run-time" RDB was produced that included such data as location, run-time, and failure mode. These

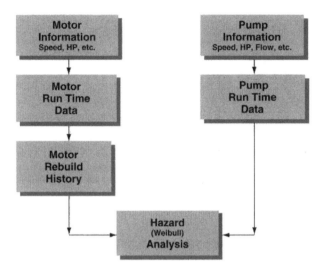

Figure 4-15. Analysis process linking relational data bases with hazard analysis.

two were linked through the location parameter to produce a file that was compatible with a Hazard Analysis Program. A third "Rebuild" RDB for motors that deals with the number of documented rebuilds was also produced. This process is illustrated in Figure 4-15.

To perform an analysis, the data base with the particular characteristics of interest and the data base with the failure information are linked through the equipment location number to extract the running hours and failure mode. This data is exported as a file to the Hazard Analysis Program that produces the plot. This method allows for the investigation of a wide variety of questions on equipment reliability and is limited only by the amount of data. For example, the analysis can determine the life and predominant failure mode of all crude oil pumps at a particular site. Another very powerful use of this analysis is the capability of determining the results of design modifications.

Analysis of Pumps

To evaluate the rebuild criteria for pumps, the Pump RDB and the Pump Run-time RDB were linked and the data extracted using a select criterion of pump service. Pumps in water and crude oil services were selected for this study.

The Hazard Analysis Plot for the water pumps is given in Figure 4-16. This figure indicates that the water pumps have a median life of

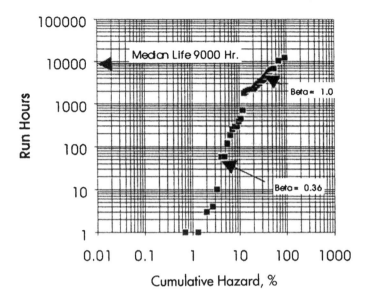

Figure 4-16. Hazard analysis of water pumps.

approximately 9000 hr compared to the rebuild criteria of 8000 hr. Half of the water pumps fail before 9000 hr and half after that time.

For the first portion of the curve in Figure 4-16, the slope of the data is approximately 2.7 ($\beta = 0.36$), indicating a "wear-in" type of failure mode. At about 1800 running hours, the failure mode changes to a constant failure rate mode. At no time do the data indicate that the pumps are failing in a predominantly "wear-out" mode.

The significant number of machines that fail before 1800 hr indicates that either the pumps are not being rebuilt properly or that they are being installed incorrectly. However, the lack of documentation of failures precludes an analysis that might determine the reasons for such early failure. Although a statistical analysis such as this cannot offer detailed explanations of the cause of failures, it can determine general failure modes. The number of machines that are shown to suffer from a wear-in failure mode raises a strong concern about the operating and cost-effectiveness of the current rebuild philosophy. Removing a machine from its location, disassembling, reassembling, storing, and reinstalling exposes a machine to significant risk of damage and mishap.

At about 1800 hr, the figure shows a significant change in the slope of the data. The slope becomes unity ($\beta = 1.0$), which indicates the failures become random with time. That is, the chance of a failure becomes independent of the time that the machine went into service. An example

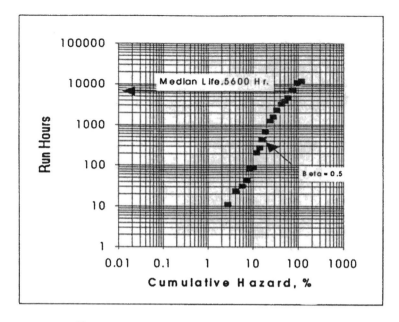

Figure 4-17. Hazard analysis of crude oil pumps.

of this mechanism would be the accidental closing of a discharge valve that caused a pump to run dead-headed and fail. The timing of the valve closing would not, in general, be a function of when the unit went into service, and thus the failure is random. Many operational and environmental causes of failure fall into this category.

The analysis of crude oil pumps shows a somewhat similar pattern to that of the water pumps. The hazard plot for these data is shown in Figure 4-17. The plot indicates that the median life of these units is only about 6600 hr. There is no indication in the data that the crude oil pumps have significant wear-out modes. This is not to say that a few individual pumps might not wear-out, only that the bulk of the population suffers from a wear-in or random failure. The longer the pumps are left in service, the lower the overall failure rate becomes. The plot of failure data in Figure 4-17 has a slope of approximately 2.0 ($\beta = 0.5$) indicating a wear-in mode over most of the life of the machines.

Analysis of Motors

The run-time data were analyzed for all critical motors. The hazard analysis plot is shown in Figure 4-18. This plot indicates that the median life of the motors is approximately 13,000 hr. Up until that time, the

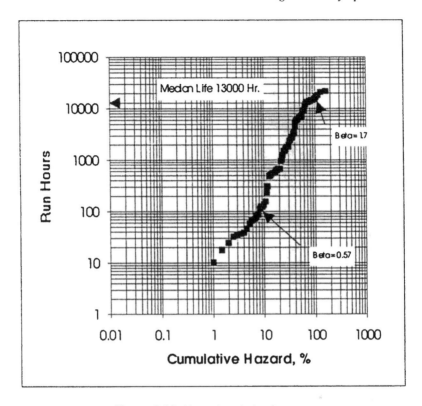

Figure 4-18. Hazard analysis of motors.

failures are predominantly a "wear-in" or "infant mortality" mode. At about 13,000 hr, the plotted data change to a β that is greater than one, indicating that the motors begin to wear out. This is seen in Figure 4-18, where the data shifts from a slope of approximately 1.7 ($\beta = 0.57$) to a slope of about 0.6 ($\beta = 1.7$). Of all of the cases that were analyzed, motors seem to be the only ones that reach a significant wear-out mode that might justify a rebuild philosophy based upon running hours. However, even this conclusion may not be valid because rebuilding the motors, like the pumps, introduces significant "infant mortality" failures.

Analysis of Rebuild Data

In addition to the run-time data that were analyzed above, there also exists information that relates the life of pumps and motors as a function of the number of times that they are rebuilt. Here, only the motors will be studied. The pump rebuild information has run-time data that includes a mix of carbon steel and stainless steel cases. The introduction of SS pumps, while significantly improving the life of the units, complicates the analysis when attempting to study the effects of pump rebuild. The rebuild data consists of a group of carbon steel pumps that includes an unknown number of rebuilds and cumulative running hours and SS pumps that are newer. Attempting to separate all of these variables to determine the effect of rebuilds on pumps was judged to be unproductive. Had the failure and run-time data been available at the beginning of the plants and been more complete, the hazard analysis tools would have been invaluable in quantifying the improvements in reliability of the pumps due to the material change.

The rebuild data for the motors were analyzed in a similar manner to that described above for the run-time data. To study the effect that rebuilding has on a machine's life, the data were organized by the number of times the motor had been rebuilt. The data are only valid since 1981; therefore, the first "documented" rebuild probably does not represent the first "actual" rebuild. For the purposes of this study, the first "documented" rebuild will be referred to simply as the "first" rebuild.

As was the case with the run-time data, the information on the rebuild does not indicate that a machine was removed because of failure. Using the same criteria as described above, it was assumed that a machine had not failed if it were near its rebuild run-time or five-year criteria or if it was on the Rebuild Schedule. Otherwise, it was assumed that a machine had failed for some reason. It is recognized that this assumption will introduce some error in the analysis; however, it is felt that this would not significantly change the conclusions as long as the population under consideration was reasonably large.

Figure 4-19 presents the Hazard Analysis of the rebuild data for the first four documented rebuilds of the motors. There were up to seven rebuilds for some motors, but it was felt that the sample population was not large enough to use without introducing significant errors. As seen in the plots, the median life decreases with the number of rebuilds. This is seen more clearly when the median life of each rebuild is plotted in Figure 4-20.

In this figure, the motor life for the first rebuild is approximately 14,000 hr and decreases to about 1600 hr for the fourth rebuild. This data indicates that every time the motors are rebuilt or reconditioned, the life decreases and motors are not brought back to "like new" condition. Figure 4-20 indicates that many of the motors may have reached the end of their economic life and should be replaced. A motor rebuild cost can be substantial, and the data shows that after about three rebuilds, this expenditure extends the life of the machine only a few months of run-time.

Examination of the individual curves in Figure 4-19 indicates that only the motors represented by the first rebuild have a significant wear-out

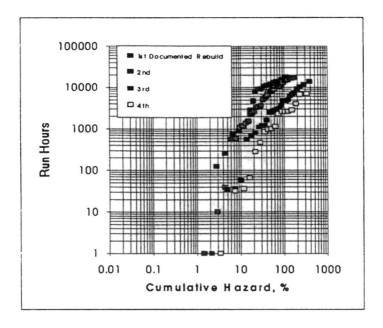

Figure 4-19. Hazard plots of rebuilt motors.

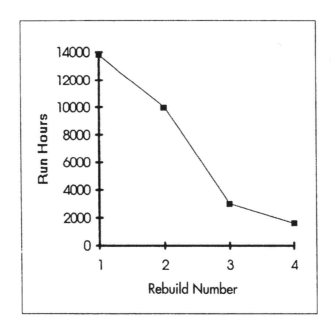

Figure 4-20. Median motor life as a function of number of rebuilds.

mode and that all of the motors are subjected to infant mortality failures. For the first rebuild case, the failures shift from a wear-in to a wear-out mode that occurs at approximately 9000 hr. After this first documented rebuild, the failures are "wear-in" changing to "random." As was concluded for the current motor population discussed earlier, a large percentage of machines suffer from early failures after they are reinstalled. Because the failure documentation system was incomplete, the reasons or causes for these failures could not be determined. However, common causes of a high infant mortality in machines are generally recognized as due to poor or marginal design, poor assembly, or bad installation. For any of these conditions, a machine can fail soon after installation.

The current state of the failure history documentation does not permit a determination of the exact causes for the decrease in life of the motors because they are repeatedly rebuilt. However, some general conclusions can be made. For pumps, a rebuild that replaces all of the internal components can restore it to a "like new" condition, at least in theory. This is not the case for a motor. The normal reconditioning of a motor does not replace the stator windings and insulation, which will continue to degrade with time. Thus, the insulation on a motor that has been rebuilt several times may be over 15–20 years old and approaching the end of its useful life. When the motor is rewound, it still is not totally restored. To remove the old winding and insulation, the stator is heated in an oven to burn out the insulation. Although care is taken to avoid any damage to the rest of the stator, some deterioration is unavoidable in the thin varnish insulation between the stator iron laminations. When this insulation is damaged, eddy current losses increase and the motor will run hotter than before the rewind.

Another element of motors that is not corrected by a rebuild is rotor bar thermal fatigue. Each motor start will subject the rotor bars to high temperature and cyclic stresses. Eventually this can cause rotor bar cracking and failure.

The fact that motors are never fully reconditioned is probably reflected in the decrease in life shown in Figure 4-20. However, because the rebuild data does not include the early portion of a motor's life, it was not always possible to separate the effects of age from the number of rebuilds.

The availability estimates currently used to establish sparing levels was based upon the assumption of a constant failure rate. However, the rebuild philosophy had the effect of resetting the clock on the machines and keeping them in a wear-in mode where the actual failure rate is higher than that assumed. Because the failure rate is shown to be a function of time, an accurate assessment of the impact on machinery availability must consider the installed time for individual machines except in the random regime.

Life Comparison With Other Industrial Experience

The current life of the motors in the study does not compare favorably with the experience of industry. Two industrial surveys of motor life and the factors that influence life are presented in Institute of Electrical and Electronic Engineers (IEEE) studies [16, 17, 18] and Electrical Power Research Institute (EPRI) studies [19, 20]. The IEEE study includes reliability data for a population of 1141 motors, 200 horsepower (hp) and above. This study covers a wide range of different types of industrial users. The EPRI study addresses only the utility industry, but covers a large population of 6312 motors. The EPRI study considers only motors of 200 hp and above. Of the two, the IEEE study, with its greater diversity of industrial applications and exposed equipment, is probably more representative of the type of service and environment that is found in the facilities of this study.

A comparison of the average life of motors in industry as compared to motor life in the study is shown in Figure 4-21. In this figure, the life of motors in the IEEE study is 14 years, and the life in the EPRI study is 31 years. This compares with the life of the study motors of 1.4 years, and for motors that have been rebuilt at least 4 times of 0.18 years. It should be noted that the IEEE and the EPRI studies deal with average or mean life, and this data is for median life. For wear-out types of failures, the mean and the median are the same. However, for populations such as those of this study, where the predominant failures are wear-in or random modes, the median life is less than the mean or average.

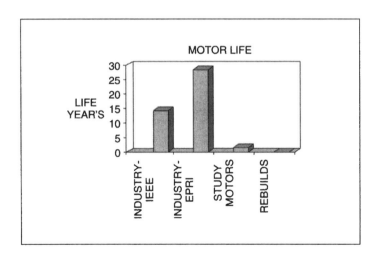

Figure 4-21. Comparison of life of study motors with industrial experience.

Conclusion

The statistical analysis used in this study indicated that the rebuild philosophy used to maintain the population of pumps was detrimental to the reliability of the units. Removing the machines from service based upon an arbitrary schedule of running time or time in service introduced failure modes that resulted in an increase in infant mortality. It was recommended to the client that a predictive maintenance program be implemented in the plants so that maintenance would be performed based upon the measured condition of the machinery. Also, early detection of faults with a monitoring program would allow for repairs to be performed *in situ* at much less expense than totally rebuilding a unit at an outside facility.

The failure analysis and maintenance documentation system for the facilities did not permit a detailed analysis of the failure modes of the equipment. It was recommended to the client that this system be upgraded so that the specific problem areas could be addressed in the future.

References

1. Czichos, H., *Tribology—A Systems Approach to the Science and Technology of Friction, Lubrication and Wear*. New York: Elsevier, 1978.
2. Yoshikawa, H. and Taniguchi, N., Fundamentals of mechanical reliability and its application to computer-aided machine design. *Annals of the CIRP*, **24** (1), 1975, p. 300.
3. Green, A. E. and Bourne, A. J., *Reliability Technology*. New York: Wiley-Interscience, 1972.
4. Henley, E. J. and Kumamoto, H., *Reliability Engineering and Risk Assessment*. Englewood Cliffs: Prentice-Hall, 1981.
5. BSI BS 5760: Part 2:1981, *Reliability of Systems, Equipments and Components. Part 2. Guide to the Assessment of Reliability*. London: British Standards Institution, 1981, pp. 9–11.
6. Fleischer, G., Probleme der Zuverlässigkeit von Maschinen. *Wiss. Z. TH Magdeburg*, **16**, 1972, p. 289.
7. British Standards Institution, op. cit., p. 12.
8. Nelson, N. W. and Hayashi, K., Reliability and economic analysis applied to mechanical equipment. *Journal of Engineering for Industry*, February 1974, pp. 311–316.
9. Sidall, J. N., *Probabilistic Engineering Design*. New York: Marcel Dekker, 1983.
10. Ascher, H. and Feingold, H., *Repairable Systems Reliability*. New York and Basle: Marcel Dekker, 1984.
11. Nelson, W., *Applied Life Data Analysis*. New York: John Wiley, 1982.
12. Nelson, W. B., *A Method for Statistical Hazard Plotting of Incomplete Failure Data That Are Arbitrarily Censored*. Schenectady: GE R&D Center TIS Report 68-C-007, January 1968.
13. King, J. R., *Probability Charts for Decision Making*. New York: Industrial Press, 1971.

14. Jardine, A. K. S., *Maintenance, Replacement and Reliability*. Bath, UK: Pitman Press, 1973.
15. Carter, A. D. S., *Mechanical Reliability*. New York: John Wiley, 1972.
16. IEEE Committee Report, *Report of Large Motor Reliability Survey of Industrial and Commercial Installations, Part I*, IEEE Transactions on Industry Applications, Vol. 1A-21, No. 4, July/August 1985.
17. IEEE Committee Report, *Report of Large Motor Reliability Survey of Industrial and Commercial Installations, Part II*, IEEE Transactions on Industry Applications, Vol. 1A-21, No. 4, July/August 1985.
18. IEEE Committee Report, *Report of Large Motor Reliability Survey of Industrial and Commercial Installations, Part III*, IEEE Transactions on Industry Applications, Vol. 1A-23, No. 1, January/February 1987.
19. Albrecht, P. F., Appiarius, J. C., McCoy, R. M., Owen E. L. and Sharma, D. K., *Assessment of the Reliability of Motors in Utility Applications*, IEEE Transactions on Energy Conversion, Vol. EC-2, No. 3, September 1987, March 1986.
20. Albrecht, P. F., Appiarius, J. C., McCoy, R. M., Owen E. L. and Sharma, D. K., *Assessment of the Reliability of Motors in Utility Applications—Update*, IEEE Transactions on Energy Conversion, Vol. EC-1, No. 1, March 1986.

Chapter 5
Is there a universal approach to predicting machinery uptime?

In the preceding chapter we showed the usefulness of hazard functions in estimating machinery reliability. Frequently, it is not possible to arrive at an appropriate distribution function due to a lack of specific data and the need for complicated calculations. In many cases, and especially when comparing competing solutions to a technical problem (i.e., *relative* reliability), a constant failure rate for machinery components may be assumed and judiciously applied.

A *constant failure rate* assumption does not deviate too much from the real world for at least two reasons. First, different distribution functions for a variety of components when combined produce a random failure pattern. Second, repair at failure tends to produce a constant failure rate when the population is large. This has been demonstrated in the literature [1].

With a constant failure rate the reliability of components or systems follows the exponential distribution:

$$R(t) = \exp(-\lambda t) \tag{5.1}$$

We have already seen that the reciprocal of failure rate is called Mean Time Between Failure (MTBF), or μ, the mean of the distribution. For example, small electric motors have typical failure rates of $\lambda = 14.3 \times 10^{-6}$/h. What is the MTBF of the motor and what is its reliability for a 8000-h operating period?

$$\text{MTBF} = \frac{1}{\lambda} = 8 \text{ years}$$

$$\text{Reliability, } R(t) = \exp(-14.3 \times 8000 \times 10^{-6})$$

$$= 0.891 \quad \text{or} \quad 89.1\%$$

We assume that these motors cannot be repaired and have to be scrapped when they fail. We would like to determine the operating time after which these motors have to be exchanged to assure a survival probability of $R(t_r) = 80\%$ based on a yearly run-time of 8000 hr.

From Equation 5.1 follows:

$$t_r = \text{MTBF} \times \ln(1/R(t_r)) = 8 \times 8760/8000 \times \ln(1/0.80) = 1.95 \text{ years}$$

The motors should be exchanged after approximately 2 years time.

Reliability of Parts In Series

The reliability of parts or components in series is:

$$R_s = R_1 \times R_2 \times \cdots \times R_n \tag{5.2}$$

$$= \exp[-(\lambda_1 + \lambda_2 + \cdots + \lambda_n)t] \tag{5.3}$$

If the components have identical failure rates, then

$$R_s = \exp(-n\lambda t) \tag{5.4}$$

Usually, this approach leads to a demand for very high component reliability in any system consisting of many parts (Fig. 5-1). For instance, we see that in order to obtain an 80% reliability in a unit with 50 components in series, an average component reliability of 99.4% is required.

A. S. Carter [2] describes a simple everyday experience of automotive transport that suggests that this approach is oversimplified:

At peak hour traffic conditions 20 to 30 vehicles may be held up at a traffic light. Each vehicle has at least 100 components in series in its transmission system, giving some 2000 components in series at each traffic light. Yet how often does the queue fail to move when the traffic lights change to green due to mechanical failure? Chaddock [3] has carried out a more scientific investigation of the supposed correlation between reliability and number of components, studying a number of weapons for which accurate data existed. He concludes there is no such correlation.

The truth is that success is achieved when the weakest or least adequate individual component of a system is capable of coping with the most severe loading or environment to be encountered, that is the strength of the chain equals that of its weakest link. This has been emphasized by other researchers who at the same time recognize the fact of variability, or scatter, both in the capability or strength of the product. It has been further emphasized in the duty it will have to face, that is the load which will be imposed on it.

The author then goes on to explain this phenomenon by a model in which both load and strength are distributed and where the strength distribution, due to some form of progressive weakening, invades the load distribution causing more and more of the population to fail (see Fig. 5-2). Carter's work also shows that where a component proves inadequate in changed duty or environment, it has only to be strengthened a little to restore the failure rate to an acceptable level. This goes to show that a "management-by-exception" approach to machinery reliability assessment is justified, that is vulnerabilities, as we will see later, have to be exposed.

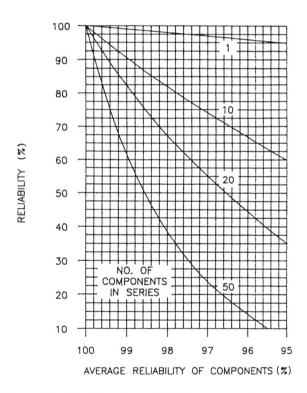

Figure 5-1. Overall system realiability: components in series.

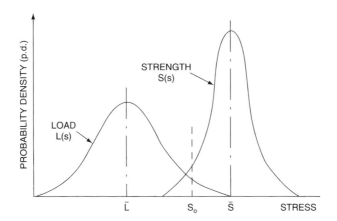

Figure 5-2. Distribution of load and strength [2]. (Reprinted from Carter, A.D.S., *Mechanical Reliability*, 1972, p. 5, by courtesy of Macmillan Press Ltd.)

Concepts as shown in Figure 5-1 have nevertheless sometimes led to unjustified waste in the process industries by providing spares, for instance, that are poorly or not at all utilized. We are alluding to cases where spares are almost "automatically" furnished without prior evaluation of the alternatives. The alternatives are to procure machinery reliable enough so that spares are not required, or to weigh the risks of not furnishing spares against the incentives of providing them [4].

Two Components In Parallel

The combined reliability of two identical components in parallel depends on the system requirement (Fig. 5-3). Two cases are possible. First, the failure of either component disables the system. Both *A* and *B* must survive. They are in series from the reliability point of view:

$$R_s = R_A \times R_B = \exp(-2\lambda t) \tag{5.5}$$

Second, survival of one component is sufficient. Here, system reliability (R_s) is the probability that *A* or *B* or both survive:

$$R_s = R_A \times R_B - R_A R_B \tag{5.6}$$

$$= 2\exp(-\lambda t) - \exp(-2\lambda t) \tag{5.7}$$

which is valid for identical components.

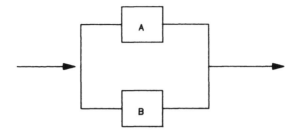

Figure 5-3. Diagram of two components in parallel.

An example for this case can be found in petrochemical pumping services. Here the need for parallel redundancy is based on cost of lost production, cost of unscheduled versus emergency repairs, and capital cost [5]. In order to determine the need for a spare or standby pump, one would first evaluate equation 5.1. With a failure rate of $\lambda = 1.5$/year, the resulting reliability referred to one year would be:

$$R_s = \exp(-1.5 \times 1) = 0.22 \text{ or } 22\%$$

That is, the probability of failure (P_f) would be:

$$P_f = 1 - R = 0.78 \text{ or } 78\%$$

Obviously, this is an unacceptable proposition.
We will now evaluate equation 5.7:

$$R_s = 2\exp(-1.5) - \exp(-3) = 0.446 - 0.050$$
$$R_s = 0.40 \text{ or } 40\%$$

This represents an improvement by almost a factor of 2. However, the result of equation 5.7 does not tell the whole story. Remember the definition – "Probability of survival of A or B or both." Obviously, we have to consider the fact that the system tends to fail only if the operating pump fails while the spare is out for repair. For $\lambda = 1.5$ per year, $\mu = 8$ months MTBF, $t = 5$ days repair time $(= 0.0137$ years$)$:

$$R_s = \exp(-1.5 \times 0.0137)$$
$$= 0.98 \text{ or } 98\%$$

or

$$P_f = 1 - 0.98 = 0.02$$

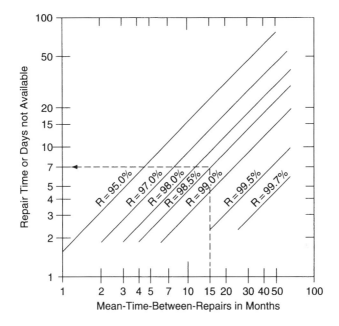

Figure 5-4. Reliability versus mean-time-between-failure and repair time-spared service.

Installing a spare pump in our system reduces probability of failure of the system during one operating year from 78 to 2%.

Often the repair quality of spared machinery is unduly compromised by shortening repair times as much as possible. Obviously, this is done intuitively in order to maintain reliability of the system. Figure 5-4 explains the relationship between MTBF of a spared machinery installation, time to repair the spare, and the resulting reliability factor. It assumes a "mature" machinery population, meaning that failures occur mutually independent of each other or perhaps as the result of some random outside influence such as the result of a unit startup or upset. We assume that no common failure causes exist, such as suction system or shared utility service problems.

Suppose we wanted to know how long a spared pump can be out for repair without endangering the process unit reliability goal which has been decreed to be 98.5%. The presently unspared pump was started and is running satisfactorily. It belongs to a population of similar pumps in similar service with an MTBF of 15 months. We move vertically from 15 on the horizontal axis and intersect the reliability line of 98.5% at a horizontal line corresponding to an allowable spare pump outage of 7 days. We conclude that there should be no need to rush the repair of the spare pump.

Three Identical Components In Parallel

1. Failure of any component disables the system illustrated in Figure 5-5:

$$R_s = R_A \times R_B \times R_C = \exp(-3\lambda t) \tag{5.8}$$

2. The system can stand failure of one component. Two or more components must survive.

$$R_s = 3R^2 - 2R^3 \tag{5.9}$$
$$= 3\exp(-2\lambda t) - 2\exp(-3\lambda t) \tag{5.10}$$

3. The system can stand failure of any two components. One of the three must survive.

$$R_s = 1 - (1 - R)^3 \tag{5.11}$$
$$= 1 - [1 - \exp(-\lambda t)]^3 \tag{5.12}$$
$$= 3\exp(-\lambda t) - 3\exp(-2\lambda t) + \exp(-3\lambda t) \tag{5.13}$$

4. Consider now a system as shown in Figure 5-6, where the failure of any one component can cause a system failure. Assume that the failure of any one part in this series system is independent of the

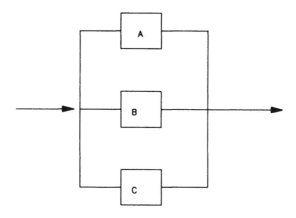

Figure 5-5. Diagram of three components in parallel.

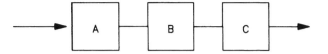

Figure 5-6. Diagram of components in series.

failure of another. The probability that the system will survive is, according to equation 5.2:

$$R_s(t) = R_A(t) \times R_B(t) \times R_C(t) \tag{5.14}$$

For an exponential time-to-failure density of each individual part, we can write:

$$R_s(t) = \exp(-\lambda_A t) \times \exp(-\lambda_B t) \times \exp(-\lambda_C t) \tag{5.15}$$

$$= \exp[-(\lambda_A + \lambda_B + \lambda_C) \times t] = \exp(-\lambda_s t) \tag{5.16}$$

Prediction Procedures

Most reliability engineering prediction procedures are based upon the above described exponential time-to-failure density. This permits simple addition of average component failure rates in order to arrive at the equipment or system failure rate from which the MTBF or reliability function may be obtained. The mathematical basis for this approach was demonstrated by equations 5.2 and 5.16. In this technique, we merely add the number of indispensable or non-redundant components of each type, multiply this by the basic average failure rate for each type of component, and add these figures to obtain the machinery unit failure rate. The MTBF is then the reciprocal of that failure rate.

As a more sophisticated approach for electronic systems MIL-HBK-217D [6] advocates the above method for predicting electronic equipment failure rates using data contained in it. The method depends on the quality of generic failure rates. These are derived from the equation:

$$\lambda_P = \lambda_b(\pi_E \times \pi_Q) \tag{5.17}$$

where λ_P = predicted failure rate,
$\quad \lambda_b$ = base failure rate for the generic part,
$\quad \pi_E$ = an environmental factor,
$\quad \pi_Q$ = a quality factor.

π_Q relates to sources and specified quality, π_E to the general environment in which the part will be used. The generic part values themselves are based on a large amount of data from laboratory, development, and field sources.

A similar source exists for mechanical generic data [7]. It subdivides the data by source so that the environment is known, but does not classify by quality. With its aid, however, a parts count for a piece of mechanical equipment can be performed, taking into account the environmental effects, providing that all the required data are in the lists.

Other methods take part stress levels into account. This is done, for example, by determining ratios of operating versus design pressure, operating versus design temperature, design size versus median size, or other important parameters. Relevant stress ratios are then weighted and the result applied to a suitable distribution function from which failure rates are determined [8].

Since our purpose is to make *relative* machinery reliability assessments we feel that a judicious application of the parts count method is justified.

We have already used the concept of expressing failure rates in terms of failures per 1 million hours. This would amount to 150 years if we assume continuous around-the-clock operation. This seems like quite a long time. However, we can look at it in another way. An equivalent experience would be if a process unit with 150 different kinds of failures had one outage a year.

Expressing failure rates per million hours is convenient because many failures occur at a rate of 1 per million operating hours. Machinery component failures would lie mostly between 1 and 100 failures per 1 million hours or $1-100 \times 10^{-6}$ hr. Table 5-1 illustrates how these failure rates relate subjectively to various levels of reliability.

Table 5-1
Numerical interpretation of subjective reliability terms

Reliability	$\lambda \times 10^{-6}$
Extremely reliable	0.01
Highly reliable, OK in large numbers	0.01–0.1
Good reliability for moderate numbers	0.1–1.0
Average reliability, OK in small numbers	1.0–10
Very unreliable	10–100
Intolerable	>100

Source: Atomic Energy of Canada Ltd.

Failure Rate Data

Failure rate data is best obtained from operating experience. Table 5-2 illustrates how failure rate data for machinery components can be obtained from field statistics. Column 2 shows the actual service experience of reciprocating compressors based on a company's experience in several plants. Column 3, the failure rates, are obtained by first postulating two incidents per year on these particular machines at a given plant site. The failure rates are then calculated by multiplying the field data percentages by the failure rate equivalent of two incidents per year (i.e., 228×10^{-6} hr).

Another important aspect of reliability prediction using failure rates is the consideration of failure modes. Failure modes have distinct failure rates and the component or part failure rate is the sum of its mode failure rate.

Failure modes are typically first a description of loss of function or malfunction and then a more detailed expansion in terms of the basic failure mode, namely the appearance of the failure (see Table 7-1). Earlier we looked at some basic failure modes in connection with failure distributions. We refer our readers to Table 4-1. Basic failure modes and the failure mechanisms associated with them play a central role in machinery failure analysis [9].

In using failure rate data for machinery reliability assessment it is a good idea to work with "worst," "best," and "expected" concepts. This reflects the fact that machinery parts and components can have different qualities. In order to make things less complicated we will calculate reliability based on two qualities – best and worst. We will then investigate if the worst case is viable. If that is the case, we need not worry because the actual quality will be closer to the expected value. If, however, our reliability based on the best case scenario is unacceptable, we have to take corrective action by looking for improved designs.

Table 5-3 lists failure rates for machinery components and parts as well as failure modes compiled from various literature sources and the

---

Table 5-2
Failure rate statistics: Reciprocating compressor

Elements	Failures (%)	Rate per 1×10^6 hr
Valves	43.0	98.4
Pistons and cylinders	19.0	43.0
Lube systems	18.0	41.0
Piston rods	10.0	22.8
Packings	10.0	22.8
Total	100.0	228.0[a]

[a] Equivalent to two incidents per year.

Table 5-3
Failure rates for machinery components [10–13]

	$\lambda(10^{-6})$	
	Best	*Worst*
1.0 Transmitting elements		
1.1 Couplings		
1.1.1 Elastomeric	20.0	30.0
1.1.2 Gear	8.0	20.0
1.1.3 Disc/diaphragm	0.01	0.1
1.2 Gear sets		
1.2.1 General purpose	8.0	50.0
1.2.2 High-speed helical	0.5	15.0
1.3 Shafts		
1.3.1 Lightly stressed	0.02	0.1
1.3.2 Heavily stressed	0.1	0.5
1.3.3 Crankshafts (R.C.)	5.0	8.0
1.4 Clutches		
1.4.1 Friction	2.0	8.0
1.4.2 Magnetic	4.0	10.0
1.5 Drive belts		
1.5.1 V-belts	20.0	80.0
1.5.2 Timing belts	40.0	80.0
1.6 Springs		
1.6.1 Lightly stressed	0.01	0.1
1.6.2 Heavily stressed	0.8	2.5
2.0 Constraining, confining, containing elements		
2.1 Bearings		
2.1.1 Sleeve bearings	4.0	10.0
2.1.2 Ball bearings	5.0	50.0
2.1.3 Roller bearings	3.0	10.0
2.2 Seals		
2.2.1 O-rings	0.1	0.7
2.2.2 Oil seals	8.0	10.0
2.2.3 Mechanical seals	25.0	200.0
2.3 Valves		
2.3.1 R.C. (Recip. comp.)	50.0	150.0
2.3.2 Check valves	0.8	10.0
2.3.3 Manual valves	0.4	6.0
2.3.4 Relief valves	1.0	10.0
3.0 Fixing elements		
3.1 Threaded fasteners		
3.1.1 Bolts	0.001	0.007
3.1.2 Pins	8.0	40.0
3.1.3 Set screws	0.03	1.0
3.1.4 Rivets	0.001	0.01

Table 5-3
Failure rates for machinery components–cont'd

	$\lambda(10^{-6})$	
	Best	*Worst*
4.0 Support elements		
4.1 Casings		
4.1.1 R.C. cylinder jackets	0.01	0.1
4.1.2 R.C. cylinder liners	10.0	30.0
4.1.3 Pump casings	0.01	1.0
4.2 Vibration mounts		
4.2.1 Elastomeric	6.0	20.0
4.2.2 Wire rope coils	0.1	1.0
4.3 Motor windings		
4.3.1 Small motors <250	10.0	20.0
4.3.2 Large motors >250	5.0	10.0
5.0 Basic failure modes		
5.1 Force/stress/impact		
5.1.1 Deformation	0.01	0.1
5.1.2 Fracture	0.001	0.01
5.1.3 Binding/seizure	0.1	1.0
5.1.4 Misalignment	0.1	1.0
5.1.5 Displacement	0.01	0.1
5.1.6 Loosening (fastener)	0.1	1.0
5.2 Reactive environment		
5.2.1 Corrosion		
1. Accessible parts	0.01	0.1
2. Inaccessible parts	0.1	1.0
5.2.2 Fretting		
1. Mostly stationary	0.1	1.0
2. Exposed to dirt	1.0	10.0
5.3 Temperature effects (see under aging)		
5.4 Time effects		
5.4.1 Wear/relative motion		
1. Non-lubricated	0.1	1.0
2. Lubricated	0.01	0.1
5.4.2 Erosion		
1. Accessible parts	0.01	0.1
2. Inaccessible parts	0.1	1.0
5.4.3 Aging		
1. Lubricants	0.01	0.1
2. Rubber	0.01	0.1
3. Metals, thermally stressed	0.1	1.0
5.4.4 Contamination		
1. Accessible parts	0.01	0.1
2. Inaccessible parts	0.1	1.0
5.4.5 Fouling/plugging		
1. High-velocity areas	0.01	0.1
2. Low-velocity areas	0.1	1.0

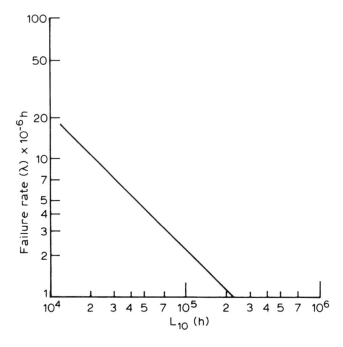

Figure 5-7. Failure rate for anti-friction bearings versus L_{10} life (adapted from [13]).

authors' experience. Figure 5-7 may be used to obtain failure rates for anti-friction bearings based on known or assumed L_{10} lives.

The Procedure

The procedure to calculate reliability based on failure rates is simple. It can be applied to predict reliability of machinery on the assembly, hierarchy, and system level within the limits of underlying assumptions. Figure 5-8 shows the form used in this effort.

The first step is to list all parts essential to the successful functioning of the system under study. The second step is to determine the quantity of parts. Third, after checking Table 5-3 for failure rate information on the specific component under consideration, determine the most probable failure mode for it; typically, fractures with shafts, wear or contamination with simple oil seals, and bearing or winding failures with motors, and so forth. If more than one failure mode is expected consider the mode with the highest failure rate. Compare this with the failure rate for the part if it is available. Multiply the highest failure rate in terms of best and worst and enter the values in the appropriate columns. It stands to reason that

Non-redundant components	Quantity	0.01 Best	0.01 Worst	0.1 Best	0.1 Worst	1.0 Best	1.0 Worst	10.0 Best	10.0 Worst	100.0 Best	100.0 Worst
1. Clutch cplg (3)											
Bearing	1			1			1				
Oil seals	2					2	2				
Coupling	1					1	1				
2. Gear Box (4)											
Bearing	2			2			2				
Oil seals	2					2	2				
Gear set	1	1			1						
3. Coupling (5)											
Elastomer	1					1	1				
4. Motor (6)											
Bearing	2			2			2				
Winding	1			1	1						
5. Gear box (7)											
Bearing	6			6			6				
Oil seals	2					2	2				
Gear set	3	3			3						
Sum best 9.24		0.04		1.20		8.00					
Sum worst 19.50					0.50		19.00				

Failure rate per 10^6 h

Figure 5-8. Calculating machinery failure rates.

this analysis will be as accurate as one is able to recognize the elements of a part and their corresponding failure modes.

The fourth step is to add the best and worst values separately. One has to determine now whether or not the sum of the worst values is tolerable. If the conclusion is affirmative no action is necessary. If the "worst quality" assumption is not tolerable we have to look for improvements. Usually, the individual failure rate values will be a clue! If there are some particularly high values try to substitute their "best quality" value and see how this affects the overall failure rate. If this does not satisfy our expectations we have to embark on a design change.

The example illustrated in Figure 5-8 pertains to a drive critical to the operation of a rotary furnace air pre-heater in a large process plant. The drive consisted of eight machinery components schematically shown in Figure 5-9. The reliability analysis was made in order to determine whether or not the drive was the weak element in an otherwise highly reliable and cost-effective scheme to recover waste heat.

Finally, how can we reduce component failure rates? The following questions need to be asked:

- Can the component be replaced by a known improved component?
- Has a review of the component design been done?
- Have all known weaknesses been eliminated?
- Have all uncertainties been identified?
- Are the uncertainties being eliminated by analysis or test?

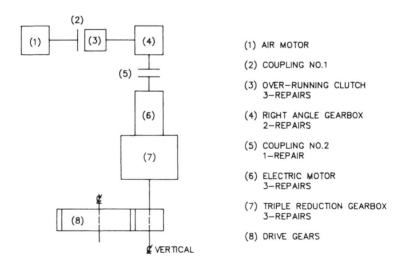

Figure 5-9. Rotary air pre-heater drive train.

- Can the design be simplified

    - by reducing the number of parts?
    - by eliminating need for high precision?
    - by requiring less maintenance skill?

- Have new features and new materials been proven by analysis or test?
- Can components tolerate abnormal conditions?

We have chosen to lump abnormal conditions under the acronym "FRETT." The letters F-R-E-T-T stand for:

F: Forces, mechanical loads, deflections, and pressure
R: Reactive agents
E: Environment
T: Temperature
T: Time, exposure to long-term and short-term loads, and deflections, i.e. vibration and shock.

Whenever F-R-E-T-T are outside the as-designed or anticipated values or quantities, the part, machine, or component will be prone to fail prematurely.

Current Methods of Predicting Reliability*

A reliability prediction is performed in the early stages of a development program to support the design process. Performing a reliability prediction provides for visibility of reliability requirements in the early development phase and an awareness of potential degradation of the equipment during its life cycle. As a result of performing a reliability prediction, equipment designs can be improved, costly over-designs prevented and development testing time optimized.

---

* By permission. From *The Handbook of Reliability Prediction Procedures for Mechanical Equipment*. It has been developed by the Logistics Technology support Group, Carderock Division, Naval Surface Warfare Center (CDNSWC) in Bethesda, Maryland. The handbook presents a new approach for determining the reliability and maintainability (R&M) characteristics of mechanical equipment. It has been developed to help the user identify equipment failure modes and potential causes of unreliability in the early design phases of equipment development, and then to quantitatively evaluate the design for R&M and determine logistics support requirements.

A software program called "MechRel" has also been developed by the Logistics Technology Support Group to automate the Handbook procedures and equations. The Handbook and MechRel software program are available free of charge and can be downloaded at http://wwwMechReLcom.

Performance of a reliability prediction for electronic equipment is well established by research and development. For example, MIL-HDBK-217 has been developed for predicting the reliability of electronic equipment. Development of this document was made possible because the standardization and mass production of electronic parts has permitted the creation of valid failure rate data banks for high population electronic devices. Such extensive sources of quality and reliability information can be used directly to predict operational reliability while the electronic design is still on the drawing board.

A commonly accepted method for predicting the reliability of mechanical equipment based on a data bank has not been possible because of the wide dispersion of failure rates which occur for apparently similar components. Inconsistencies in failure rates for mechanical equipment are the result of several basic characteristics of mechanical components:

a. Individual mechanical components such as valves and gearboxes often perform more than one function and failure data for specific applications of non-standard components are seldom available. A hydraulic valve, for example, may contain a manual shut-off feature as well as an automatic control mechanism on the same valve structure.

b. Failure rates of mechanical components are not usually described by a constant failure rate distribution because of wear, fatigue and other stress-related failure mechanisms resulting in equipment degradation. Data gathering is complicated when the constant failure rate distribution cannot be assumed and individual times to failure must be recorded in addition to total operating hours and total failures.

c. Mechanical equipment reliability is more sensitive to loading, operating mode, and utilization rate than electronic equipment reliability. Failure rate data based on operating time alone are usually inadequate for a reliability prediction of mechanical equipment.

d. Definition of failure for mechanical equipment depends upon its application. For example, failure due to excessive noise or leakage cannot be universally established. Lack of such information in a failure rate data bank limits its usefulness.

The above deficiencies in a failure rate database result in problems in applying published failure rates to an actual design analysis. The most commonly used tools for determining the reliability characteristics of a mechanical design can result in a useful listing of component failure modes, system level effects, critical safety related issues, and projected maintenance actions. However, estimating the design life of mechanical equipment is a difficult task for the design engineer. Many life-limiting failure modes such as corrosion, erosion, creep, and fatigue operate on the component at the same time and have a synergistic effect on reliability. Also, the loading on the component may be static, cyclic, or dynamic at different points during the life cycle and the severity of loading may also be a variable. Material variability and the inability to establish an effective database of historical operating conditions such as operating pressure, temperature, and vibration further complicate life estimates.

Although several analytical tools such as the failure mode, effect and criticality analysis (FMECA) are available to the engineer, they have been developed primarily for electronic equipment evaluations, and their application to mechanical equipment has had limited success. The FMECA, for example, is a very powerful technique for identifying equipment failure modes, their causes, and the effect each failure mode will have on system performance. Results of the FMECA provide the engineer with a valuable insight as to how the equipment will fail; however, the problem in completing a quantitative FMECA for mechanical components is determining the probability of occurrence for each identified failure mode.

The above-listed problems associated with acquiring failure rate data for mechanical components demonstrates the need for reliability prediction models that do not rely solely on existing failure rate data banks. Predicting the reliability of mechanical equipment requires the consideration of its exposure to the environment and subjection to a wide range of stress levels such as impact loading. The approach to predicting reliability of mechanical equipment presented in the handbook considers the intended operating environment and determines the effect of that environment at the lowest part level where the material properties can also be considered. The combination of these factors permits the use of engineering design parameters to determine the design life of the equipment in its intended operating environment and the rate and pattern of failures during the design life.

Development of the Handbook

Useful models must provide the capability of predicting the reliability of all types of mechanical equipment by specific failure mode considering the operating environment, the effects of wear, and other potential causes of degradation. The models developed for the handbook are based upon identified failure modes and their causes. The first step in developing the models was the derivation of equations for each failure mode from design information and experimental data as contained in published technical reports and journals. These equations were simplified to retain those variables affecting reliability as indicated from field experience data. The failure rate models utilize the resulting parameters in the equations, and modification factors were compiled for each variable to reflect its quantitative impact on the failure rate of individual component parts. The total failure rate of the component is the sum of the failure rates for the component parts for a particular time period in question. Failure rate equations for each component part, the methods used to generate the models in terms of failures per hour or failures per cycle, and the limitations of the models are presented. The models were validated to the extent possible with laboratory testing or engineering analysis.

The objective of the handbook and MechRel® software program is to provide procedures which can be used for the following elements of a reliability program:

- Evaluate designs for reliability in the early stages of development.
- Provide management emphasis on reliability with standardized evaluation procedures.
- Provide an early estimate of potential spare parts requirements.
- Quantify critical failure modes for initiation of specific stress or design analyses.
- Provide a relative indication of reliability for performing trade-off studies, selecting an optimum design concept or evaluating a proposed design change.
- Determine the degree of degradation with time for a particular component or potential failure mode.
- Design accelerated testing procedures for verification of reliability performance.

One of the problems any engineer can have in evaluating a design for reliability is attempting to predict performance at the system level. The problem of predicting the reliability of mechanical equipment is easier at the lower indenture levels where a clearer understanding of design details affecting reliability can be achieved. Predicting the life of a mechanical component, for example, can be accomplished by considering the specific wear, erosion, fatigue and other deteriorating failure mechanism, the lubrication being used, contaminants which may be present, loading between the surfaces in contact, sliding velocity, area of contact, hardness of the surfaces, and material properties. All of these variables would be difficult to record in a failure rate data bank; however, the derivation of such data can be achieved for individual designs and the potential operating environment can be brought down through the system level and the effects of the environmental conditions determined at the part level.

The development of design evaluation procedures for mechanical equipment includes mathematical equations to estimate the design life of mechanical components. These reliability equations consider the design parameters, environmental extremes, and operational stresses to predict the reliability parameters. The equations rely on a base failure rate derived from laboratory test data where the exact stress levels are known, and engineering equations are used to modify this failure rate to the appropriate stress/strength and environmental relationships for the equipment application.

As part of the effort to develop a new methodology for predicting the reliability of mechanical components, Figure 5-10 illustrates the method of considering the effects of the environment and the operating stresses at the lowest indenture level.

A component such as a valve assembly may consist of seals, springs, fittings, and the valve housing. The design life of the entire mechanical system is accomplished by evaluating the design at the component and part levels considering the material properties of each part. The operating environment of the system is included in the equations by determining its impact at the part level. Some of the component parts may not have a constant failure rate as a function of time and the total system failure rate of the system can be obtained by adding part failure rates for the time period in question.

Many of the parts are subject to wear and other deteriorating type failure mechanisms and the reliability equations must include the parameters

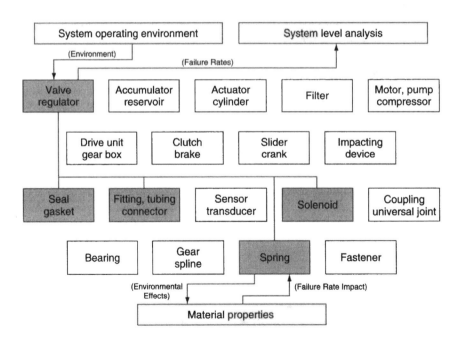

Figure 5-10. Mechanical Components and Parts.

which are readily accessible to the equipment designer. As part of this research project, Louisiana Tech University was tasked with establishing an engineering model for mechanical wear which correlated to the material strength and stress imposed on the part. This model for predicting wear considers the materials involved, the lubrication properties, the stress imposed on the part, and other aspects of the wear process [14]. The relationship between the material properties and the wear rate was used to establish generalized wear life equations for actuator assemblies and other components subject to surface wear.

In another research project, lubricated and unlubricated spline couplings were operated under controlled angular misalignment and loading conditions to provide empirical data to verify spline coupling life prediction models. This research effort was conducted at the Naval Air Warfare Center in Patuxent River, Maryland [15]. A special rotating mechanical coupling test machine was developed for use in generating reliability data under controlled operating conditions. This high-speed closed-loop test bed was used to establish the relationships between the type and volume of lubricating grease employed in the spline coupling and gear life. Additional tests determined the effects of material hardness, torque, rotational speed, and angular misalignment on gear life.

Results of these wear research projects were used to develop and refine the reliability equations for those components subject to wear.

Example Design Evaluation Procedure

A clutch assembly will be used to illustrate the Handbook approach to predicting the reliability of mechanical equipment.

Clutch Varieties

Clutches are made up of two basic components – the pressure plate and disc. The pressure plate supplies sufficient force or pressure to the disc so enough friction is developed to transmit torque to the driveline.

Friction clutches, although available in many different forms tend to be of the axial or rim type. Axial clutches operate where the movement is parallel to the axis of the shaft. Rim types operate where the movement is radial. Examples of the former are the plate and cone clutches. Examples of the latter include coil or wrap spring and chain clutches.

Plate clutches are divided into two designs – single and multi-plate. The single plate design is the type favored by automotive designers for transmission and light-to-medium power applications. The single plate is normally provided with a friction lining on each side of the disc. Multi-plate designs employ a number of discs lined on both sides, which serve to distribute the load over a large area. These types are used for high torque and high load applications. They required only moderate clamping pressures and are suitable for high-speed operation because their relatively small size generates lower centrifugal forces.

Cone clutches are used for smaller, medium power, low-speed transmission systems which may be subjected to rough usage. These devices cope well with such treatment because of their simple robust construction, and due to the fact that heat is dissipated more readily than with plate clutches.

Rim and block clutches employ various means of engaging the stationary half of the assembly through radial movement against the rim of the driving member. The action is similar to that of an internally expanding brake shoe.

Centrifugal clutches are often used with squirrel cage motors. The fabric facing may be fitted to shoes or blocks mounted on a spider which is keyed onto the driving shaft. The shoes or blocks are thrown outward by centrifugal force, engagement being automatic when a predetermined speed is reached from starting.

Coil or wrap spring clutches operate on the principle of a spring mounted on a drum being tightened. The action is much like that of a rope tightening around a revolving capstan. The design is compact, simple in construction, and is used where high torques are required from low power. For this reason, the clutches have found applications in small equipment such as plain paper copiers and, in their larger versions, for haulage gears and rolling mills and presses.

Chain clutches employ inner and outer friction rings in an oil-filled housing actuated by cams bearing on chain toggles which force the rings together.

Sprag clutches consist of a number of specially shaped steel springs or wedges which jam inner and outer races in one direction only. This action leads to their use for applications in overrunning (where the clutch acts as a free-wheel) and back-stopping. This design is particularly useful for intermittent rotary motion involving, for example, indexing or inching [16].

Materials classification divides the friction materials into organic and metallic groups. The organic group includes all materials composed of both asbestos and non-asbestos fibers and bound by some resin binder. The metallic group consists of all friction materials containing iron, copper, ceramic bronze, graphite, carbon, or other metallic material as the base material.

Clutch Failure Rate Model

The clutch system reliability model will contain the following component parts:

- actuators
- bearings
- clutch friction linings
- seals
- springs.

The total clutch system failure rate is the sum of the failure rates of each of the above component parts in the system:

$$\lambda_{CL} = \lambda_{AC} + \lambda_{BE} + \lambda_{CF} + \lambda_{SE} + \lambda_{SP} \qquad (5.18)$$

where λ_{CL} = total failure rate for the clutch system, failures/million hours

λ_{AC} = total failure rate for actuators, failures/million hours (see HB*)

λ_{BE} = total failure rate for bearings, failures/million hours (see HB†)

λ_{CF} = total failure rate for clutch friction materials, failures/million hours

λ_{SE} = total failure rate for seals, failures/million (see HB‡)

λ_{SP} = total failure rate for springs, failures per million hours (see HB¶)

* Chapter 9, † Chapter 7, ‡ Chapter 3, ¶ Chapter 4

Clutch Friction Material Reliability Model

In the following we are going to show an example of the principle of failure rate development for just one component of equation (5.18), namely λ_{CF}.

A list of failure modes for clutch friction materials is shown in Table 5-4. By using the clutch system beyond the life of the friction material, a drastic reduction of friction coefficient can occur. This rapid deterioration can result in a catastrophic failure of the clutch.

Under normal operating conditions, the friction materials used in clutches are reliable mechanical components. Like brake friction

Table 5-4
Clutch friction surface failure modes [17]

Problem	Characteristics	Causes
Dishing	Clutch plates distorted into a conical shape	Lack of conformability. The temperature of the outer region of the plate is higher than the inner region
Waviness or buckling	Clutch plates become buckled into a wavy platter	Lack of conformability. The inner area is hotter than the outer area
Banding or crushing	Loss of friction material at the ends of a band	Crushing and excessive wear of the friction material
Material transfer	Friction material adhering to opposing plate, often giving rise to excessive wear	Overheating and unsuitable friction material
Bond failure	Material parting at the bond to the core plate causing loss of performance	Poor bonding or overheating, the high temperature affecting bonding agent
Burst failure	Material splitting and removed from the spinner plate	High stresses on facings when working at high speeds
Grooving	Grooving of the facing material on the line of movement	Material transfer to opposing plate
Reduced performance	Decrease in coefficient of friction giving a permanent loss in performance	Excess oil or grease on friction material or on the opposing surface
Distortion	Facings out of flatness after high operating temperature	Unsuitable friction material

materials, the wear of clutch materials is dependent on the amount of accumulated energy dissipated by the mechanical component.

$$h = k \times p \times s \qquad (5.19)$$

where $h =$ change in thickness of the clutch friction material caused by wear in inches

$k =$ wear coefficient, $(\text{lb/in.}^2)^{-1} = k_o k_t,$

$k_o =$ wear coefficient at ambient temperature,

k_t = temperature influence factor for lining material,
p = nominal pressure between the clutch wear plates, psi $= P/A$,
s = sliding distance during clutch actuation, in $= v_s/t_a$,
v_s = sliding velocity in seconds,
t_a = actuation time in seconds

If the effective thickness of the clutch lining is d (inches), life of the clutch friction material is given by the following equation:

$$Life = \frac{d}{W_p} \tag{5.20}$$

where *Life* = number of applications before friction material is completely worn
d = lining thickness in inches
W_p = friction material wear per application, in

$$W_p = \frac{k_o \times P \times v_s \times t_b}{A} \tag{5.21}$$

and

$$\lambda_{CF,\ B} = \frac{1}{Life} = \frac{W_p}{d} = \frac{k_o \times P \times v_s \times t_b}{d \times A} \tag{5.22}$$

By normalizing equation (5.22) to those values for which historical failure rate data is available, the following failure rate model can be derived:

$$\lambda_{CF} = \lambda_{CF,\ B} \times C_{NP} \times C_T \tag{5.23}$$

where λ_{CF} = failure rate of the clutch friction material in failures/million hours
$\lambda_{CF,\ B}$ = base failure rate of the clutch friction material in failures/million hours
C_{NP} = multiplying factor which considers the effect of multiple plates on the base failure rate
C_T = multiplying factor which considers the effect of ambient temperature on the base failure rate

Base Failure Rate for Clutch Lining/Disk Material

The clutch friction material base failure rate, $\lambda_{CF,B}$, may be provided by the manufacturer of the clutch assembly. If not, then the base rate can be calculated from equation (5.22).

Clutch Plate Quantity Multiplying Factor

The correction factor for the number of plates is given by:

$$C_{NP} = \text{number of disks in the clutch}$$

Temperature Multiplying Factor

Because the temperature of the friction material affects the wear of the material, the ambient temperature to which the clutch is exposed will affect the wear of the friction lining [18]. As a result:

$$C_T = 1.42 - 1.54E - 3X + 1.38E - 6X^2 \tag{5.24}$$

(for sintered metallic linings)

$$C_T = 2.79 - 1.09E - 2X + 1.24E - 5X^2 \tag{5.25}$$

(for resin-asbestos linings used in light-duty automotive and moderate-duty industrial brakes)

$$C_T = 3.80 - 7.58E - 3X + 5.07E - 6X^2 \tag{5.26}$$

(for carbon-carbon linings)

$$C_T = 17.59 - 6.03E - 2X + 5.43E - 5X^2 \tag{5.27}$$

(for resin-asbestos truck linings)

where $X = 590 + T$
       $T = $ Ambient temperature in °F

Similar procedures are used to work out the remaining components of equation (5.18).

Validation of Reliability Prediction Equations

A very limited budget during the development of the handbook prevented the procurement of a sufficiently large number of components to perform the necessary failure rate tests for all the possible combinations of loading roughness, operational environments, and design parameters to reach statistical conclusions as to the accuracy of the reliability equations. Instead, several test programs were conducted to verify the identity of failure modes and validate the engineering approach being taken to develop the reliability equations. For example, valve assemblies were procured and tested at the Belvoir Research, Development and Engineering Center in Ft. Belvoir, Virginia. The number of failures for each test were predicted using the equations presented in the handbook. Failure rate tests were performed for several combinations of stress levels and results compared to predictions. Typical results are shown in Table 5-5.

Another example of reliability tests performed during development of the handbook is the testing of gearbox assemblies at the Naval Air Warfare Center in Patuxent River, Maryland [19]. A spiral-bevel right angle reducer type gearbox with 3/8 in. steel shaft was selected for the test. Two models having different speed ratios were chosen, one gearbox rated at 12 in.-lb torque at 3600 rpm and the other gearbox rated at 9.5 in.-lb torque. Prior to testing the gearboxes, failure rate calculations were made using the reliability equations from this handbook. Test results were

Table 5-5
Sample test data for validation of reliability equations for valve assemblies

Test series	Valve number	Test cycles to failure	Actual failures/10^6 cycles	Average failures/10^6 cycles	Predicted failures/10^6 Cycles	Failure mode
15	11	68,322	14.64	14.64	18.02	3
24	8	257,827				1
24	9	131,126	7.63	10.15	10.82	1
24	10	81,113	12.33			1
24	11	104				2
24	12	110,488	9.05			1
24	13	86,285	11.59			1
25	14	46,879	21.33	19.67	8.45	2
25	15	300				3
25	19	55,545	18.00			1

Test parameters – System pressure: 3500 psi; Fluid flow: 100% rated; Fluid temperature: 90 °C; Fluid: Hydraulic, MIL-H-83282.
Failure mode: 1 – Spring fatigue; 2 – No apparent; 3 – accumulated Debris.

compared with failure rate calculations and conclusions made concerning the ability of the equations to be used in calculating failure rates.

Similarly, other reliability tests have been performed pertaining to stock hydraulic actuators using a special-purpose actuator wear test apparatus [14], air compressors for 4000 hr under six different environmental conditions to correlate the effect of the environment on mechanical reliability [20], gear pumps and centrifugal pumps [21, 22], impact wrenches [23], brakes and clutches, a diesel engine-driven rotary vane compressor mounted on a housed mobile trailer [24, 25], and a commercial actuator assembly [26].

Summary

The procedures presented in the handbook should not be considered as the only methods for a design analysis. An engineer needs many evaluation tools in his toolbox and new methods of performing dynamic modeling, finite element analysis and other stress/strength evaluation methods must be used in combination to arrive at the best possible reliability prediction for mechanical equipment.

The examples included here are intended to illustrate the point that there are no simplistic approaches to predicting the reliability of mechanical equipment. Accurate predictions of reliability are best achieved by considering the effects of the operating environment of the system at the part level. The failure rates derived from equations as tailored to the individual application then permits an estimation of design life for any mechanical system. It is important to realize that the failure rates estimated using the equations in the handbook are time dependent and that failure rates for mechanical components must be combined for the time period in question to achieve a total equipment failure rate.

It will be noted upon review of the equations that some of the parameters are very sensitive in terms of life expectancy. The equations and prediction procedures were developed using all known data resources. There is of course additional research required to obtain needed information on some of the "cause-effect" relationships for use in continual improvement to the handbook. In the meantime, the value of the handbook lies in understanding "cause-effect" relationships so that when a discrepancy does occur between predicted and actual failure rate, the cause is immediately recognized. It is hoped that users of the handbook and the MechRel® software program will communicate observed discrepancies in the handbook and suggestions for improvement to the Naval Surface Warfare Center.

References

1. Henley, E. J. and Kumamoto, H., *Reliability Engineering and Risk Assessment.* Englewood Cliffs: Prentice-Hall, 1981, pp. 198–205.
2. Carter, A. D. S., *Mechanical Reliability.* New York: John Wiley, 1972, pp. 5–11.
3. Chaddock, D. M., The reliability of complicated machines. *Inter Services Symposium on the Reliability of Service Equipment.* London: Institute of Mechanical Engineers, 1960.
4. Simmons, P. E., The optimum provision of installed spares. *Hydrocarbon Processing,* April 1982.
5. Davis, G. O., How to make the correct economic decision on spare equipment. *Chemical Engineering,* November 1977.
6. MIL-HBK-217D, *Reliability Prediction of Electronic Equipment,* ANSI, 1982.
7. *Guided Weapon System Reliability Prediction Manual,* MOD (PE) Report DX/99/013-00.
8. Venton, A. O. F., *Component-based Prediction for Mechanical Reliability. Mechanical Reliability in the Process Industries.* London: MEP, 1984.
9. Bloch, H. P. and Geitner, F. K., *Machinery Failure Analysis and Troubleshooting.* Houston: Gulf Publishing, 1983, pp. 527–530.
10. Hauck, D., *A Literature Survey.* AECL Report No. CRNL-739, 1973.
11. Grothus, D., *Die Total Vorbeugende Instandhaltung.* Dorsten, Germany: Grothus Verlag, 1974.
12. Giacomelli, E., Agostini, M., and Cappelli, M., *Availability and Reliability in Reciprocating Compressors Evaluated to Modern Criteria.* Florence, Italy: Quaderni Pignone, 42/12/1986.
13. *Standard Handbook of Machine Design.* Edited by J. E. Shigley and C. R. Mischke. New York, Toronto: McGraw-Hill, 1986, pp. 27–1 to 27–17.
14. Newcomb, T. P., *Thermal Aspects of Vehicle Braking,* Automobile Engineering, August 1960.
15. Pratt, D., *Results of Air Compressor Reliability Investigation,* Report No. TM 88-38 SY, January 1989, Naval Air Warfare Center, Patuxent River, Maryland.
16. Pratt, D., *Results of Gear Pump Reliability Investigation,* Report No. TM 89-24 SY, February 1990, Naval Air Warfare Center, Patuxent River, Maryland.
17. Pratt, D., *Results of Centrifugal Pump Reliability Investigation,* Report No. TM 89-69 SY, February 1990, Naval Air Warfare Center, Patuxent River, Maryland.
18. Pratt, D., *Results of Pneumatic Impact Wrench Reliability Investigation,* Report No. TM 90-88 SY, December 1990, Naval Air Warfare Center, Patuxent River, Maryland.
19. CDNSWC, *Interim Reliability Report on the MC-2A Compressor Unit,* January 1992.
20. Pratt, D., *Results of Air Force MC-2A Air Compressor Unit Reliability Investigation,* Report No. TM 92-89 SY, March 1993, Naval Air Warfare Center, Patuxent River, Maryland.
21. Pratt, D., *Results of Dayton 5A701 Linear Actuator Reliability Investigation,* Report No. TM 93-89 SY, Naval Air Warfare Center, Patuxent River, Maryland, 1994.
22. Neale, M. J., Tribology Handbook, Butterworths, London, 1985.
23. Stankovich, I., Modern pneumatic handling. *Bulk Solids Hardware,* **4 (1)**, March 1984, pp. 183–188.

24. Anderson, A. E., *Wear of Brake Materials*, Wear Control Handbook, M. B. Peterson and W. O. Winer, eds, *Am. Soc. Mech. Eng.* New York, 1983, pp. 843–857.

25. **API**Standard 617, *Centrifugal Compressors for General Refinery Service*, 5th Edition, April 1988.

Chapter 6
Predicting uptime of turbomachinery

As the preceding chapters showed, numerical and statistical methods can be used to identify areas of vulnerability before the analyst is actually confronted by the hardware and its potential problems. A similar systematic reliability evaluation of major turbomachinery and centrifugal pump components, for example, can warn of potential problems in future or existing installations.

This chapter presents structured approaches to predicting the reliability of such major turbomachines as centrifugal compressors and steam turbines, as well as general purpose equipment such as centrifugal pumps. The major turbomachinery train shown in Figure 6-1 features a steam turbine driving a low-pressure (LP) and a high-pressure (HP) turbocompressor.

A procedure and a set of curves can be developed to coordinate the major factors influencing reliability of this type of equipment. These factors are type of machine, unit size, speed, pressures and temperatures, coupling effects, number of start–stop cycles, starting cycle time, characteristics of supports, foundation, piping, and the effects of operating practices and maintenance provisions.

Reliability factors were established to improve the accuracy of equipment evaluations, and to make sure a maximum number of remedies can be considered quickly. Reliability factor curves presented in the following pages are based on personal experience and extensive use of references. Such curves can never be highly accurate, and they can never cover all possible types of installation. Common sense must be used in their application. The curves are given more to outline a systematic procedure than to provide numbers ready for use.

Figure 6-1. Two-casing oxygen compressor with main and intermediate gearing. *Source*: Mannesmann-Demag, Duisburg, West Germany.

Interpretation of Reliability Factors (RF)*

- RF = 2.0 or above: Excellent probability of trouble-free operation. Breakdown rate is estimated about half that of normal.
- RF = 1.0: Average installation with normal probability of failures and breakdowns.
- RF = 0.5: Probability of problems is about twice that of normal.
- RF = 0.1: Probability of problems is about ten times normal. In other words, chances of trouble-free operation will be very poor, and time between breakdowns will be short. Usually, basic changes will be required to correct the situation.

The interpretation of reliability factors should be valid for individual components as well as for the overall installation. Table 6-1 gives a random example to illustrate the procedure. This assumed example indicates a good overall plant design but poor installation and facilities. Unusual trouble is not likely to occur. But if it should happen, significant improvement could be obtained quickly by correcting the piping and foundation rather than by looking into couplings, bearings, or other basic equipment details. Unusual troubles could just as easily develop the other way around. It depends on where the weak spots of an installation are located. This same installation could become marginal if it were started and stopped every day or if it ran at higher speeds or if it were to be quick-started, etc.

---

* Adapted from [13], with the kind permission of the author. John S. Sohre. Turbomachinery Consultant Ware, Massachusetts 01082.

Table 6-1
Example of overall reliability determination

Equipment: 10,000 rpm, turbine driven compressors
From the design curves (Figure 6-2)

Type of equipment	Turbine	RF = 1.0
	Compressor	1.2
Equipment size	Turbine, 5-ft bearing span	1.0
	Compressor, 1st body 3-ft bearing span	1.5
	Compressor, 2nd body 4-ft bearing span	1.2
Number of bearings in train	6	0.9
Startup time	40 min	1.0
Maximum pressures	Turbine: 600 psi	1.6
	No. 1 compressor: 75 psi	1.9
	No. 2 compressor: 300 psi	1.8
Maximum temperature	Turbine: 750 °F	0.8
	Compressor No. 1: 150 °F	2.0
	Compressor No. 2: 250 °F	1.9
Coupling	Gear, curved teeth	1.0
Casing support	Turbine, flexplates, centerline	1.1
	Compressor No. 1, non-centerline	0.8
	Compressor No. 2, centerline, sliding	1.0
Starting frequency	One per year	1.8
Multiplied subtotal, design features		
		$\overline{51.3}$

Installation (from Figs 6-3 and 6-4)

Piping strains	Turbine: 150% NEMA and API	0.7
	No. 1 compressor: 100%	1.00
	No. 2 compressor: 100%	1.00
Pipe supports	Turbine: springs	1.00
	Compressor: rack and rod	0.7
Expansion joints	One poorly restrained joint on compressor	0.6

Foundation

Rigidity	Mat and slab weak	0.7
Vibration	Non-resonant (weight1 × unit weight)	0.8
Vibration isolation	Not isolated, significant non-resonant transmission	0.8
Installation, multiplied subtotal		$\overline{0.132}$

Operation

Operators	Average	1.0
Maintenance personnel	Good	1.5
Maintenance facilities	Poor	0.5
Operation, multiplied subtotal		$\overline{0.75}$

Total overall reliability of installations: RF = 51.4(0.132)(0.75) = 5.1

The example shown would be a relatively easy project to work on. It would be far more difficult to come up with a solution where all the factors run close to 1.0. When all factors are 1.0, a random breakdown is probably involved. Troubleshooting would then require a well-planned and coordinated effort. Many different symptoms would have to be analyzed before an improvement could be made.

Factors Influencing Reliability

Type of Equipment *(Fig. 6-1)*

Electric motor driven units are highly reliable at low speeds. As speeds and gear ratios increase, double increaser gear trains become necessary. As units become more and more sophisticated, the reliability drops off rather sharply. Reliability drop is especially a problem with long equipment trains.

Because of the extremely short starting time with motor driven high-speed trains, it is practically impossible to supervise the unit during startup. The short starting time may result in a very high damage level if trouble occurs. Frequent starting aggravates the situation.

Synchronous motor drives introduce the additional risk of torsional failure when passing through slip-frequency resonance. Heavy torsional shock and vibration can be the result of relatively minor malfunctions during the synchronizing cycle. Other shock and vibration problems are caused by short-circuit and phase faults. Particularly susceptible are long trains exposed to frequent starting.

Steam turbines require a considerable amount of auxiliary equipment such as boilers, piping, or condensers. Auxiliaries affect overall reliability, but reliability is very high once the unit is running. High speeds present no more problems than would be expected with a centrifugal compressor, allowing for some effects of temperature, pressure, and auxiliaries.

Equipment Size *(Fig. 6-2)*

The faster a machine runs, the smaller it must be. Otherwise problems of stress, and especially vibration problems, will develop. Critical speeds and other rotor instabilities become a dominating factor at high speeds. Situations can arise where it becomes impossible to pass through a critical phase without risking destruction of the machine. Such conditions are mainly affected by bearing span. Therefore, the reliability curves are plotted for various spans.

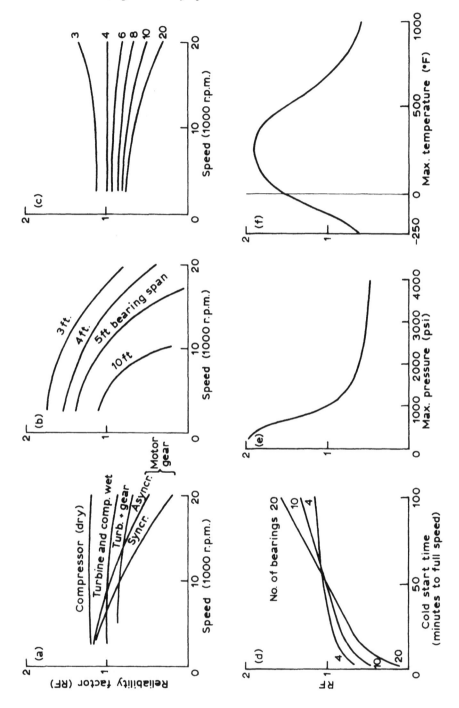

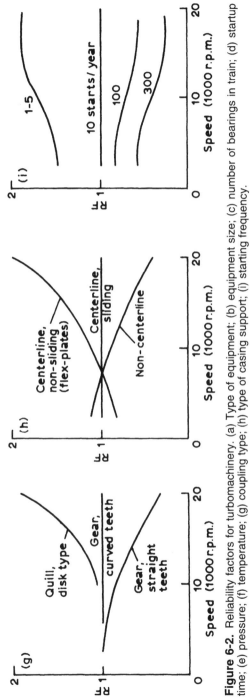

Figure 6-2. Reliability factors for turbomachinery. (a) Type of equipment; (b) equipment size; (c) number of bearings in train; (d) startup time; (e) pressure; (f) temperature; (g) coupling type; (h) type of casing support; (i) starting frequency.

For three-bearing machines, the stability of the rotor improves considerably and longer spans can be used. For a given reliability, the longest span between adjacent bearings can be increased by 10–50%. Span depends on speed and design features. An increase in span can more than double the shaft length without loss in reliability. Obviously, there must be provisions to hold the three bearings lined up nearly perfect under all normal and abnormal conditions. Otherwise, the reliability can go down rather than up. Alignment is the major problem with three-bearing machines.

Number of Bearings in the Train

The reliability of a long train with many couplings will be less than that of a simple, short unit. Reliability can be expressed by the number of bearings in the train. If there are gears in the train, multiply the reliabilities of the low-speed section with the one for the high-speed section to get an overall reliability.

Startup Time

Quick starts (motor drive) are more likely to damage long trains than short trains. There is no time to supervise the long unit during the few seconds it takes to come to full speed. A curve has been included to show the effect of startup time. Much depends on supervisory instrumentation, protective devices, starting shock severity, temperature, pressures, surging, switch gear operation, etc. Individual estimates must be made to include these factors.

Pressure

The pressure factors are reflected in the reliability curves. Pressures shown are maximum pressures on the unit. The advantages and disadvantages of high-pressure machines are:

Advantages	*Disadvantages*
Compact design	Small machines, sensitive to pipe strain
Small distortions	Heavy pipe walls
Small piping	Small supports, close together
Small internals	High-impact loads on internals
	High-thrust load and thrust load variations
	Long seals
	Thick casing walls
	Tight clearances

Temperature

Temperature is the main offender where reliability is concerned. Most compressors, gears, and motors are only exposed to moderate temperatures, as compared to turbines. Lower temperatures compensate for some other shortcomings of motors such as short starting cycle, torsional vibrations, electrical problems, etc.

Temperature can cause distortion of the casing, foot, and foundations as well as misalignment and problems with pipe expansion. Temperature inflicts restrictions on materials. Seals are only one example. Material restrictions affect the entire design philosophy, as well as the efficiency and life-expectancy of the unit.

Coupling Types

Gear couplings are often considered standard for large, high-speed equipment. At high speeds straight teeth can contribute to certain rotor instabilities. Curved or barreled teeth often give smaller exciting forces on the rotor. Much depends on the design and coupling quality.

Someone once said: "You can never waste money buying the best coupling you can get." This statement is especially true for large, fast machines and long trains. The problem is not so much that the coupling breaks down – although this may also happen – but that the coupling excites the rotor system into vibrations and instabilities which can be very violent. This excitation is caused by the interaction of periodic tooth friction forces with the rotor-stator damping system. Other problems, such as those caused by misalignment or rotor critical speeds, may also be emphasized or de-emphasized by variations of coupling design and quality.

Well-designed quill shafts or flexible disk-type couplings are much lighter. They do not generate the instabilities caused by looseness, friction, and lubricant contamination or lubricant breakdown, which are inherent in gear couplings to a greater or lesser degree. This makes flexible, dry-disk couplings more reliable, especially at high speeds.

Improper installation or poor maintenance can easily cause failures. Damage to highly stressed quills and membranes is one reason why these couplings are sometimes not used. Couplings of this type have been used successfully in aircraft engines, where a very high level of maintenance control is standard.

Casing Support

A casing support structure is sometimes suspected of causing trouble in an area where it cannot do much harm or even where it is the best type to apply. Each type has its advantages and disadvantages. A rugged sliding foot support is often best for large, low-speed machines such as large turbines, gears, motors, generators, and compressors. Centerline supports become necessary at higher speeds and temperatures. If supports are of the sliding type, they may bind or lift under pipe forces and thermal distortion. Also, sliding supports can cause serious vibration. Flexible plates avoid these problems and are especially advantageous for relatively small machines running at high and very high speeds. Small, slow machines are often simply bolted down, because thermal expansions are small and can be absorbed with little distortion. A bolted machine has the advantage of ruggedness and insensitivity to piping strain.

Piping Strain *(Fig. 6-3)*

The effect of piping strain on a machine reduces reliability by:

- causing misalignment and subsequent vibration;
- causing case distortion and subsequent vibration, rubs, case leakage, and possible cracking;
- causing foundation or base deflection, which may result in misalignment, case distortions, and subsequent vibrations or rubs.

Excessive piping strain may be the result of:

- Thermal expansion and contraction of the pipe, boiler, and machine. This indicates faulty pipe design. Expansion joints or loops may have to be installed.
- Improper pipe support. Frequent problems arise from indiscriminate use of rod hangers – instead of spring hangers – anchors, and other non-elastic restraints and supports. For correction, disconnect the piping at both ends and support it on spring hangers, except where anchors or restraints are required by the pipe design.

Improper pipe installation is very frequently a source of trouble and hard to find once the pipe is installed. Usually, piping is not properly lined up at the flanges. If flanges are not parallel when lined up, very large moments and forces can occur in the casing and at the case supports. To identify strains caused by flange misalignment, mount dial indicators at the coupling and supports, disconnect the pipes, and observe the movements. These installation strains are superimposed on thermal expansion strains.

Cold spring is one source of piping strains. The cold spring usually encountered is provided by cutting the pipe short by about half the anticipated thermal expansion. The pipe ends are pulled together and

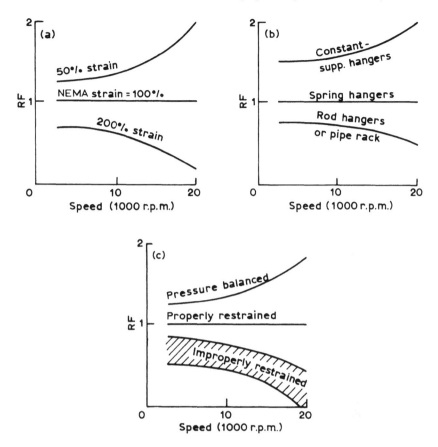

Figure 6-3. Effect of piping on reliability. (a) Pipe strain; (b) pipe supports; (c) expansion joints.

welded. Cold spring strain is practically unpredictable, especially the resultant moments, and quite often the equipment suffers from it.

Expansion Joints

Expansion joints are often useful in low-pressure lines, but they are not anywhere nearly as flexible as many engineers believe. If expansion joints are not lined up properly, or indiscriminately exposed to shear or torsion, the strains on the machinery can cause serious problems. One must also consider the thrust caused by an unstrained expansion joint. It is equal to the cross-sectional area at the largest bellows diameter multiplied by the internal pressure, in psig. Tie rods often used on expansion joints to absorb the thrust are only effective and harmless when the joint is used in shear. If tie rods are used on a joint which is meant to move in tension-compression, they bypass the joint and make it virtually useless, because pipe forces are then transmitted through the rods. Or, if restrained only in one direction, the rods may become loose. Then, the pressure thrust will act on the machine again, as if the rods were not there.

Settling foundations of machinery, boilers or condensers, can cause serious pipe strain. Often involved are metal expansion joints between equipment and condensers (or coolers) and large, low-pressure piping with little flexibility. Concrete shrinkage and creep also belong in this category. A 10-ft column shrinks 0.06 in. during the first 6 years. Creep during the first 2 years is three to four times the original static deflection.

Size and Speed

The faster a machine runs, the more sensitive it will be to pipe strain.

- A high-speed machine is smaller than a low-speed machine.
- Piping is normally sized for flow, regardless of speed. It is therefore often large compared to the casing size and support strength. The result is more severe distortion and misalignment for the faster machine. Tolerance of a machine for distortion and misalignment decreases as speed increases.
- As speed increases, the tendency for a rotor to become unstable also increases. Instability is caused by oil whirl and certain friction-induced and load-induced whirls. Therefore, a given displacement which is harmless at low speed can cause instability at high speeds.
- Bearing clearances are small (smaller journals) for high-speed machines. Thus, bearings are less tolerant of distortion and displacement.

The above factors are reflected in the curves showing the effect of piping strain on reliability (Fig. 6-3). The piping is assumed to be in accordance with API and NEMA standards. That is, allowable strains are a function of casing weight and size and therefore allowable forces and moments are smaller for fast running machines. The curves show the effect of excessive strain. This includes all strain regardless of the source such as installation, support settling, or expansion joints.

Pipe Supports

Pipe supports are shown separately because their effect is pronounced during startup, shutdown, and load changes. Also, pipe supports are a significant factor in long-term reliability due to settling, jamming of springs and slides, or plain aging effects.

It is unrealistic to base allowable pipe reactions on pipe size only, disregarding the size, mass, and speed of the equipment. Such a design allows the same pipe strain no matter whether the machine is large or small.

Foundation *(Fig. 6-4)*

The foundation is one of the most influential factors where overall reliability of a unit is concerned. A foundation must:

1. Maintain alignment under all normal and abnormal conditions. The conditions include soil settling, thermal distortion, piping forces, vacuum pull, or pressure forces in expansion joints. A heavy and

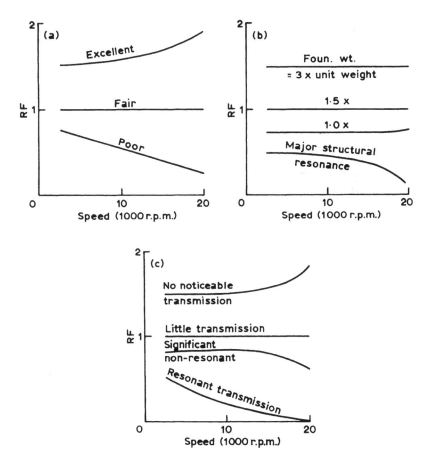

Figure 6-4. Effect of foundation characteristics on reliability. (a) Rigidity to maintain alignment; (b) vibration characteristics; (c) isolation from surroundings.

rigid mat, the portion resting on the soil, is a key to good alignment. Other aids to alignment are equal deflections of all columns under load, as well as mass, continuity, symmetry, and rigidity of the top slab on which the unit rests. The way the foundation is supported on the soil, as well as soil characteristics and soil resonances, deserve special attention.

2. Minimize vibration. The foundation must be as heavy as possible and non-resonant. If a foundation is resonant, it does not matter much whether it is a light structure or a heavy one; reliability will be greatly reduced in either instance.

3. Isolate the unit from external vibrations. For larger or more critical units, one should provide an air gap filled with mastic sealer all around the slab and mat. Vibration transmission may be from the unit to the surroundings or vice versa, and it may be aggravated by resonance at transmission frequencies. Piping, stairways, and ducts may also transmit vibration, which should be prevented by proper isolation. Ground water transmission is often serious. Reliability is reduced when units, especially large ones, are mounted on baseplates which are then mounted on top of the foundation. Baseplates introduce an additional member in the system which increases deflections and vibrations. Usually, deflections and resonant frequencies become unpredictable and have a way of showing up at the wrong place and at the wrong time. Besides, the base usually interferes with proper foundation design. Therefore, cost savings of a unit mounted on a steel frame base should be evaluated against reduced reliability.

Operation *(Fig. 6-5)*

The larger and/or faster the unit, the more influence operators will have upon reliability. One must use one's own judgment in rating an operating crew. The main factors include training, intelligence, cooperation, but especially organization and leadership.

Operating personnel as a factor in reliability may not seem important at first. One is usually stuck with a given crew when a troublesome job comes up. But evaluation of operating personnel will tell us how reliable or foolproof we must make a unit if it is to be operated successfully by such a crew.

Maintenance Personnel

Maintenance crew evaluation is essentially the same as for an operating crew. The same factors must be considered together with their economic

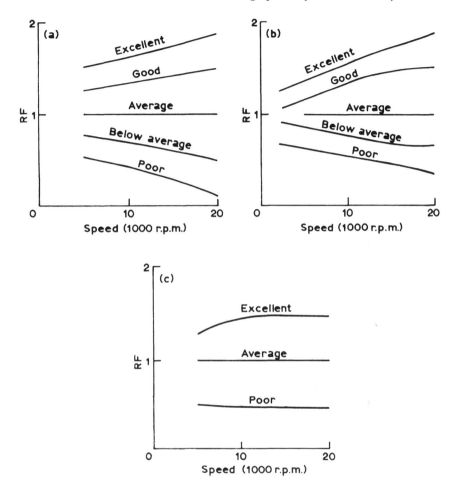

Figure 6-5. Effect of operations and maintenance on reliability. (a) Operating personnel; (b) maintenance personnel; (c) maintenance facilities.

effects and calculated risks. To illustrate the effect of maintenance, consider as an example the internal inspection of a high-speed compressor which may improve the chances of successful operation. With a good crew one would make the inspection; with a poor crew one would rather take a calculated risk of a failure up to a certain level of severity. One can reason that the machine is likely to be worse off rather than better, after a poor maintenance crew inspects it. Evidently, such a unit will be considerably less reliable, whether the crew is put to work on it or not.

Maintenance facilities include working conditions with the process unit as well as shops, tools, availability of spare parts and, last but not least, availability of instruction books, drawings, and technical data.

Ruggedness of Turbomachinery

Many people feel that a more massive construction provides higher reliability. Others question this point, believing that one can build a very light machine with the same reliability as a design weighing many times as much. Aircraft engines are usually referred to, and one can hardly argue the point that they are highly reliable. The question seems to be mainly whether or not the necessary sophistication went into a lightweight machine, to make up for the obvious advantages of mass and rigidity. This can only be decided by looking at the respective designs.

However, this item seems to receive increasing attention and it was suggested by a specialist that the massiveness metal content of a machine can be expressed in some comparative form, to allow evaluation. To do this, the weight not contributing to ruggedness must be disregarded.

Length is perhaps the most critical dimension of high-speed machines. A short machine is more reliable than a long one. If we calculate the weight per inch of turbine, this should tell us a good deal about its construction. For example, a turbine for the same speed, efficiency, and conditions can be built with, say, six or nine stages, depending on the thermodynamics and hardware sophistication. The shorter machine will be more compact and will have a greater average weight per inch of length, although the total weight may actually be less than that of the longer unit. Design sturdiness experienced by parameters such as wall thickness or mass will also be reflected in this number. Speed has an effect because as speeds go up machines get smaller in diameter.

Weight per inch has been plotted for several units in Figure 6-6. This figure covers several turbine generators from 3000 to 22,000 kW and several high-speed compressor drives in the 2000–15,000 hp range, both condensing and non-condensing. Five machine types of different manufacture are included. Most machines are of average to heavy design. Therefore, the curves indicate fairly heavy construction.

Weight used in the weight per inch ratio is overall weight of the installed turbine, including control valves, trip, and throttle valve, but not baseplates, oil tanks, and the like. Length used is overall body length. But the length does not include small protrusions such as protruding shaft ends, valves, or flanges.

To get a somewhat better feel of the actual metal content of a turbine, the equivalent solid diameter (steel) has been plotted in Figure 6-7. Equivalent solid diameter is the diameter if the whole turbine were compacted until no air remained inside and then made into a round bar of the same length and weight as the turbine. Equivalent solid diameters are surprisingly large. Another way of looking at this is to multiply the weight per inch with the rated speed as a parameter of diameter. This

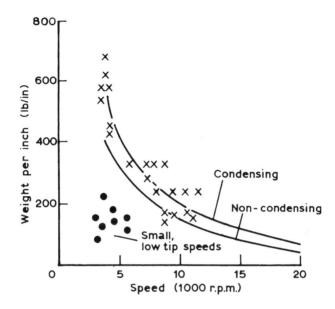

Figure 6-6. Weight per inch of turbine (overall length).

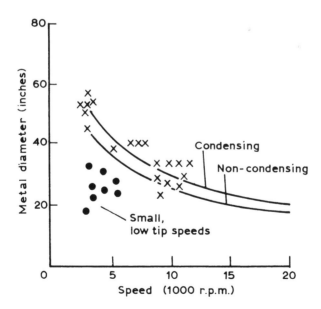

Figure 6-7. Equivalent solid metal diameter.

gives us a factor which is independent of speed (since weight per inch appears to vary as a linear function of speed). We can call this factor a Mass Concentration Factor (MCF). Then:

$$\text{MCF} = \frac{\text{Total turbine weight (lb)}}{\text{Total turbine length (in.)}} \times (\text{rated speed, rpm})$$

For machines of comparable ruggedness of construction this factor remains surprisingly constant for a wide variety of designs and speeds. The upper curves (condensing turbines) are plotted for an MCF of 2.0×10^6. Non-condensing turbines appear to be about 10–20% lighter, with an MCF of $1.6 - 1.8 \times 10^6$.

Examples

A condensing turbine compressor driver with a 6000-hp rating operates at 8000 rpm with $W = 256,000\,\text{lb}$ and $L = 76\,\text{in}$. MCF = $(26,000/76)(8000) = 2.74 \times 10^6$. This is an unusually heavy machine.

Another condensing turbine compressor drive has a 5000-hp rating and operates at 8300 rpm with $W = 20,000\,\text{lb}$ and $L = 108\,\text{in}$. MCF = $(20,000/108)(8300) = 1.54 \times 10^6$. This is a long, relatively lightly constructed machine.

The group of light machines in the low-speed area represents small units with many stages, which are often used for this type of service.

We should use the description above only as a guide to help assemble meaningful data which can then be interpreted to suit individual preferences and requirements. Compressors and other machinery equipment can be evaluated in a similar manner. Reliability curves can then be plotted to include these factors in the overall evaluation of a unit. A similar approach can be employed to determine the probable reliability of centrifugal pumps.*

Application Issues for Centrifugal Pumps

Application issues may be divided into two categories: optimum selection of a pump size and optimum selection of auxiliary equipment (i.e., mechanical seals, lubrication methods, bearings, couplings, etc.). In this section we will focus on size selection when applied to pumps of

---

*Contributed by Maurice Jackson (Tennessee Eastman Company) and Barry Erickson (then with Durco Pumps).

SELECTED PUMPS

Pump	RPM	Impeller (in.)	TDH (ft)	Horsepower	NPSH Req'd (ft)	%Eff.
3 × 2-8	3550	7	158	18.2	11.1	66
3 × 2-13	1750	12 $^5/_8$	153	18.5	5.6	62
4 × 3-13 HH	1750	12 $^1/_8$	154	19.1	4.0	61
3 × 2-10 A	3550	7 $^1/_8$	157	19.8	15.0	60
4 × 3-13	1750	12 $^1/_2$	153	19.8	4.5	58

Figure 6-8. Pump selections.

a given design (i.e., from a single manufacturer) operating on a given service. We will not consider the effects of different services at this time, nor will we discuss selection of auxiliary equipment.

When selecting a pump, one of the first things a user does is to determine the head and capacity required. After deciding on a supplier and a product line, the user must still select the pump size that will handle the duty. As an example, Figure 6-8 lists five ANSI pump sizes from a single manufacturer that could be selected to handle a duty of 300 gpm at 150 ft. Review of the options indicates that the first pump, a size $3 \times 2 - 8$, would probably be the least expensive because it is smaller. Because it draws the least horsepower (18.2), it would have the lowest operating cost. Figure 6-8 does not provide any information regarding the relative reliability of the five selections.

Customer	Chemical Company A
Pump Service	Solution × Circulator
Flow	300 gpm
TDH	150 ft
NPSH Available	17 ft
Specific Gravity	1.0
Viscosity	1.0 cp

This is an interesting situation because maintenance expense can be the major cost item in the life-cycle cost of a pump. Several surveys have shown that the average mean time between repairs for an ANSI pump is 15 months. The average repair cost cited by users is $2500 per repair. This figure does not include burden, overhead expenses, and lost production. Because the average cost of a small ANSI pump is in the $4000 range, the repair costs will exceed the initial cost in considerably less than 3 years.

Pump Selection Reliability Factors

From a reliability point of view there are three major factors that affect the selection: operating speed, impeller diameter, and flow rate. As in the case of major turbomachinery, method of assigning a numerical value for each factor is proposed. This value allows ranking the relative reliability of alternative pumps on each factor. The numerical values range between zero and one, higher values indicating more-reliable selections. Because a poor ranking on any one factor can significantly affect the reliability of the pump, an overall reliability index is formed by taking the product of the three individual factors. This product will be referred to as the Reliability Index (RI).

RPM (F_R). The operating speed affects reliability directly through wear in rubbing contact surfaces (mechanical seals and shaft seals), bearing life, heat generated by the bearings and lubricants, and wear caused by abrasives in the pumpage. For most of these items the rate of wear has a linear relationship to the pump RPM. Thus the RPM factor is taken as a linear function of operating speed. Figure 6-9 illustrates this factor. The starting and ending points for the relationship are set as zero RPM and the maximum RPM for which the pump is designed, because reliability is a function of the basic design.

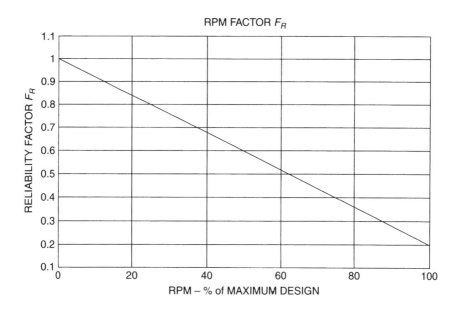

Figure 6-9. RPM factor.

A value of 0.2 is assigned to the RPM factor when application is at maximum design RPM. This value was arrived at by recognizing that the index is a comparative rating. The maximum and minimum values of the parameter affect how each parameter is weighted relative to the other parameters. Although the value of 0.2 is somewhat arbitrary, it does ensure that RPM is weighted equally with the other parameters. It was also found that the final index values are not very sensitive to this value.

For example, if a given pump was designed to operate at a maximum speed of 3500 rpm, an application at 3500 rpm would be assigned $F_R = 0.2$. If the same pump was applied at 1750 rpm, the value of F_R would be assigned a value halfway between 0.2 and 1.0, or 0.6 (the speed is one-half the maximum).

Impeller diameter (F_D). The impeller affects reliability through the loads it imposes on the shaft and bearings. Impellers produce two types of loads: one that is relatively steady in both magnitude and direction, and a second which is variable in both magnitude and direction. The first is a result of non-uniform pressure distribution in the casing. It produces a shaft deflection in one direction that causes the mechanical seal faces to run off-center but not wipe radially (for most seal designs). The second load is a result of the interaction between the impeller vanes and the casing discharge tongue. It produces a deflection as each vane passes the tongue or cutwater. This second effect can be very damaging because it continually causes the seal faces to move radially relative to each other. The magnitude of this movement may be greater than the steady deflection.

Both loads are related to the impeller diameter in a cubic manner; thus they decrease rapidly as the impeller diameter is reduced, and reliability increases equally rapidly. But, as the diameter is further reduced, the possibility of encountering suction recirculation and resulting random loads increases. Because suction recirculation occurs at the pump inlet where fluid energy levels are lower than at the exit, the loads produced by recirculation are not as great as those produced by the impeller/discharge interaction. Consequently, there is an optimum diameter that is closer to the maximum diameter than to the minimum. An optimum diameter maximizes reliability. Because the loads produced by recirculation are less severe at lower RPM, F_D is made a function of RPM. Figure 6-10 illustrates the variation of the diameter factor. The optimum diameter is taken as 75% of the trim range (25% from maximum).

Thus a pump with an impeller diameter trim range of 10–6 in. would be assigned a value of $F_D = 1.0$ when trimmed to 9 in. at any speed. When trimmed to a maximum diameter (10 in.), F_D would be assigned a value of 0.0 if operation was at the maximum design RPM, and 0.5 when operating at one-half of maximum RPM.

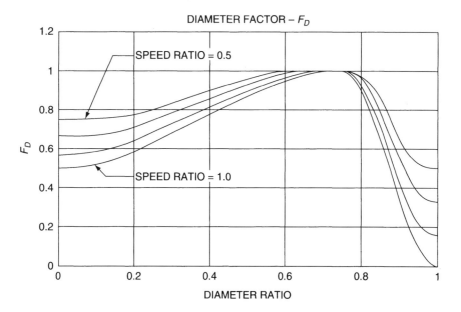

Figure 6-10. Diameter factor.

Flow rate (F_Q). A centrifugal pump is designed to operate most reliably at one capacity for a given RPM and impeller diameter. This flow rate is called the best efficiency point (BEP). At this flow, hydraulic loads imposed on the impeller are minimized and are steady. At flows greater than or less than the BEP, the hydraulic loads increase in intensity and become unsteady because of turbulence in the casing and impeller. These unsteady loads have the same effect on reliability as the impeller/discharge loads discussed above. In order to measure the effect of these loads, a series of tests were conducted on a pump. The tests involved varying the following parameters:

- RPM
- Impeller diameter
- Flow rate
- Pump shaft to motor shaft alignment
- NPSH margin

Vibration at the bearings was selected as a convenient direct indication of relative shaft motion. Figure 6-11 presents the vibration levels averaged over the range of the parameters as a function of flow rate. This figure shows that the vibration at BEP is 60% of the level at 10% of BEP, and is 45% of the level at 120% of BEP. Thus, if a reliability factor for flow

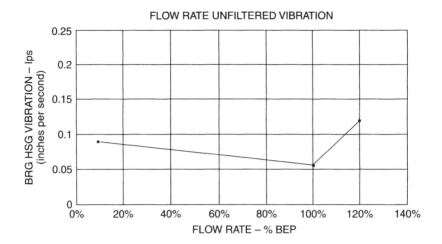

Figure 6-11. Bearing housing vibration versus flow rate.

is assigned a value of 1.0 at BEP, then values of 0.60 at 10% of BEP and 0.45 at 120% of BEP are appropriate for this pump.

Experience with pumps of a variety of sizes has shown that smaller pumps vibrate less when throttled back on their curves than do larger pumps. This is probably attributable to smaller pumps being more rugged relative to the imposed loads than larger pumps. Thus the reliability factor for flow rate was made dependent on BEP capacity. Figure 6-12 illustrates the F_Q function.

A pump selected at BEP capacity is assigned $F_Q = 1$. A small pump (BEP <50 gpm) is assigned $F_Q = 0.5$ when operated near shutoff. A large pump (BEP >3000 gpm) is assigned $F_Q = 0$ when operated near shutoff. For all pumps, F_Q is assigned a value of zero when applied at flows greater than 125% of BEP. This is done in recognition of rapidly increasing $NPSH_R$ as well as high impeller loading.

Reliability index (RI). The Reliability Index is formed by the product of the three individual factors:

$$RI = F_R \times F_D \times F_Q$$

Values will range from 0 to 1, with higher values indicating greater reliability. Because this factor does not take into account design characteristics, it cannot be used to compare pumps of different designs. Its value is in assisting in the selection of the most reliable pump of a given design.

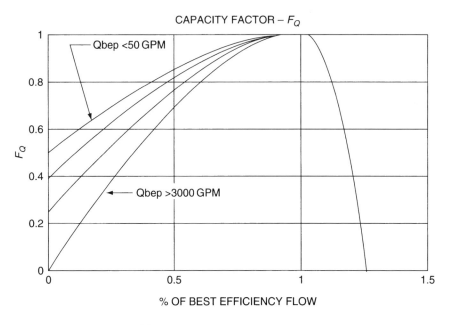

Figure 6-12. Capacity factor, F_Q.

An example of the use of the Reliability Index is given in Figure 6-13. This figure lists the five pumps identified in Figure 6-8 that could be selected for a duty of 300 gpm at 150 ft. In Figure 6-13, these pumps are ranked using the Reliability Index. The fourth column lists the RI for each pump. The size 3 × 2-13 pump has the highest index value, primarily because it operates at less than its maximum design speed. The 3 × 2-10 A pump has an RI of zero because of its selection at 125% BEP. It can be seen in the next to last column in Figure 6-13 that the $NPSH_R$ for this pump is considerably higher than the others.

Columns labeled "Cost Factor" and "Energy Factor" are provided to assist in making the final selection. The Cost Factor is the ratio of the cost

Pump Size	D	RPM	RI Factor	F_R	F_D	F_Q	Dia. Ratio	Q/Qbep	Cost Factor	Energy Factor	$NPSH_R$ (ft)	$NPSH_A$ (ft)
3 × 2-8	7.0	3550	0.24	0.4	0.96	0.63	0.63	1.15	1.00	1.00	11.1	17
3 × 2-13	12.6	1750	0.44	0.7	0.63	1.00	0.91	1.00	1.49	1.02	5.6	17
4 × 3-13 HH	12.1	1750	0.33	0.4	0.96	0.87	0.78	0.58	1.97	1.05	4.0	17
3 × 2-10 A	7.1	3550	−0.00	0.4	0.66	−0.00	0.28	1.25	1.11	1.09	15.0	17
433-13	12.5	1750	0.17	0.4	0.50	0.83	0.88	0.52	1.86	1.09	4.5	17

Figure 6-13. Reliability optimized pump selection.

of a given pump to the cost of the least expensive one. The Energy Factor is the ratio of the operating horsepower to that of the pump with the least operating horsepower. These two columns indicate that the size 3×2-8 pump has the least initial cost and the least operating expense. The most reliable pump, size 3×2-13, would cost 49% more initially than the 3×2-8 and would have an annual operating cost 2% greater. The user now has significant additional information upon which to base his selection.

Comparison with Field Experience

In order to make full use of this additional reliability information, it is necessary to develop a relationship between RI and the mean time between failure (MTBF). This could possibly be done through laboratory testing, although simulating field operation in a laboratory environment is often difficult. The alternative is to use field reliability data. While obviously preferable, this requires not only good information on actual field experience, but also a field data base that is not affected by other poor application practices.

In the industrial case history given here, a search was made for suitable field data; it yielded two sets of usable data. In both instances, all pumps involved were of the same manufacture and design; they were all installed similarly and used the same mechanical seal. The fluids pumped were not the same, and the duty cycles are not fully known. Figure 6-14 is a scatter

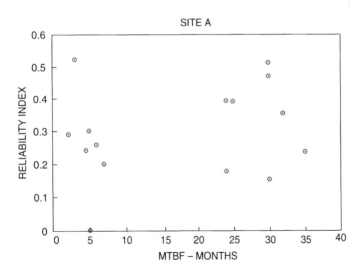

Figure 6-14. Field reliability data.

plot of the measured MTBF versus RI for Site A. While at first glance the data appear to exhibit random scatter, a detailed investigation of the raw data reveals otherwise. The following significant factors should be recognized:

- The data represents history on pumps that were installed within a 3-year period, but not all had been in service for 3 years. None of the pumps on the right-hand side of the figure (those with an MTBF greater than 20 months) had experienced a failure. Thus these data points may begin to show scatter as operating time increases.
- Those pumps with a MTBF less than 10 months and with Reliability Indices above 0.2 were in a variety of slurry services. The remainder of the pumps were in less severe services. Because slurry service will affect MTBF significantly, those data points are not comparable with the remainder.

Despite these shortcomings in the data, there is a general correlation between MTBF and the Reliability Index. If the slurry service duties are ignored, then a much clearer correlation exists. Note that the only pump with a zero value for RI exhibits frequent failure, whereas the more-reliable pumps have higher Reliability Indices.

Figure 6-15 is a scatter plot from a second field site. Again, there is considerable scatter, but the general trend is apparent. It is significant that the two pumps for which RI is zero also exhibit very poor reliability.

Effective testing of the proposed index against field data requires better control of the field operating parameters than is usually possible. Parameters such as NPSH margin, presence of entrained gases, and operating excursions away from the nominal duty point are not addressed in the index but are important factors in determining reliability.

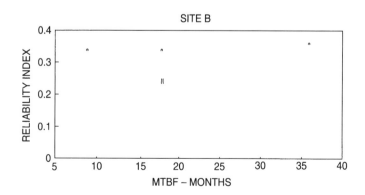

Figure 6-15. Field reliability data

Although these field data are insufficient to permit developing an absolute relationship between RI and MTBF, the correlation is encouraging, considering the different services. At present, the proposed Reliability Index is restricted to comparing the relative reliability of different pumps applied in the same service. Despite this restriction, the index does address an important aspect of pump application – optimizing the choice from multiple offerings. This index has been used to select pumps on at least two large projects.

Summary

A mathematical formulation of a Reliability Index is a reasonable indicator of the relative reliability of different pump sizes operating on the same service. The index accounts for the effects of operating speed, impeller diameter, and flow rate. Laboratory vibration data were used to correlate the effects of flow rate. Theoretical correlations were applied for operating speed and diameter.

The index was tested against field data from two sites. Although a strong correlation was not demonstrated, the following conclusions can be drawn:

- Pumps for which the Reliability Index is zero demonstrated very poor reliability.
- Barring severe operational parameters, reliability indices above 0.2 correlated with longer MTBF (20 months).
- Reliability is influenced by operational and installation parameters as much as by application factors.

Bibliography

1. Sohre, J. S., *Transient Torsional Criticals of Synchronous Motor Driven, High Speed Compressor Units.* ASME paper 65-FE-22.
2. Gunter, E. J., Jr., *Dynamic Stability of Rotor-Bearing Systems.* NASA SP-113.4.5. Washington, D.C.: US Government Printing Office.
3. *American Standard Code for Pressure Piping.* ASME, 345 East 47th Street, New York 10017.
4. *NEMA Approved Standards, Piping for Turbine-Generator Units, and for Mechanical-Driven Steam Turbines.* NEMA, 155 East 44th Street, New York 10017.
5. *API Standards 615 (Turbines) and 617 (Compressors).* American Petroleum Institute, 50 West 50th Street, New York 10017.
6. Sohre, J. S., *Foundations for High-Speed Machinery.* ASME paper 62-WA-250.
7. Sohre, J. S., *Operating Problems with High-Speed Turbomachinery, Causes and Correction*, 3rd revision. Originally presented at ASME Petroleum Mechanical Conference, 1968.

8. Pollard, E. I., *Torsional Response of Systems*. ASME paper 66-WA/Pwr-5.

9. Naughton, D. A., *Preventable Accidents to Turbines and Speed Increasing Gear Sets*. Hartford, Connecticut: The Hartford Steam Boiler Inspection and Insurance Co.

10. Huppman, H., *Das Hydrodynamische Gleitlager im Grossmaschinenbau*. Referat #11, Allianz Versicherungs AG "Der Maschinenschaden," 8 München 22, Postfach 220, Germany.

11. Bahr, H. C., *Recent Improvements in Load Capacity of Large Steam Turbine Thrust Bearings*. ASME paper 59-A-139.

12. Schmitt-Thomas, K. G., *Das Zusammenwirken von Korrosion und mechanischer Beanspruchung an metallischen Werkstoffen bei der Auslösung von Schäden*. Allianz Versicherungs AG.

13. Sohre, J. S., *Turbomachinery Analysis and Protection*. Proceedings of the 1st Turbomachinery Symposium, Texas A&M University, College Station, Texas, 1972.

Chapter 7
Failure mode and effect analysis*

Failure Mode and Effect Analysis (FMEA) is a name given to a group of activities which are performed to ensure that all that could potentially go wrong with a product has been recognized and that actions are taken to prevent things from going wrong.

In the 1960s, the moon flight program engineers, faced with staggering consequences of malfunctioning space vehicles, devised a method of forecasting the problems that could occur with every component. They were thinking beyond the normal design considerations, all the way to the most bizarre situations one could devise. They did this in long and concentrated brainstorming sessions. The result of this approach contributed to the success of the moon landing in 1969.

With the decline of the space program in the early 1970s, many NASA engineers found jobs in other industries and brought failure forecasting with them. The technique became known eventually as the FMEA. In 1972, NAAO, a Quality Assurance organization, developed the original reliability training program which included a module on the execution of FMEA.

Although good engineers have always performed an FMEA type of analysis on their designs, most of their efforts were documented only in the form of their final parts and assembly drawings. Repetition of past mistakes, however, was possible, because people were assigned to other tasks, left the company, etc. With liability insurance besieging, for instance, the automotive industry in the 1970s, FMEA became a natural tool to lower the occurrence of failures. Since that time, the discipline has been spreading among the multibillion dollar companies. In turn, these large companies have been pressing their suppliers to adopt FMEA to improve the reliability of their products.

* Adapted from a paper presented by S. R. Jakuba, S. R. Jakuba FMEA Consultants, at the 1987 Spring National Design Engineering Conference, ASME. By permission of the author.

The Process

Fully implemented, the FMEA process is applied to each new product and to any major change of an existing product. The FMEA documents become an integral part of the product design documentation and are as such continuously updated.

The FMEA process is used to ensure that all problems that could possibly occur in the design, procurement, and servicing of a product have been considered, documented, and analyzed.

In this regard, the Product Engineering organization has the responsibility for product performance criteria establishment, and product design and development, which includes consideration of manufacturability, service-ability, and the user's potential misuse. The Manufacturing organization has the responsibility for the fabrication or purchase of a product to an engineering drawing and specification. The Marketing, Sales, and Service organization has the responsibility for technical support of the product after sale.

Accordingly, there are three independent FMEA documents dealing with the three different aspects of the process. The Design FMEA lists and evaluates the failures which could be experienced with a product and the effects these failures could have in the hands of an end user. The Manufacturing FMEA lists and evaluates the variables that could influence the quality of a particular process. The Service FMEA evaluates service tools and manuals to ensure that they cannot be misused or misrepresented. In the following, we deal with the Design FMEA procedure only.

Design FMEA

The Design FMEA identifies areas that may require further consideration of design and/or test. It captures and implements design inputs, some of which might otherwise not be made, and, if made, might get lost. They include inputs from other departments such as Manufacturing, Sales, Purchasing, Service, Reliability, and Quality Assurance. Combining the different viewpoints and experience not only improves the design of a machinery product but also improves the acceptance of it throughout the company and in the field.

The Design FMEA, further referred to as FMEA, is initiated after a conceptual design has been finalized. It should be substantially completed before the production hardware is made, to ensure that the Production Release documentation includes the FMEA inputs and that the potential benefit of the FMEA information is fully utilized. Subsequent changes to a product should also be incorporated. FMEA documentation should be periodically updated to record the changes and their impact on reliability and risk.

The FMEA is initiated by the engineer responsible for release of the product for manufacturing after he has determined that the product as designed will perform to the performance specification, can be made and assembled to print, and is serviceable and "foolproof."

The objective in performing the Design FMEA is to:

- find whether the performance specification is proper and complete;
- find if and how the design could be inadequate both to the design intention as well as for reasons of overload, contamination, weather extremes, manufacturing variations, serviceability, customer misuse, or negligence, etc.;
- evaluate the consequences of a marginal product reaching a customer;
- quantify risk;
- identify the need for corrective actions and to assign priorities for their execution;
- implement and follow agreed-upon actions.

Definitions and FMEA Forms

The following definitions apply:

Failure mode	The manner in which a part or system fails to meet the design intent
Effect of failure	The experience the owner encounters as a result of a failure mode
Cause of failure	An indication of a design weakness
Cause prevention	The in place and scheduled design verifications and quality assurance inspections
Severity ranking	A subjective evaluation of the consequence of a failure mode on the end user
Occurrence ranking	A subjective estimate of the likelihood that if a defective part is installed it will cause the failure mode with its particular effect
Detection ranking	A subjective estimate of the probability that a cause of a potential failure will be detected and corrected before reaching the end user
Risk Priority Number (RPN)	The product of severity, occurrence, and detection rankings

Failure mode, effect of failure, and cause of failure serve to document all that could fail, how the failure would be perceived if it happened, and what could cause it. The cause prevention serves to document all existing and firmly scheduled measures intended to assure that the cause of a failure has been eliminated. Severity ranking, occurrence ranking, and detection ranking provide a numerical means of stating a subjective estimate of the respective parameters. Typically, on a scale of 1–10, the rankings represent a number which reflects how severe the effect of a failure is, how likely the failure is to happen, and how unlikely the cause of failure is to pass undetected.

The Risk Priority Number (RPN) is the number resulting from the multiplication of the three rankings. Risk Priority Number allows prioritization of the actions that need to be performed to lessen the risk.

Tables 7-1a and b illustrate a typical FMEA form. There is no one FMEA form that suits all companies and all applications. The first four of the above categories are, however, almost always present. Also, the form always contains a space for information needed to identify the product, such as drawing number, product application(s), where it is made, and its function. There should also be a space for the listing of corrective actions. The corrective actions are recommended by the FMEA participants, and regardless whether they will be pursued or not, they should all be recorded on the form.

Finally, there should be a space for the name of a person responsible for the implementation of a corrective action.

Procedure

The Design FMEA procedure is an integral part of a product development. The engineer responsible for the product should make entries on the FMEA forms, listing his thoughts and reasoning concurrently with performing the other design and test activities. It is important that information written on the forms is concise, clear, and systematically arranged, because people unfamiliar with both the product and FMEA will later read and evaluate the entries. If the entries are vague or incomplete, the potential of the FMEA effort will not be realized; not only will the time of several people be wasted but also potentially dangerous problems may be overlooked.

Experience indicates that it is more cost-effective not to perform an FMEA at all than to produce a vague, half-hearted one. A certain writing and organizational talent is needed to produce the FMEA document. Not every engineer has the talent, and not every engineer is willing to devote the time needed for researching all the information, and write and rewrite it until it conveys the relevant message in just three or four words.

Table 7-1
An FMEA form – Sheet 1 of 2

Failure Mode and Effect Analysis (FMEA)								Sheet No. _____ of _____
System:	Component:							
System status:	Component status:			Operating conditions:			Documentation:	
1	2	3	4	5	6	7	8 9 10 11	12
No.	Part/ function	Failure mode	Basic failure mode/ possible cause	Failure detection (surveillability)	Available countermeasures	Failure effects	Occurrence Severity Detection RPN	Failure assessment and recommended action

Table 7-1
An FMEA form – Sheet 2 of 2

Sheet No. _____ of _____

Failure Mode and Effect Analysis (FMEA)

1	2	3	4	5	6	7	8 9 10 11	12
No.	Part/ function	Failure mode	Basic failure mode/ possible cause	Failure detection (surveillability)	Available countermeasures	Failure effects	Occurrence Severity Detection RPN	Failure assessment and recommended action

When production drawings are available the engineer contacts the person responsible for the FMEA activities. Together they select the people who should review the FMEA drafts, and amend and rank the entries. The selection of the reviewers is done on the basis of their qualification both with respect to their knowledge of the product and their ability to contribute to the FMEA process. The selected participants are briefed on the product and on the duties expected of them in the FMEA process.

After they have had a chance to study the FMEA documents, gather information related to their involvement with the product and call an FMEA meeting to amend and rank the entries. The meeting may last several days, so its timing must be planned. During the meeting all the entries on the form are reviewed, recommended actions confirmed, and priorities assigned.

When managerial approval of the recommended actions is obtained, the actions are given deadlines. The control of the completion is assured by entries on the FMEA form, usually on a separate, shorter form, which lists the approved actions only.

It can generally be said that the training of the FMEA participants, and the effort involved in performing FMEA is substantial. The benefits of an FMEA are reliability enhancement and cost avoidance, not a measurable saving in the bottom line. Therefore, to be carried through in an effective way, the FMEA activities require the unconditional commitment of the management and a dedicated leadership.

The FMEA technique provides the means of presenting one's thoughts in a methodical way. The objective is to document all potential flaws of a product, evaluate the risks associated with each, and prevent the occurrence of high risks.

The benefits of the FMEA process extend clearly beyond the design aspects of a machinery product. It enables the designer and owner to gain a deeper knowledge of the product. Further, it increases the awareness of the product features by all involved parties and it provides a basis for an assessment of reliability, maintainability, and safety of similar or newly designed products.

Examples

In the preceding paragraphs we have seen that an FMEA produces the following results:

- It identifies potential and known failure modes.
- It identifies the causes and the effects of each failure mode.

- It prioritizes identified failure modes according to frequency of occurrence, severity, and defect formation.
- It allows to plan for problem follow-up and corrective action.

An effective FMEA depends on certain key steps. (We refer our readers to completed examples on pp. 151 and 153). The essential steps are as follows:

1. *Describe the anticipated failure mode.* The analyst must ask the question: "How could this part, system or process fail? Could it break, deform, wear, corrode, bind, leak, short, open, etc.?" Table 7-2 and the following list of failure mode functions may serve as a guide:

   1.1 Fails to open – complete or partial
   1.2 Fails to remain – in position
   1.3 Fails to close – complete or partial
   1.4 Fails to open
   1.5 Fails to close
   1.6 Internal leakage
   1.7 External leakage
   1.8 Fails out of tolerance
   1.9 Erroneous output
   1.10 Reduced output
   1.11 Loss of output
      – thrust
      – indication
      – partial
      – false
   1.12 Erroneous indication
   1.13 Excessive flow
   1.14 Restricted flow
   1.15 Fails to stop
   1.16 Fails to start
   1.17 Fails to switch
   1.18 Premature operation
   1.19 Delayed operation
   1.20 Erratic operation
   1.21 Instability
   1.22 Intermittent operation
   1.23 Inadvertent operation
   1.24 Rupture
   1.25 Excessive vibration

Table 7-2
Failure mode: Basic

2.1 Part/element level	2.2 Assembly level
2.1.1 Force/stress/impact	2.2.1 Force/stress/impact
1. Deformation	1. Binding
2. Fracture	2. Seizure
3. Yielding	3. Misalignment
4. Insulation rupture	4. Displacement
2.1.2 Reactive environment	5. Loosening
1. Corrosion	2.2.2 Reactive environment
2. Rusting	1. Fretting
3. Staining	2. Fit corrosion
4. Cold embrittlement	2.2.3 Temperature
5. Corrosion fatigue	1. Thermal growth/contraction
6. Swelling	2. Thermal misalignment
7. Softening	2.2.4 Time
2.1.3 Thermal	1. Cycle life attainment
1. Creep	2. Relative wear
2. Cold embrittlement	3. Aging
3. Insulation breakthrough	4. Degradation
4. Overheating	5. Fouling/contamination
2.1.4 Time	6. Plugging
1. Fatigue	
2. Erosion	
3. Wear	
4. Degradation	

The investigator is trying to anticipate how the design being considered could possibly fail. At this point, he should not make the judgment as to whether or not it *will* fail, but concentrate on how it *could* fail.

2. *Describe the effect of the failure.* The analyst must describe the effect of the failure in terms of owner reaction. In other words "What does the operator experience as a result of the failure mode described?" For example, in considering the failure mode of a diaphragm coupling in a high-speed turbine-driven process compressor application (Fig. 7-1a), the analyst would have to determine how this would affect the operation. Would there be a sudden acceleration of the turbine and would its overspeed protection device properly respond by activating the steam shut-off valve? Is there a need for a redundant emergency drive for safe run-down?

3. *Describe the cause of the failure.* The analyst will now anticipate the cause of the failure. Would temporary overload cause the coupling diaphragm failure? Would environmental conditions cause a

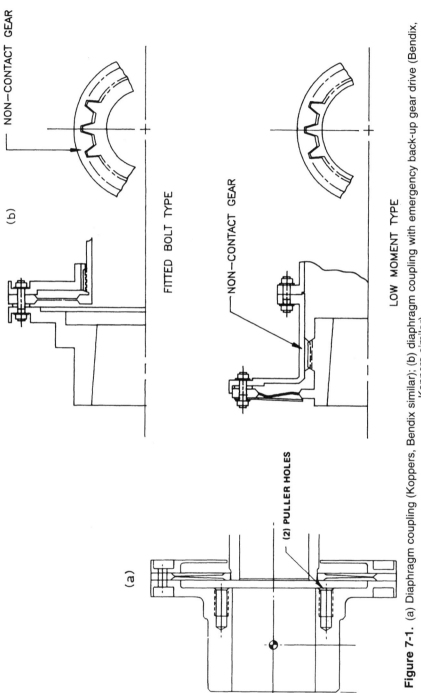

Figure 7-1. (a) Diaphragm coupling (Koppers, Bendix similar); (b) diaphragm coupling with emergency back-up gear drive (Bendix, Koppers similar).

problem? In short, the analyst investigates what conditions could bring about the failure mode. He concentrates on "FRETT," the possible effects of excessive Force, a Reactive Environment, abnormal Temperature and excessive Time.

4. *Estimate the frequency of occurrence of failure.* The analyst must estimate the probability that the given failure mode will occur. He assesses the likelihood of occurrence, based on his knowledge of the system, using an evaluation scale of 1–10. An 1 would indicate a low probability of occurrence, whereas a 10 would indicate a near certainty of occurrence.

5. *Estimate the severity of the failure.* In estimating the severity of the failure, the investigator weighs the consequence of the failure. An 1 here would indicate a minor nuisance, whereas a 10 would indicate a severe consequence such as "turbine run-away" or "stuck at wide open governor valve."

6. *Estimate failure detection.* The investigator will now proceed to estimate the probability that a potential failure will be detected before it can have any consequences. He will again use a 1–10 evaluation scale. An 1 would signal a very high probability that a failure would be detected before serious consequences would arise. A 10 would indicate a very low probability that the failure would be detected and consequences therefore would be appreciable. For instance, a failure of the above described diaphragm coupling might be assigned a detection probability of 10 because it would happen suddenly, without any detection possibilities. Similarly, a diaphragm coupling with an emergency run-down feature (Fig. 7-1b) would be assigned a detection probability of 4, because upon diaphragm failure there would be a detectable noise to allow initiation of contingency measures. Finally, the failure of the auxiliary resetting lever of the steam turbine overspeed trip system (Fig. 7-2) might be assigned a detection number of 1 for obvious reasons.

7. *Calculate the risk priority number.* The RPN obviously provides a relative priority of the anticipated failure mode. A high number indicates a serious failure mode. Using the risk priority numbers, a critical items summary can be developed to highlight the top priority areas that will require action.

8. *Recommended corrective action.* It is vital that the analyst takes sound corrective actions, or sees that others do the same. The follow-up aspect of the exercise is clearly critical to the success of this analytical tool. Responsible parties and timing for completion

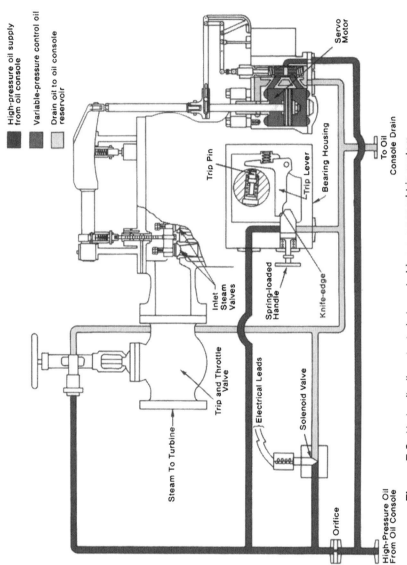

Figure 7-2. Hydraulically actuated steam turbine overspeed trip system. *Source:* United Technologies Elliott, Jeannette, PA.

should be determined for all corrective actions. The decision tree in Figure 7-3 can be used when deciding on corrective actions as a result of an FMEA.

The FMEA Form

The FMEA form (Tables 7-1a, b) may be used for machinery parts or assembly and systems failure mode and effect analysis. In order to complete the form the analyst needs the following information:

- system specifications;
- description of function, flow sheets, and drawings;
- description of operating conditions.

This information is entered into the appropriate rows. Components and their failure modes are numbered for identification (see column 1). Part, system, or process *function* are entered into column 2; *failure mode* into column 3; failure mechanisms and possible *causes* into column 4. *Failure mechanisms* in this context are more detailed explanations of the failure mode in terms of expanding on the mechanical, physical, or chemical mechanisms leading to the anticipated failure mode.

Failure causes should be listed as far as they are assignable to each failure mode. It would be well to assure that the list is all-inclusive so that remedial action can be directed at all pertinent causes. Examples of causes are as follows:

1.0 Design stage

    1.1 Wrong material selection, i.e. brittle when cold
    1.2 Wrong design assumptions, i.e. design temperature too low
    1.3 Design error

2.0 Materials, manufacturing, testing, and shipping

    2.1 Material flaw, i.e. inadequate plating thickness
    2.2 Improper fabrication, i.e. inferior welding quality
    2.3 Improper assembly, i.e. insufficient torque (fastener)
    2.4 Inadequate testing, i.e. not tested at operating conditions
    2.5 Improper preparation for shipment, i.e. part allowed to rust
    2.6 Physical damage, i.e. damaged in transit
    2.7 Insufficient protection, i.e. part or assembly dirty

3.0 Installation, commissioning, and operation

    3.1 Improper foundations, i.e. foundation sagging
    3.2 Inadequate piping support, i.e. piping deflects machinery

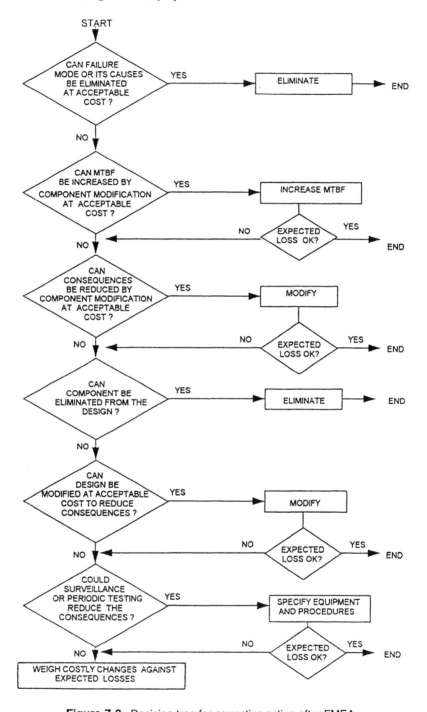

Figure 7-3. Decision tree for corrective action after FMEA.

3.3 Wrong final assembly, i.e. built-in misalignment
3.4 Improper startup, i.e. shaft bow in steam turbines
3.5 Inadequate maintenance, i.e. build-up of dirt
3.6 Improper operation, i.e. no lubrication

4.0 Basic failure modes (FRETT)

4.1 Failure due to high forces, stresses, and impact, i.e. broken stem on gate valve
4.2 Failure due to reactive environment, i.e. corroded casing on pump
4.3 Failure due to thermal problems, i.e. thermal rise causes misalignment
4.4 Time-dependent failures, i.e. aging causes O-ring leak

Column 5 shows the possibilities of *failure detection*, such as for instance automatic annunciation, inspections, and functional tests. Column 5 provides information on *surveillability*. Column 6 may contain information about appropriate countermeasures already available by design. These would be all measures and features contributing to limiting or avoiding the consequences of an anticipated failure mode. Examples are spare devices, redundancy designs, switch-over features, and devices which will limit consequential damage. When entering *failure effects* into column 7 we assume that the countermeasures listed in column 6 are effective. Effects of failure should be expressed in terms of operator reaction. The following will serve as a guide:

1. No effect.
2. Loss of redundancy, i.e. failure of one of dual shaft seals.
3. Functional degradation, i.e. excessive operating effort.
4. Loss of function, i.e. pump does not deliver.
5. Liquid/fumes/gas leakage/release, i.e. failing joint gasket.
6. Excessive noise/vibration, i.e. internal rub due to thermal expansion.
7. Violation of rules and safety standards, i.e. blocked safety valve.
8. Fails to indicate.
9. Fails to alarm.
10. Fails to trip.
11. Fails to start.
12. Fails to stop.

Column 8 evaluates the probability of *occurrence* on a scale of 1–10. For example, 10 would indicate an extremely probable occurrence, whereas 1 would signify a very improbable occurrence.

Column 9 estimates the *severity* or consequence of the failure on a 1–10 scale.

The number assigned to *detection* in column 10 is based on the probability that the anticipated failure mode will be detected before it becomes a problem. Again, 10 indicates a low probability that the failure would be detected before consequences occur. An 1 means high detection probability. Column 5 will help in the evaluation of column 10.

Column 11 contains the *risk priority number* (RPN) and is calculated by multiplying the numbers in columns 8–10, inclusive. The RPN number is an indicator of relative priority.

Finally, column 12 contains the anticipated failure assessment together with a brief description of the corrective actions recommended. Under remarks one would find the persons or departments responsible

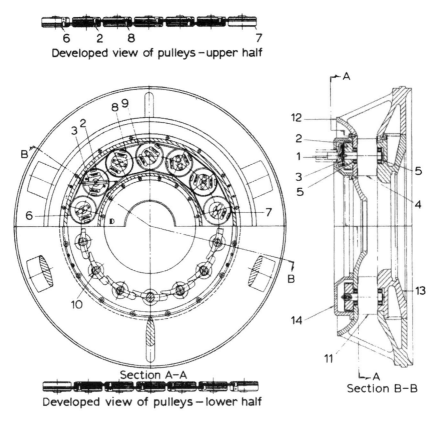

Figure 7-4. Adjustable inlet guide vane assembly. 1, Upper drive shaft; 2, drive pulley; 3, key; 4, guide vane; 5, ball bearing; 6, end pulley; 7, end pulley; 8, pulley; 9, 1/8″ diameter aircraft type cable; 10, lower drive shaft; 11, guide vane; 12, intake wall; 13, inboard support; 14, cover.

Table 7-3
FMEA example: Adjustable inlet guide vane assembly

No.	Part/function	Failure mode	Basic failure mode/ possible cause	Failure detection (surveillability)	Available countermeasures	Failure effects	Occurrence	Severity	Detection	RPN	Failure assessment and recommended action
1	2	3	4	5	6	7	8	9	10	11	12
1	Upper Drive Shaft	Fracture	Jamming	When making adjustments	Manual intervention	Reduced compressor output	2	3	6	36	Test periodically, i.e. Preventive maintenance
2	Drive Pulley	Loosening	Key sheared	Visible erratic movement	Manual intervention	Reduced compressor output	1	3	6	18	Test/exercise periodically
3	Shaft Key	Shearing/ Fracture	Improper fitting procedure	See above	As above	As above	1	3	6	18	See above
9	Cable	Breakage	Breakage								

Failure Mode and Effect Analysis (FMEA)

System: Variable Inlet Guide Vane Assembly

Component: As per Bill of Material or Assembly Drawing

System status: Inlet GV Operating

Component status: Controlling Compressor Inlet Flow

Operating conditions: Ambient Pressure 60–100°F Oily and dusty

Documentation: Comp. Outline Drawing Inlet GV Assembly Drawing Operation and Maintenance Manual

Sheet No. _____ of _____

for corrective actions as well as their status in terms of progress
and timing.

Evaluation

The evaluation of the effects of component failures may be done according to different criteria. Examples of assessment criteria are:

- The maintenance case, i.e. the failure effect does not lead to system
  failure.
- System failure.
- Inadmissible system status, i.e. the failure effect results in a system
  status which violates safety rules.
- Danger status, i.e. the system's risk potential is being liberated.

Examples

Examples are presented in the form of an assembly or part FMEA covering the adjustable inlet guide vane assembly of a critical process gas
compressor (Fig. 7-4). Table 7-3 shows the completed FMEA analysis.
Another example is an analysis of a compressed air system (Fig. 7-5 and
Table 7-4).

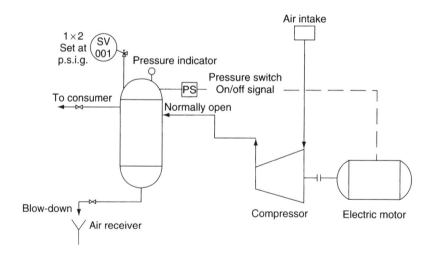

Figure 7-5. Schematic diagram of a compressed air system.

Table 7-4
FMEA example: Air compression system component [1, 2]

Failure Mode and Effect Analysis (FMEA)

Sheet No. 2 of 3

System: Air supply system		
Component: Safety value		
System status: Undisturbed design conditions, pressurized	Component status: Design conditions (closed)	Operating conditions: Room temperature: 50–100 °F, Relative humidity: <80%, Dust-free atmosphere
		Documentation: Drawings, System specifications

1	2	3	4	5	6	7	8	9	10	11	12
No.	Part/function	Failure mode	Basic failure mode/possible cause	Failure detection (surveillability)	Available countermeasures	Failure effects	Occurrence	Severity	Detection	RPN	Failure assessment and recommended action
2.1	Staying closed	Leakage	Spring fatiguing	Compressor kicking off/on more often. Field inspection will detect increased noise level	Compressor keeps up supply of air	Compressor makes up for pressure loss	5	2	3	30	Preventive maintenance case, shutdown of system for repair
2.2	Staying closed	Fails to open	Spring fracture	Pressure indication, field inspection	None	Rapid loss of pressure	6	7	1	42	System outage

Table 7-4
FMEA example: Air compression system component–cont'd

Sheet No. _____ of _____ 12

No.	Part/function	Failure mode	Basic failure mode/ possible cause	Failure detection (surveillability)	Available countermeasures	Failure effects	Occurrence	Severity	Detection	RPN	Failure assessment and recommended action
1	2	3	4	5	6	7	8	9	10	11	
2.3	Opening at $110 < P < 120\,\text{psi}$	Fails to close	Corrosion, dirt, wrong setting	None, by valve inspection and test	None	No immediate effects, loss of safety function at over-pressuring	2	8	10	160	* Intolerable system condition * Enforce safety valve inspection and test program * Provide indicators that safety valve was activated

Failure Mode and Effect Analysis (FMEA)

References

1. Browning, R. L., Analyzing industrial risks. *Chemical Engineering*, October 20, 1969, pp. 100–114.
2. DIN 25448, *Ausfalleffektanalyse*. Berlin: Beuth Verlag, 1980.

Chapter 8
Fault tree analysis

Fault Tree Analysis (FTA) is a deductive method in which a hazardous end result is postulated and the possible events, faults, and occurrences which might lead to that end event are determined. Fault Tree Analysis also overlaps Sneak Circuit Analysis (SCA) because the FTA is concerned with all possible faults, including component failures as well as operator errors.

Sneak Circuit Analysis is used to troubleshoot and improve hydraulic, electronic, shutdown instrumentation and other control interfaces around process machinery [1].

Fault Tree Analysis is a "top–down" analysis that is basically deductive in nature. The analyst identifies failure paths by use of a fault tree drawing. A fault tree is a graphical representation of a thought process. It is constructed from events and logical operators. An event is either a component failure or system operation. The events and their graphical representation are given in Table 8-1.

A fault tree commences by selecting a top event. This event is the undesired event or ultimate disaster. From there, the analyst endeavors to find the immediate events that can, in some logical combination, cause the top event. These lower events are examined, in turn, for causes and the process is repeated to levels of greater detail. Ideally, the lowest level events will be all basic events and represented by a circle.

Fault trees provide a method for determining the logical causes of a given event. It illustrates all of the ways an undesired event can occur. It helps determine the critical components and the need for other analytical efforts. Numerical computations indicating the probability of occurrence for the top event and intermediate events can be obtained. The major drawback of the fault tree is that *there is no way to ensure that all causes have been evaluated consistently*: large fault trees are difficult

Table 8-1
Fault tree analysis symbols

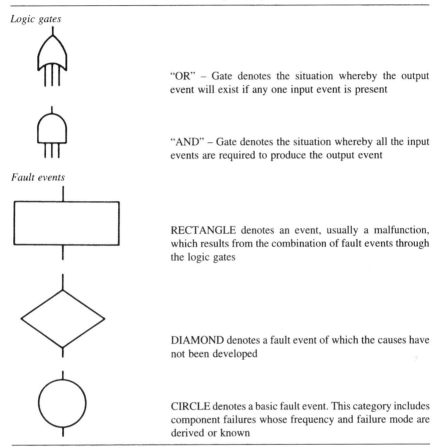

Logic gates

"OR" – Gate denotes the situation whereby the output event will exist if any one input event is present

"AND" – Gate denotes the situation whereby all the input events are required to produce the output event

Fault events

RECTANGLE denotes an event, usually a malfunction, which results from the combination of fault events through the logic gates

DIAMOND denotes a fault event of which the causes have not been developed

CIRCLE denotes a basic fault event. This category includes component failures whose frequency and failure mode are derived or known

to understand. On the system level, they do not resemble the system flowsheet. Complex logic is frequently involved.

Fault Tree Analysis is performed on the system configuration, determined by the analyst. Determining the configuration of a system is generally central to all analyses.

Although this concept, like the FMEA earlier, has been intuitively used by engineers for a long time, its systematic and formal application in reliability analysis is relatively recent. Events which could cause the top event are generated and connected by logic operators AND, OR, and EOR. The AND gate provides a TRUE output if and only if all the inputs are TRUE. The OR gate provides a TRUE output if and only if one or more inputs are TRUE. The EOR, exclusive OR, gate provides a TRUE output if and only if one but not more than one input is TRUE [2, 3]. The analysis proceeds by generating events in a successive manner until the events need not be developed further. Those events are called primary events. The fault tree itself is the logic structure relating the top event to the primary events.

The linking of events according to logical rules is shown in Figure 8-1. Fault Tree Analysis may be applied at any level from component part to full system. General applications of FTA are:

- reliability assessment of machinery parts (see the compressor rotor example on p. 163);
- reliability assessment of simple subsystems;
- probability assessment of specific failure events in complex systems;
- critical failures identified by FMEA.

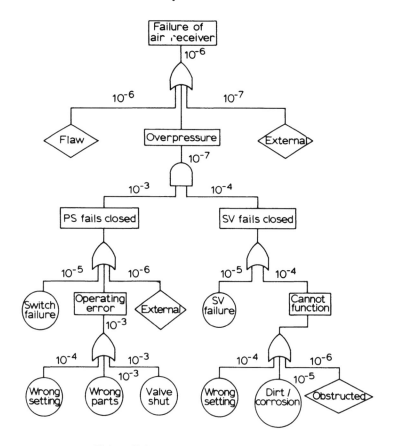

Figure 8-1. Fault tree (adapted from [7]).

Procedure

Although the fault tree can become complex, each part is simple. (For reasons of simplicity the reader is referred to the compressed air system in Fig. 7-5, which is used to illustrate the procedure.) Figure 8-1 depicts the fault tree. The following eight steps should be followed:

1. *Define the undesired event.* "Failure of Air Receiver" has been chosen. The system can fail in other ways as shown in the preceding FMEA. A tree can be drawn for each defined event.
2. *Identify the possible prime causes of failure.* These are shown on the fault tree as "defect," "overpressure," and "external events."
3. *Identify conditions which could contribute to the prime causes.* Both "defect" and "external events" are shown as "undeveloped events." The reader should imagine that these events could of course be developed downwards to pinpoint design deficiencies for instance, or aging effects such as fatigue and corrosion. Further, external events could be shown as earthquakes or fire.

   All three conditions could result in the undesired or top event. They are therefore connected through an OR gate.
4. *Repeat step (3) at the next lower level.* Only the "overpressure" condition remains. Two causes are shown – the overpressure cut-out switch does not open when required and the relief valve fails to blow. Both conditions are required to produce the undesired event. They are therefore connected by AND gates.
5. *Continue to the required level.* Four events were left undeveloped in Figure 8-1. They reflect the designers judgment that their probabilities were small enough to be ignored.

   We are now left with five basic inputs to which probabilities may be assigned. In a critical case, each of these could be developed further to eliminate inherent weaknesses or to reduce the likelihood of human error.
6. *Determine whether or not quantitative analysis is required.* The usefulness of the fault tree technique can be enhanced by the use of quantitative data. In this way not only can the fault paths be identified, but their probability of occurrence may be established [4].

   The decision to employ quantitative analysis should be made on the basis of experiences, system complexity, and severity of consequences. One should ask the following questions:

   • What is the severity of the undesired event?
   • Are quantitative data in terms of failure rates available, meaningful, and relevant?

- Does the fault tree contain many AND gates, expressing degrees of redundancy?
- Does a particular branch of the fault tree appear marginal?
- Do we want to commit resources to this tedious task?

7. *Allocate a probability value to each event.* A failure rate can be assigned to each input event. The probability of the undesired event, or top event, can then be calculated.

Failure rate data can come from experience, test data, published data as shown in Table 5-3, or engineering judgment. The latter is applied as a first approach. Here the analyst makes use of arbitrary relative probabilities which can be selected from Figure 8-2.

We proceed to assign probabilities from Figure 8-2 to each event in Figure 8-1.

It would be well to consider maintainability and surveillability factors at this time. For instance, a high probability has been assigned to "valve shut" as the valve could be left closed after it has been maintained. As during an FMEA, we should ask ourselves the following questions in this context:

- What is the general maintenance and operational environment?
- Can an item be overlooked; can it be maintained?
- Is the part unusual or non-standard?
- Can status be easily ascertained, i.e. an open or closed gate valve?
- Is it difficult to assemble?
- Can it be installed incorrectly?
- What is the in-service failure or deterioration mode, i.e. how is it influenced by FRETT:
  - force
  - reactive environment
  - temperature
  - time

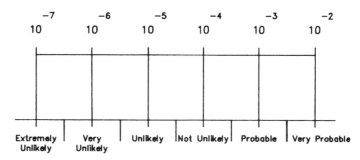

Figure 8-2. Relative probabilities. *Source*: Atomic Energy of Canada Ltd.

8. *Connect the input values*. The rules for each type of gate as shown in Table 8-1 might not be rigorously correct. Errors, however, are negligible for inputs smaller than 10^{-1}. The reader is referred to the list of references at the end of this chapter for further information on the subject. Two rules have to be observed:

- The output of an AND gate is the product of the input.
- The output of an OR gate is the sum of the inputs.

Because powers of ten are used we only have to add or subtract. Low probabilities do not have to be accumulated. For example:

The AND gate leading to "overpressure" in Figure 8.1:

$$10^{-3} \times 10^{-4} = 10^{-7}$$

The OR gate leading to "PS fails to Open":

$$10^{-4} + 10^{-3} + 10^{-6} \approx 10^{-3}$$

If only a few data points are available for a quantitative FTA, one might want to resort to a method proposed in [5]. In order to arrive at a failure prediction at the unit level, this approach combines subjective weighting of part failures and failure modes with objective data for a small number of failures or failure modes. To do this the complete tree is developed down to a part or part failure mode level. At each gate subjective probability estimates are made by using service engineers with relevant experience [6].

With a fully weighted tree it is possible to take one piece of hard data relating to one part failure or part failure mode and use the subjective weightings to propagate this upwards through the tree to arrive at an equipment failure rate estimate.

Examples

Three examples are shown. Example 1 demonstrates how fault trees have been used to explain failure events logically. Figure 8-4 is an expansion of one of the events in Figure 8-3. It illustrates the events leading to mechanical bearing failure. The example conveys that fault trees can be used effectively without necessarily assigning probabilities or failure rates.

Example 2 deals with an investigation made in connection with a rotor (Fig. 8-5) for a process gas compressor owned by a major petrochemical company. The effort is depicted in Figure 8-6 and had to be undertaken in order to justify the replacement of a spare rotor that had been damaged during repair and overhaul.

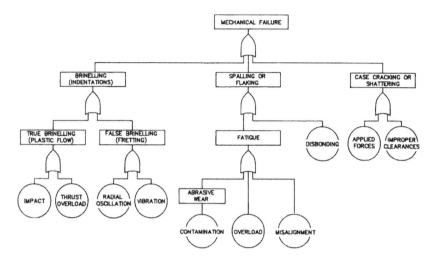

Figure 8-3. Fault tree of mechanical bearing failure [6].

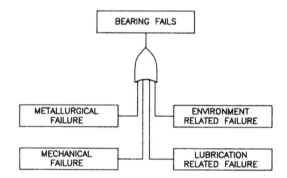

Figure 8-4. Types of bearing failure shown as intermediate events on a fault tree [6].

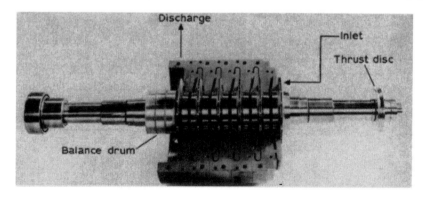

Figure 8-5. Centrifugal compressor rotor.

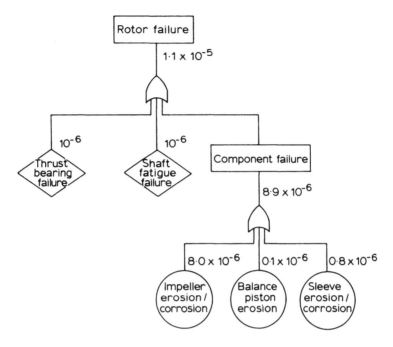

Figure 8-6. Fault tree: centrifugal compressor rotor.

Example 3 explains the failure events leading to the no-flow or low-flow initiated mechanical failure of a multistage deepwell pump (Fig. 8-7). The fault tree is shown in Figure 8-8.

Assessment and Evaluation

Assessment of the FTA results should lead to an action plan. Several questions should therefore be asked:

- Is the overall reliability acceptable? In our last example the relative probability of the undesired event is 0.00001/h, or "unlikely."
- What inputs are subject to large uncertainty? In our examples, there are no appreciable uncertainties.
- Is there substantial redundancy? The distribution of AND gates is an indication of the degree of redundancy.
- Could loss of redundancy go undetected? Essentially, we will ask the same questions covered in our FMEA procedures.
- Do we need help? If the risk seems high and it cannot be lowered without a major system modification, it would be a good idea to summon help from a specialist.

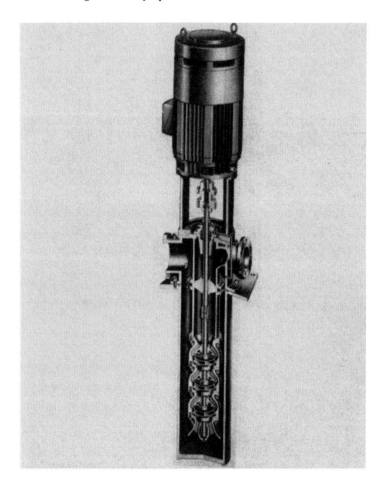

Figure 8-7. Multistage deepwell pump (Goulds).

Review

FTA has advantages and disadvantages. Listing the advantages motivates purpose and application of the method:

- The fault tree can serve in all phases of the machinery life cycle because it can help to determine possible causes of undesirable events.
- FTA may be used to evaluate competing designs by revealing qualitative and quantitative event interdependencies.

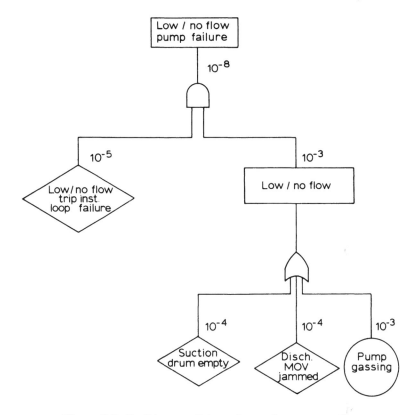

Figure 8-8. Fault tree: multistage deepwell pump operation.

The disadvantages of FTA can be explained by its design principle. It attempts to build a mathematical model of a complex physical condition by logical linking of events. If all peripheral, environmental, and operating conditions are not defined, then the method depends on the judgment of the analyst.

- One chief disadvantage is that there is no effective formal control against overlooking of events or the neglect of operating or environmental conditions. The best preventive measure would be to have several analysts make independent analyses.
- One main difficulty with a quantitative FTA exists in the lack of reliable and relevant failure rate data as well as the probabilities of events.
- Finally, the construction of fault trees can demand a lot of effort and may become expensive.

References

1. Bloch, H. P. and Geitner, F. K., *Machinery Failure Analysis and Troubleshooting.* Houston: Gulf Publishing, 1983, pp. 560–579.
2. Dhillon, B. S., *On Reliability and Maintainability Engineering.* Advance paper, University of Ottawa, 1980.
3. Dhillon, B. S., Bibliography of literature on fault trees. *Microelectron. Reliab.* **17**, 1978, pp. 501–503.
4. Scerbo, F. A. and Pritchard, J. J., *Fault Tree Analysis: A Technique for Product Safety Evaluation.* New York: ASME publication 75-SAF-3.
5. Innes, C. L. and Hammond, T., Predicting mechanical design reliability using weighted fault trees. *Failure Prevention and Reliability Conference*, Chicago, September 1977, pp. 213–228.
6. Strauss, B. M., Fault tree analysis of bearing failures. *Lubrication Engineering*, November 1984, pp. 674–679.
7. AECL-4607, *Reliability and Maintainability Manual*, edited by J. G. Melvin and R. B. Maxwell. Atomic Energy of Canada Ltd. 1974.

Chapter 9
Machinery risk and hazard assessment

The general objective of a hazard and risk assessment is the identification of machinery features which could threaten the safety of personnel, property, or the environment. Hazard and risk assessment methods are evaluated in Table 9-1. *Hazard* is defined as the source of harm, and *risk* is the possibility of experiencing this harm. For example, the hazard around a pumping service for toxic material could be the failure of the shaft seal. Two designs are suggested to prevent a leakage of the toxic material. Design B uses multiple mechanical seals (Fig. 9-1B), whereas design A calls for a single mechanical seal (Fig. 9-1A). Both designs may fail during operation of the pump. However, the probability of a toxic release for the multiple seal design (Fig. 9-1B) is much less than for the design incorporating only a single seal. Consequently, for the same hazard level, design B poses less risk for the plant operators and the public than design A.

From this, risk can be defined [2] as the answer to three questions:

- What can go wrong?
- How likely is it to go wrong?
- What are the effects and consequences?

The answer to the first question is a series of accident or incident scenarios. The answer to the second question is the probability of any given scenario. The answer to the third question lies in arriving at a measure of the extent of damage. This can be, as in our example, the number of people affected by the toxic release, the extent of damage to the environment, or the amount of business losses.

167

Table 9-1
Risk and hazard assessment methods [1]

Method	Characteristic	Advantages	Disadvantages
Preliminary hazards analysis	Defines the system hazards and identifies elements for FMEA and fault tree analysis; overlaps with FMEA and criticality analysis	A required first step	None
Failure mode and effect analysis (FMEA)	Examines all failure modes of every component. Hardware oriented	Easily understood. Well accepted, standardized approach, non-controversial, non-mathematical	Examines non-dangerous failures. Time-consuming. Often combinations of failures not considered
Criticality analysis	Identifies and ranks components for system upgrades. May be part of FMEA	Well-standardized technique. Easy to apply and understand. Non-mathematical	Follows FMEA. Frequently does not take into account human factors, common cause failures, system interactions
Fault tree analysis	Starts with "top event" and finds the combination of failures which cause it	Well accepted technique. Very good for finding failure relationships. Fault oriented; we look for ways system can fail	Large fault trees are difficult to understand, bear no resemblance to system flowsheet, and are not mathematically unique. Complex logic is involved
Event tree analysis	Starts with initiating events and examines alternative event sequences	Can identify (gross) effect sequences, and alternative consequence of failure	Fails in case of parallel sequences. Not suitable for detailed analysis
Cause–consequence analysis	Starts at a critical event and works forward using consequence tree; backwards using fault tree	Extremely flexible. All-encompassing. Well documented. Sequential paths clearly shown	Cause–consequence diagrams can become too large very quickly. They have many of the disadvantages of fault trees
Hazards and operability studies (HAZOP)	An extended FMEA which includes cause and effect of changes in major plant variables	Suitable for large plants	Technique is not well standardized

Single Cartridge Seal   **Double Cartridge Seal**

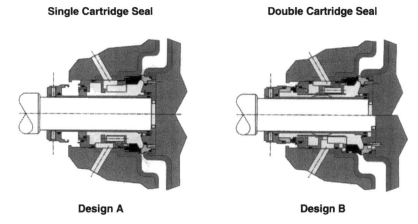

Design A   **Design B**

Figure 9-1. Cross-sections of typical single (left) and double (right) mechanical cartridge seals for chemical process pumps (courtesy of AESSEAL, pty, Rotherham, UK, Knoxville, Tennessee).

Assessing Risk

The methodology for assessing risks from rare events has gone through more than two decades of development. It started in the defense industry and is practiced in the nuclear industry. Today, many different industries, such as the hydrocarbon processing industry (HPI), are using and modifying the basic methods to match their needs.

An example is the quantitative risk analysis developed by a major HPI company [3]. The primary goal was to determine investment levels as a consequence of safety considerations. The company first analyzed their past experience with various types of process machinery and equipment. It then used the statistical data to establish a computerized data bank. The data bank allowed them to prepare a curve similar to that in Figure 9-2 for each type of equipment. Actual graphs might also show upper and lower 95% confidence limits as dashed lines above and below the main curve. Property damage is on the *y*-axis, and the probability of occurrence per unit-year for a dollar loss equal or greater than a given value is on the *x*-axis.

The curves can be used in two ways. First, they reflect historical experience and make estimates of probabilities possible. As an example, the probability of a $500,000 or greater loss for the particular equipment represented by Figure 9-2 is about one incident per 1000 unit-years of service. These curves, then, are tools for quantitative risk assessment. They are primarily used for screening purposes and are one factor in a decision-making process. Other factors are, for instance, public relations aspects, government regulations, personnel exposures, and so forth.

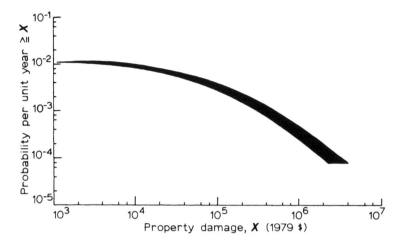

Figure 9-2. Fires/explosions, 1961–75, in hydrocarbon processing plants [2].

A second way of applying the graphs is as part of a risk analysis process. It is really a reversal of the procedure mentioned above. After the potential losses for equipment and business interruption are estimated, the probability of their occurrence is determined by finding the point of intercept of the loss curve with the *y*-axis. Multiplying the total loss by the probability results in the annual loss costs. Here is where the risk is quantified in dollars.

Assessing Hazards

An important first step in evaluating the risk associated with a particular machinery system is to establish the source of hazard. Often it will not be necessary to employ the whole range of methods shown in Table 9-1. A first approach could be experience-based observation of the facts. Other simple methods are checking existing designs against basic technical rules (see Fig. 9-3), internal company standards for new equipment, local and national codes, or industry standards such as issued by ANSI, API, and so forth. The next step would perhaps be a preliminary hazard analysis.

Preliminary Hazard Analysis (PHA)

This is the first systematic analysis of the machinery system and is designed to identify gross system hazards as the basis for more rigorous and detailed analysis later.

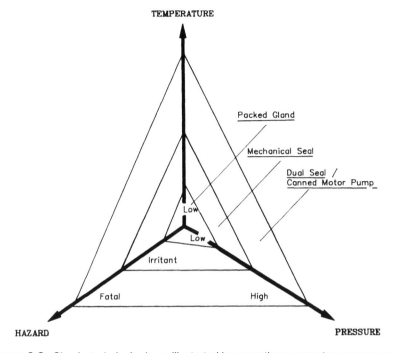

Figure 9-3. Simple technical rule as illustrated by operating parameters versus pump shaft seal selection.

It can be stated that PHA is an examination of the generic hazards known to be associated with a system at its conceptual phase of development. The purpose of this analysis is to:

1. Identify hazards.
2. Determine the effects of the hazards.
3. Establish initial safety requirements.
4. Determine areas to monitor for safety problems.
5. Initiate the planning of a safety program.
6. Establish safety scheduling priority.
7. Identify areas for testing.
8. Identify the need for additional analyses.

The PHA determines the recognized and anticipated design safety pitfalls and provides the method by which these pitfalls may be avoided. When this analysis is undertaken, there is little information on design details and less on procedures.

The PHA is usually a top-level review for safety problems. In most instances, the following basic steps are undertaken for a PHA:

1. Review problems known through past experience on similar machines or systems to determine whether they could also be present in the equipment under consideration.
2. Review the functional and basic performance requirements, including the environments in which operations will take place.
3. Determine the primary hazards that could cause injury, damage, loss of function, or loss of material.
4. Determine the contributory and initiating hazards that could cause or contribute to the primary hazards listed.
5. Review possible means of eliminating or controlling the hazards, compatible with functional requirements.
6. Analyze the best methods of restricting damage in case there is a hazard due to loss of control.
7. Indicate who is to take corrective action, and the actions that each will undertake.

Three basic approaches that can be used to ensure that all hazards are being covered are the columnar form, top-level fault tree, and narrative description. These methods will not in themselves find hazards. They will orient the analyst so that a thorough coverage of all aspects of the system will be performed.

The columnar form is the simplest method to implement. The chief advantage is that it is easy to review. The form has a heading that patterns questions in the mind of the analyst. The headings must at least incorporate the following terms or descriptions: Hazard, Cause, Effect, Hazard Category, Corrective or Preventive Measures.

The hazard is the generic area or condition that may influence system safety. The following is a partial list of hazards (the analyst can usually think of many more):

- acceleration
- contamination
- corrosion
- chemical dissociation
- electrical
- explosion
- fire
- heat and temperature
- leakage

- moisture
- oxidation
- pressure
- radiation
- chemical replacement
- shock (mechanical)
- stress concentrations
- stress reversals
- structural damage or failure
- toxicity
- vibration and noise
- weather and environment.

The cause column of a report is used to explain when the system is exposed to the hazard. It is here that the results of system generation are considered. Project phasing must also be considered, as well as an estimate of the percentage of system operation time that the hazard will be in effect.

An effect column is system-centered. It details the action of the hazard on system operation. In this column the possibility of causing injury or death, however remote, must be stated.

The hazard category is a numerical measure of how important the hazard is. The number of categories should be kept small, usually four or less, so that attention may be placed where it will do the most good.

The corrective or preventive measures column is almost self-explanatory. Here, methods of abating the hazard are given.

A top-level fault tree follows the method of FTA with generic events. Although this method helps define causes and effects, it does not follow that the system is checked hazard by hazard. Since the fault tree is event-oriented, it helps analyze undesired events, but does not determine that a particular event is a hazardous condition, element, or potential accident.

The narrative approach is less rigorous, and usually less complete, than the top-level fault tree and narrative approaches. Narrative writing style is a lengthy and time-consuming task. This approach is less susceptible to systematic method or technique and, therefore, the results usually have serious gaps or incomplete areas. The hazardous conditions and potential accidents are generally identified from experience, and then are explained in great depth and detail, more on the order of a final report than an analysis.

Once a PHA has been gone through, a more thorough hazard assessment may be made using the techniques of a hazard and operability study.

Hazard and Operability (HAZOP) Study

A HAZOP study is usually a systematic technique for identifying hazards or operability problems throughout an entire facility [4]. In our context, HAZOP studies have been successfully applied around major compressor installations. Here the technique provides opportunities to think of all possible ways in which hazards or operating problems might occur. In order to reduce the chance that something might be missed, this is done in a systematic way, each pipeline and each sort of hazard being considered in turn.

A pipeline for our purposes here is one that joins two pieces of equipment. Our example is a high pressure gas supply system (Fig. 9-4) consisting of two reciprocating compressors (Fig. 9-5), necessary motorized valves for remote operation, and interconnecting piping. The compressors are designed to move process gas from a common source of supply to either a customer or a high-pressure storage facility. The objective of a HAZOP review was specifically to find out whether or not the compressor piping could be simplified by eliminating some of the existing safety valves and check valves without compromising the design, safety, and operability of the system.

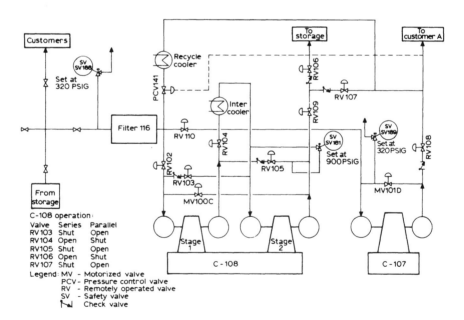

Figure 9-4. Simplified flow plan: gas compression system.

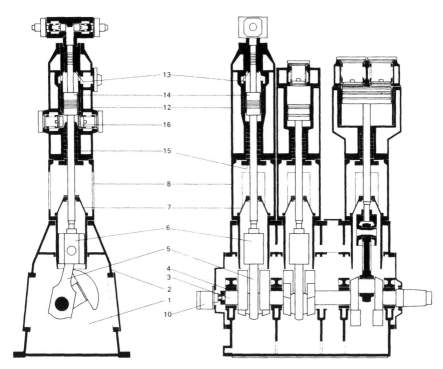

Figure 9-5. Major reciprocating compressor (Sulzer). 1, Crankcase; 2, frame; 3, crankshaft; 4, bearing; 5, connecting rod; 6, crosshead; 7, cover; 8, distance piece; 9, purge chamber; 10, lubricating group for crankcase; 12, cylinder; 13, cylinder liner; 14, piston; 15, piston rod packing; 16, valves; 17, capacity control.

Table 9-2 is a summary of the HAZOP investigation. The review items are usually dealt with by applying a series of HAZOP guide words [5]:

- none
- more of
- less of
- part of
- more than
- other than.

"None" in our example means no forward flow or, in effect, reverse flow when there should be forward flow. The questions to ask now are:

- Could there be reverse flow?
- If so, how could it happen?
- What are the consequences of reverse flow?
- Are the consequences hazardous or do they just prevent efficient operation?

Table 9-2
Summary of risks and suggested actions

Item no.	Equipment no.	Hazard	Possible cause	Probable severity	Existing indication and/or protection	Comments	Suggested actions	Follow-up by
1	C-108 Remove check valve at RV-103	Backflow from C-108 interstage to suction piping	Failure of RV-103 (i.e., valve open but limit switches show closed)		– SV188 protects suction piping at 320 psi – SV181 protects C-108 2nd stage from high P. – Customer line will go high on pressure – indication at unit – at 270 psi dump valve to storage will open. – During startup of a machine, the operator will be in the area and he will: a) Check visually the position of valves before startup b) Hear SV release if it occurs	Valve failure of RV-103 only a concern during startup Possibility of a RV valve opening unassisted is very remote. Single stage bypass has a check valve, however, overall bypass has *no* check. (Check if the bypass would be going in the wrong direction)	Remove check valve at RV-103	(Name)
5	C-107 Check valve at RV108	Backflow from Customer A to C-107 suction	During S/U of C-107 all the block valves including the bypass MV101D will be open. Only the check valve at RV108 is keeping Customer A from depressuring back		SV189 protects C-107 suction piping. Unit set to dump to storage at 270 psi	Check valve at RV108 is required for operability of C-107. This valve should be the most reliable type of check valve available	Review alternate check valve types	(Name)

- If so, can we prevent reverse flow by changing the design or the operating procedures?
- If so, does the size of the hazard, i.e. severity of consequences multiplied by the probability of occurrence, justify the additional expense?

Since our objective is the removal of redundant check valves, "none" or "reverse flow" were the only guide words used.

When looking at the lines containing safety valves we would of course invoke additional guide words such as "more of" (i.e., more pressure, meaning failure to open) or "less of" (i.e., less pressure, indicating failure to close or reseat). The meanings of the guide words are summarized below and Figure 9-6 reflects the entire HAZOP process.

HAZOP studies are now being conducted on a routine basis in many companies for all new units and major modifications. Some opinions exist in the United States that it is no longer a question of "if" the

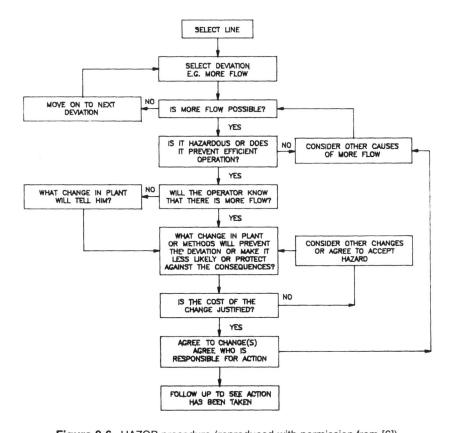

Figure 9-6. HAZOP procedure (reproduced with permission from [6]).

HAZOP guide word	*Deviations*
None	No forward flow when there should be, i.e. no flow or reverse flow
More of	More of any relevant physical property than there should be, e.g. higher flow (rate or total quantity), higher temperature, higher pressure, higher viscosity, etc.
Less of	Less of any relevant physical property than there should be, e.g. lower flow (rate or total quantity), lower temperature, lower pressure, etc.
Part of	Composition of system different from what it should be, e.g. change in ratio of components, component missing, etc.
More than	More components present in the system than there should be, e.g. extra phase present (vapor, solid), impurities (air, water, acids, corrosion products), etc.
Other than	What else can happen apart from normal operation, e.g. startup, shutdown, uprating, low running, alternative operation mode, failure of plant services, maintenance, catalyst change, etc.

government – through the Occupational Health and Safety Administration (OSHA) – will require HAZOP reviews, but "when."

There is no doubt in the authors' opinion that the result of a HAZOP review frequently has a significant impact on the cost of a project. Since large process machinery trains are considered major subsystems, they too can affect that cost.

For more information see Appendix B, "Safety Design Checklist for Reliability Professionals."

References

1. Henley, E. J., *Designing for Reliability and Safety*. Gas Processors Association, Annual meeting, San Antonio, Texas, March 10–12, 1986.
2. Kazarians, M. and Boykin, R. F., Assessing risk in the CP plant. *Chemical Processing*, October 1986, pp. 78–82.
3. Young, R. S., *Risk Analysis Applied to Refinery Safety Expenditures*. API Committee on Safety and Fire Protection, Fall meeting, Denver, Colo., September 23, 1986.

4. Lawley, H. G., *Operability Studies and Hazard Analysis*. In "Loss Prevention" series, Vol. 12. New York: AIChE, 1974.
5. Kletz, T. A., Eliminating potential process hazards. *Chemical Engineering*, April 1, 1985, pp. 48–68.
6. Kletz, T. A., *HAZOP & HAZAN: Notes on the Identification and Assessment of Hazards*. Rugby, UK: ICE, 1983.

Chapter 10
Machinery system availability analysis

Earlier, we defined availability of a system as *the fraction of time it is able to function*. System availability is clearly the consequence of subsystem availabilities which in turn are a function of assembly and part level availabilities.

Operating time can frequently be greater than available time. Available time therefore sets a limit on production.

The objectives of an availability analysis are:

1. To estimate system availability for comparison with a target value.
2. To identify low availability components, assemblies, or parts for improvement.

Availability analysis is an extension of FMEA. One specific effect, namely unavailability, is estimated. System availability analysis should therefore be performed after an FMEA has been completed.

The Prediction Approach

System availability assessment is a tool that can be applied during all life-cycle phases of a machinery system. Its results can be used by management as availability control. It can identify deficiencies in certain areas, or compare alternative designs.

The work steps are as follows. The system outage time caused by each level of the hierarchy is estimated from failure and maintenance data and recorded on a work sheet. Three types of downtime are identified: forced maintenance, predictive-scheduled maintenance, and planned (turnaround) maintenance. The sum of all types of downtime is used to compute machinery system availability.

The availability (A) of a component is expressed as:

$$A = \frac{\text{total time} - \text{maintenance time}}{\text{total time}} \qquad (10.1)$$

$$= 1 - \frac{\text{maintenance time}}{\text{total time}}$$

$$= 1 - \text{unavailability}(UA)$$

The unavailability of a unit is a function of the time required for two modes of maintenance: breakdown-corrective and predictive-preventive. Unavailability resulting from breakdown maintenance or forced unavailability is:

$$UA_F = \frac{\lambda \times t \times T_R}{T} \qquad (10.2)$$

where $t =$ operating time in hours,
$\lambda =$ failure rate per hour,
$T_R =$ average repair time in hours,
$T =$ total time in hours.
With high availability, t becomes approximately:

$$t = T$$

The error in this approximation is small compared to errors in λ and T_R. Similarly, preventive maintenance-related unavailability can be approximated to be:

$$UA_P = \frac{T_M}{T} \qquad (10.3)$$

where $T_M =$ average preventive maintenance time in h/year,
$T = 8760\,\text{h/year}$.
Total unavailability of an item under scrutiny then is:

$$UA = \lambda T_R + \frac{T_M}{T} \qquad (10.4)$$

Machinery System Unavailability

In order to be able to estimate a machinery system's unavailability, we require two pieces of information:

1. The system downtime required to perform a particular job.
2. The job priority determined by business impact.

If a machinery system must be shutdown for maintenance, repair, or overhaul (MRO) of a component, the resulting downtime will naturally be greater than the actual component MRO time. Therefore, the gross repair time (GRT)* is greater than T_R. This relationship is shown in Figure 10-1 for a typical repair cycle. GRT equals T_R only if an on-line repair is possible (Class I). The other extreme, Class III, may be many times greater than T_R. In order to arrive at a numerical value for GRT the length of each step in the repair cycle has to be estimated, that is obtain permit and access, identify the failure, and so forth.

Priorities

Machinery system unavailabilities are classified by priorities. Priorities are determined based on problem seriousness, urgency, and growth. Three types of unavailabilities can usually be identified.

Forced Unavailability (UA_F)

It is caused by work that, if it would not get done, would result in high business losses, that is high seriousness, urgency, and defect growth would be the case. Usually, this is work that has to be performed in response to failures that disable the system and call for immediate repair. System forced unavailability is the sum of component forced unavailability.

$$UA_F = \sum(\lambda \times \text{GRT}) \tag{10.5}$$

Maintenance Unavailability (UA_M)

Maintenance unavailability is caused by work that can be deferred for some time but usually not to a planned production unit shutdown. Operators can minimize maintenance unavailability in two ways. First, by doing

---

* Often referred to as stream-to-stream time.

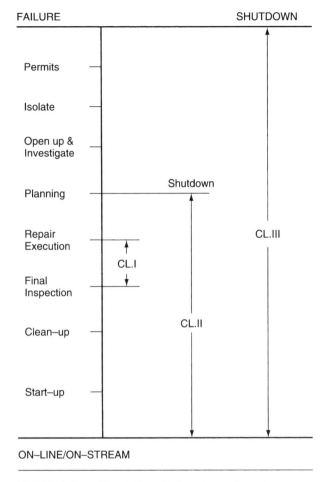

FAILURE   SHUTDOWN

Permits

Isolate

Open up &
Investigate

Planning

Repair
Execution

Final
Inspection

Clean–up

Start–up

ON–LINE/ON–STREAM

CLASS I: Gross Repair Time if failure does not require
shutdown and can be repaired on–line

CLASS II: Gross Repair Time if failure is accessible and does
not require immediate shutdown

CLASS III: Gross Repair Time if failure is inaccessible or if
failure requires immediate shutdown

Figure 10-1. Process machinery repair cycle.

the work within a suitable shutdown window (i.e., a period of time when
the system is available for maintenance due to other reasons). Second,
by performing several maintenance operations simultaneously.

It becomes apparent then that the design review can contribute to
minimizing the total work load. Consequently, even though the sum of

the component unavailabilities is not a true estimate, it is a direct estimate of the work load and a relative measure of unavailability. We already know that low availability indicates poor reliability or maintainability.

Since jobs can be done concurrently, maintenance unavailability need not include all steps in Figure 10-1. Average repair time (T_R) and maintenance time (T_M) are used rather than GRT:

$$UA_M = \sum(\text{preventive maintenance} + \text{deferrable repair}) \qquad (10.6)$$

or

$$UA_M = \sum\left(\frac{T_M}{T} + \lambda T_R\right) \qquad (10.7)$$

Planned Unavailability (UA_P)

Planned unavailability is generated by work that can be deferred from one operating period to another. This is usually referred to as scheduled maintenance, repair, overhaul, and inspection (MRO & I). For all components:

$$UA_P = \sum(\text{planned MRO \& I}) \qquad (10.8)$$

or

$$UA_P = \sum\frac{T_M}{T} \qquad (10.9)$$

Table 10-1 shows the actual procedure by way of an example. The table represents the availability assessment of three machinery subsystems as part of a process refrigeration system (Fig. 10-2).

Eight failure cases are identified for the machinery portion of the commercial refrigeration package. Two failure cases are shown for each

Table 10-1
Availability analysis: Process machinery

(A)	(B)	(C)	(D)	(E)	Reliability maintenance		Unavailability (H)		
		Rate, $\lambda \times E\text{-}6(h)$	Repair time (h)	Gross repair time (h)	(F) Predictive (h/year)	(G) Planned (h/year)	Forced[a]	Predictive[b]	Planned[c]
Component	Failure								
1.0 Refrigeration package	1.1 Comp. seal (forced)	30		30			8.E-4	0	0
	1.2 Comp. brg. (forced & RM)[d]	9	40	72			6.E-4	4.E-4	0
	1.3 Comp. rotors (RM)[d]	25	60			10	0	2.E-7	0.0011
	1.4 Motor (forced)	8		110			8.E-4	0	0
	1.5 I&E components (forced)	7	3	5			3.E-5	0	0
	1.6 I&E components (RM)[d]	10			7		0	8.E-4	0
	1.7 Oil pump (forced)	30		10			3.E-4	0	0
	1.8 Oil pump motor (forced)	12		16			2.E-4	0	0
2.0 Glycol pump	2.1 Pump (forced & RM)[d]	30	30	40	10		0.0011	0.0011	0
	2.2 Motor (forced)	8		16			1.E-4	0	0
3.0 HC pump	3.1 Pump (forced & RM)[d]	50	10	10		10	5.E-4	5.E-8	0.0011
	3.2 Motor	16		16			2.E-4	0	0
					Total Unavailability Reliability Availability		0.0046 0.0088 99.54% 99.12%	0.0019	0.0023

[a] $(C) \times (E) \times E\text{-}6 \times 8000/8760$.   [b] $(C) \times (D) \times E\text{-}6 \times 8000/8760 + (F)/8760$.   [c] $(G)/8760$.   [d] RM = reliability maintenance (i.e., predictive/preventive maintenance).

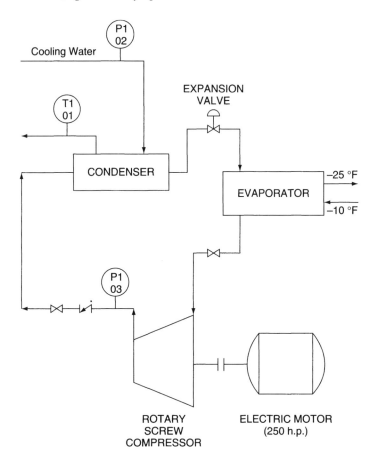

Figure 10-2. Schematic diagram of a process refrigeration system.

of the two process pumps which are in unspared service. The following points may be of interest:

1. Most individual components listed in Table 10-1 have failures which demand immediate shutdown and consequently result in forced unavailability. These failures are caused by technical life attainments, predicated by L_{10} bearing life, for example. There are three categories of these types of failures:

   • Predictable but not predicted;
   • Predicted but not acted upon;
   • Not predictable.

2. Failure 1.3 (Table 10-1), internals of the rotary screw compressor, includes both maintenance and planned unavailability. A planned overhaul or exchange of the compressor element every 5 years results in an average unavailability of 10 hr/year. In addition, random failures are expected here at a rate of 25×10^{-6} h or one every 5 years based on an 8000-h operating year. These random failures, initiated by operational accidents, are assumed to develop gradually. They would therefore respond to predictive-preventive maintenance (PM) measures resulting in a 60-h maintenance unavailability.

3. Failure 1.6 (Table 10-1) causes unavailability due to preventive maintenance actions. They have two components: preventive adjusting of malfunctioning devices and planned instrument or electrical checks. Forced outages due to instrument malfunctions are not expected.

The resulting reliability and availability values can be compared with process unit target availability. If the discrepancy is excessive, major sources of unavailability should be identified and addressed.

The Operations Management Approach

A different approach would be appropriate where machinery systems are operating and their performance has to be described and compared to other systems for management purposes. Such an approach would take the following operational states into account:

- In Operation: In service and producing
- Ready to Start: Standing by
- Forced Outage: Not in operation after a failure during operation that caused the unit to "trip" off-line; before maintenance
- Maintenance: Under preventive or corrective MRO
- Out of Service: Not required for operation during a given time period; off

The five operational states are shown in Figure 10-3. The diagram illustrates the possible changes from one state to the other. A machinery availability tracking system should describe operational states and the transition from one state to the other as accurately as possible. This is accomplished by first collecting relevant operational data and then converting them to key machinery performance indicators or management tools.

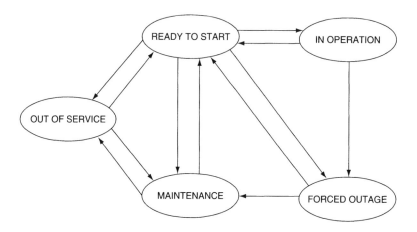

Figure 10-3. Machinery operational states.

Table 10-2
Required operational data

Available

Operating time	t	(h)	
Ready to start	r	(h)	

Unavailable

Maintenance	M	(h)	$M = T_R + T_M$
Breakdown	T_R	(h)	
Reliability (predictive, planned)	T_M	(h)	
Forced outage	F	(h)	
Out of service	O	(h)	$O = O_m + O_n + O_d + O_o$
Modification	O_m	(h)	
Not needed	O_n	(h)	
Time delay	O_d	(h)	
Other	O_o	(h)	

Calendar time	C	(h)	
Effective calendar time	E	(h)	$E = C - O$
Starting demands	S_d	(number)	
Starting successful	S_s	(number)	
Total starts	S_t	(number)	
Forced outages	n	(number)	

The required operational data is shown in Table 10-2. Key performance indicators are listed in Table 10-3. It is essential to first carefully define the machinery system with all its subsystems and components. Then each data type (Table 10-2) must be defined by considering the peculiarities of the machinery population such as service mode (continuous process,

Table 10-3
Machinery performance indicators

Index	Description	Calculation	
MTTF	Mean Time To Failure	t/n	(h)
MTTR	Mean Time To Repair	$(F + T_R)/n$	(h)
MIR	Maintenance Intensity Ratio	$M/(t + r)$	(%)
MTO	Mean Time Operating	t/S_s	(h)
S	Starting Reliability	$100 \times S_s/S_d$	(%)
A	Availability	$100 \times (t + r)/E$	(%)
R	Running Reliability	$100 \times t/(t + F + T_R)$	(%)
U	Use Factor	$100 \times t/C$	(%)

standby, peak load) and so on. This is important in those cases where company- or industry-wide comparisons of machinery performance indicators are undertaken.

Chapter 11
Practical field uptime assessment

In the absence of reliability data, we have to resort to on-site inspection, engineering judgment, and experience in order to arrive at a reasonably consistent and comparable reliability assessment. In a large plant or an organization having many plants, it would be desirable to have a numerical value established to facilitate comparisons of similar equipment and assist in the planning and budgeting of equipment maintenance, engineering manpower support, improvements, or replacements.

In the following examples, we show how a machinery index or complexity numbers may be established by actual field observation in order to assess machinery reliability management needs.

The Reliability Index Number*

This is a relative number arrived at to represent the reliability of a particular piece of equipment and to relate it to other similar pieces. This index number can be determined for each piece of critical equipment in a process plant. It also is possible to combine these pieces and express an aggregate Reliability Index Number for the system. There would be little value in doing so, however, unless there were other like systems to be compared with it.

Because it is a relative number, we must be consistent in determining the index number for each type of equipment. Some ground rules must be established to guide craftsmen or specialists in judging the factors involved. The optimum condition would be to have one individual in a plant responsible for determining the Reliability Index Number for one

---

* Courtesy of The General Electric Company, Schenectady, N.Y.

type or class of equipment. The next best condition is to have one person responsible for determining the Reliability Index Number for a class of equipment and provide time for personal communication of the guide rules or guidelines to those making the inspection of that equipment.

For those who believe that perfect should always be 100%, our Reliability Index can always have a maximum value of 100. Because of the inherent differences in designs, to use a base of 100 may require the use of guide rules (rulers) having varying graduations. Nevertheless, this may be the simplest index to apply, providing we can rely upon the use of good judgment by qualified personnel.

Determining the Reliability Index Number

There are five basic factors that must be considered in determining the reliability of an electric motor, for instance. A perfect Reliability Index Number of 100 would be made up of:

Visual inspection	40
Tests and measurements	30
Age	10
Environment	10
Duty cycle	10
Total	100

Visual Inspection

When it is made by a qualified technician, visual inspection is the most important factor in determining the reliability of critical equipment. The technician must know what to look for and how to evaluate what he sees. Critical equipment seldom fails during normal operation without giving some warning. We attempt to detect and interpret this warning before a failure occurs. The frequency of thorough visual inspections must be based upon operating experience, the recommendations of equipment manufacturers, and some consideration of the age factor. The technician should have two opportunities to view the equipment: first, in operation under load; second, when partially or completely dismantled. Also, he should have the report of the last visual inspection. A suitable checklist and report form must be used, as this enables us to determine what attention is required and to prepare a cost estimate.

Guide rules must be set up for use by the technician in evaluating the best estimate of condition versus the maximum weighted value allotted (40 in our example). If these are to be kept as simple as possible, they must be made quite broad, such as for a gear box for instance:

Power input path	10
Power conversion path	10
Power transmission path	10
Frame, housings, and base	5
Sensing, indicating, and control	5
Total (max.)	40

The above guide rules facilitate the use of the overall Reliability Index of 100 but require the inspector to be more flexible in applying his judgment (refer to Fig. 11-1 as applied to a motor). Regardless of the pattern of the guide rules used, we must always apply the same guide rules to similar equipment if our data are to have real significance.

Using checklist below, rate each item as follows:

2 = Acceptable
1 = Keep under observation
0 = Requires immediate attention

Stator
(a) .... Insulation condition
(b) .... Winding tightness
(c) .... Cleanliness
(d) .... Lamination condition
(e) .... Condition of leads
(f) .... Air gap
(g) .... Winding temperature

Rotor
(h) .... Winding tightness
(i) .... Cleanliness
(j) .... Laminations/poles
(k) .... Bearings
(l) .... Shaft-spider-coupling
(m) .... Vibration
(n) .... Lubrication

Comments (describe condition of all items rated 0): ....
....

Rating = sum of items [   ] (28 max.)

Figure 11-1. Visual checklist: AC motors.

Tests and Measurements

These are next in importance in establishing reliability. Some may question the weighted value of visual inspection versus that of tests and measurements. If you cannot make good visual inspections of equipment or if you do not have qualified personnel to make them, change the values. However, if you do lower the value of visual inspection for either of these reasons, the overall accuracy of your reliability estimates will be lowered. It might be better to hire such qualified personnel from equipment manufacturers or service contractors on a contract basis and strive for accuracy in your ratings.

Again, we must establish guide rules to help achieve uniformity in the ratings of similar equipment. In doing so, make the ratings to be applied to each subfactor as simple as possible, such as Good = 3, Fair = 2, Poor = 1, Requires Immediate Attention = 0 (see Fig. 11-2 for ground rules that have been used for large electric motors and generators typically found in large industrial plants).

It must be pointed out that in very large motors or generators of high-voltage ratings, it is often desirable to add the A-C high-potential test, even though it is a go or no-go test. However, this should be applied only after one or two of the other tests listed have been applied or when it is necessary to establish the suitability for service of the insulation system.

Electrical Tests

A Insulation Resistance – Megger

	Stator	Rotor
Megohm reading (1.0 min)	. . . . megohm	. . . .
Rated machine (kv + 1)	. . . . (kv + 1)	. . . .
Megohms (kv + 1)	. . . .	. . . .
Megohms (kv + 1)	*Stator rating*	*Rotor rating*
Over 10	5	5
2–10	4	4
1–2	3	3
1.0	2	2
Less than 1.0	0–1	0–1

(A) Rating = ... (Stator) + ... (Rotor) = ... (10 max.)

(B) Rating = ... (Stator) + ... (Rotor) = ... (10 max.)

(C) High-voltage D-C

If no discharge or rapid rise exists, rate 10; otherwise, rate 0–5

(C) Rating = ... (max.)

Total rating [   ] (30 max.)

Figure 11-2. Tests and measurements: Large motors and generators.

Likewise, a turbine-driven generator should be given other tests or measurements such as oil pressure (lubrication), bearing loading, vibration, alignment, clearance of bearings, clearance of wheels, etc. Such large machinery is usually considered individually and no attempt is made to include it here. These tests or measurements are mentioned only as examples to suggest guide rules or subfactors that might be applicable to some types of equipment. Obviously, many of our tests and measurements can be made during or at the same time as the visual inspection.

Age of Equipment

Age has a definite bearing on equipment reliability, and not just because it may be very old. Most equipment has a statistical life-expectancy curve as shown in Figure 4-8. When equipment is new, it has a higher likelihood of trouble than will be the case after it has operated for 1–2 years. This is caused by manufacturing defects, design inadequacies, shipping damage, or application unknown. As it becomes old and worn, it requires closer attention to maintenance, unless major rebuilding or upgrading has been performed, which may tend to re-establish the curve. For our use, let us pick a component of the equipment that may be most affected by age, such as the insulation system of a motor or generator, and apply a simple rating formula such as shown in Figure 11-3.

Environment and Duty Cycle

These are important factors but we rate them at only 10 points each (see Figs 11-4 and 11-5 for applications to motors and generators, as contrasted to the much higher values for visual inspection and tests and

---

Age of insulation

Stator . . . . . .   Rotor . . . . . .
(Record age of insulation)

Age (years)	Rating
0 – 2	6
2 – 12	10
13 – 15	6
16 – 20	4
Over 20	0–3

Note: If stator age differs from rotor age, rate older component

Rating [   ] (10 max.)

Figure 11-3. Age guide rules: Motors and generators.

Environment

Describe . . . .

Environment	*Open or DP	TE
(a) Warm, dry	10	10
(b) Hot (above 40 °C)	7	8
(c) Corrosive gas/vapor	5	8
(d) Moisture	0–4	3–7
(e) Abrasive dust	3–5	8
(f) Conductive dust	2–4	7

*Add 3 points to (c–f) for sealed insulation systems

Rating [   ] (10 max.)

Figure 11-4. Environment guide rules: Motors and generators.

Duty cycle

Select one condition from each of the five below:

Condition	Rate
(a) Load	
Smooth	1–2
Uneven	0–1
(b) Load	
100% NP	1–2
<100% NP	0–1
(c) Duty	
Short-time	1–2
Continuous	1
(d) Duty	
Non-reverse	2
Plug or reverse	0–1
(e) Starts	
Few (1/h)	1–2
Frequent (1/h)	0–1

Rating [   ] (10 max.)

Figure 11-5. Duty cycle guide rules: Motors and generators.

measurements). This is because the undesirable effects of the difficult environment and duty cycles are more important than the causes *per se*, and these effects are considered under visual inspection and tests and measurements.

Notice that under environment (Fig. 11-4) we allow for built-in features of the motor or generator that enable it to cope more effectively with the problem. Again, under duty cycle (Fig. 11-5), we have favored the short-time rated motor, which may never reach name-plate maximum operating temperature, and the motor which is not plugged or reversed.

However, we have penalized the motor that is plugged or reversed, or that is started and stopped frequently.

It is obvious that the last three factors – age, environment, and duty cycle – can be rated with a minimum of effort.

When all five factors have been evaluated and totaled, a single Reliability Index Number results. The reliability rating report form used to establish the rating would consist of the five factors shown in Figures 11-1–11-5. These can be incorporated on a single page with appropriate headings for equipment nomenclature, location, productivity rating, and maximum Reliability Index value (Fig. 11-6). The Reliability Index Number is not a magic number, above which all similar equipment will not fail in service and below which it will fail. One such number will have but little value; when compared with other numbers established by the same method for similar equipment, however, it can be very valuable.

Please note that our examples state that 0 = Requires Immediate Attention. If our Reliability Index report on a piece of equipment contains one or more 0 items, we must examine these before proceeding further. If they indicate that minor, or even routine, maintenance is required, it may be

(A) *Motor data*

Unit:....................................	Process number:......................
Duty:...................................	Location:.............................
Asset:...................................	
Manufacturer and type:...	
Installation date:.........................	Design:..............................
Horsepower/frame:.......................	Class of insulation:...................
Synchronous speed:......................	Full load speed:......................
Serial number:...........................	Volts/cycles/phase:...................
Full load amperes:.......................	Locked motor amperes:...............
Enclosure:..............................	
Temperature rise:........................	Maximum ambient/cooling:............
Service factor:...........................	

(B) *Reliability factors*

I.	Visual inspection:	.... of 40
II.	Tests and measurements:	.... of 30
III.	Age:	.... of 10
IV.	Environment:	.... of 10
V.	Duty cycle:	.... of 10
	Reliability index:	.... of $\overline{100}$
	Date index recorded:	....

Figure 11-6. Maintenance reliability evaluation: Large motors.

best to accomplish it right away and then correct the ratings accordingly. The resulting Reliability Index Numbers will be more accurate and of more value to us in planning and budgeting for equipment maintenance, rebuilding, or replacement.

Equipment Replacement and Rebuilding

Our Reliability Index Numbers will be significant when compared with similar equipment within the same productivity rating or classification. From this comparison, we can establish and assign priorities for equipment maintenance, rebuilding and upgrading, or replacement. We can expend our maintenance effort where it is most needed. By referring to the reliability rating report forms, we can determine the action that is required to maintain operation at the normal level. An estimate of the cost of such maintenance can be established based upon our past experience or quotations from equipment builders or maintenance contractors.

If we have used the variable base Reliability Index Numbers, we can convert them into percentages to be used as a guide in evaluating priority ratings to be assigned to different kinds of equipment within productivity ratings or classifications. An overall average Reliability Index Number can also be established for a process or an operation.

If we calculate the anticipated Reliability Index Numbers that will result from the indicated maintenance actions, we can advise management of the existing level and the anticipated level that the execution of our maintenance budget will accomplish. This can be done by individual pieces of equipment, productivity ratings, or processes.

The Machinery Complexity Number

Any assessment of machinery reliability would logically give consideration also to machinery complexity. This is sometimes recognized in engineering manpower studies for major petrochemical plants which would obviously need more personnel for large, multi-casing, or old machinery trains than for smaller, less complex, or perhaps new machinery trains.

In the mid- to late-1980s, the authors investigated the merits of categorizing rotating machinery complexity on the basis of machine type, train configuration, size, and age. Five numerical gradations were proposed (Table 11-1).

Using the numbering system described in Table 11-1, a given plant would be in a position to assess its rotating equipment complexity and

---

Table 11-1
Rotating equipment complexity assessment

---

1. Complexity by train configuration

Gas turbine-driven compressors – highest complexity	5
Steam turbine-driven compressors (turbo and/or recips)	4
Reciprocating compressors	3
Motor-gear compressors	2
GT and ST generators	2
Motor-driven compressors (direct drive)	1

2. Complexity by size, special-purpose equipment

Over 15,000 hp	4
5001–15,000 hp	3
501–5000 hp	2
1–500 hp	1

3. Complexity by size, pumps (rating given to entire pump population in a given plant)

Predominantly large pumps – least complex, over 100 hp	10
Predominantly medium size, 25–100 hp	20
Predominantly small pumps, less than 25 hp	30

4. Complexity by age

Over 10 years	3
5–10 years	2
Less than 5 years	1

5. Complexity by counting driven casings
This is a straightforward summation of casings, exclusive of drivers

---

to make a somewhat more objective judgment than pure guessing in attempting to compare various plants.

If, for example, a plant were comprised of three major machinery trains aged 4, 7, and 16 years, the age complexity numbers would be $1 \times 1 = 1$, $1 \times 2 = 2$, and $1 \times 3 = 3$, for a total of 6 complexity points. Let us assume one of the trains to be a gas turbine-compressor–gear-compressor type rated at 25,000 hp. Its "type complexity" would rank a 5, its casing complexity number would be 3. The remaining trains would likewise be ranked on the basis of these complexity numbers.

The relative complexity ranking for nine North American petrochemical plants is shown in Table 11-2. The purpose of this 1986 exercise was to divide the complexity numbers of various plants (e.g., CHET's 84 or DORA's 128) by the number of machinery support engineers entrusted with maintenance and surveillance. It was theorized that the resulting number might show some plants to have excessive and other plants to have insufficient staffing.

Table 11-2
Turbomachinery complexity ranking for several North American petrochemical plants

Plant	Type	Age	Size	Casings	Pumps	Total
AMOS	2 × 3	3 × 2	1 × 3	2 × 2		
	2 × 4	1 × 3	3 × 2	2 × 3	—	52
	14	9	9	10	10	
BRIT	4 × 2	2 × 3	1 × 4	3 × 4		
	3 × 3	3 × 1	4 × 3	3 × 3		
	1 × 1	3 × 2	3 × 2	2 × 2	—	100
	18	15	22	25	20	
CHET		2 × 3	1 × 4	3 × 4		
	2 × 4	1 × 2	2 × 3	1 × 2		
	3 × 3	2 × 1	2 × 1	1 × 1	—	84
	17	10	12	15	30	
DORA				7 × 2		
		4 × 3	2 × 3	2 × 4		
	8 × 3	3 × 2	3 × 4	1 × 3		
	3 × 2	4 × 1	6 × 2	1 × 1	—	128
	30	22	30	26	20	
ERIC	4 × 5	2 × 2	4 × 3			
	3 × 4	2 × 3	1 × 2	7 × 4		
	1 × 3	4 × 1	3 × 1	1 × 3	—	107
	35	14	17	31	10	
FRAN	2 × 4	3 × 3	2 × 4	1 × 5		
	2 × 3	1 × 2	2 × 2	2 × 4		
	1 × 2	1 × 1	1 × 1	2 × 2	—	88
	16	12	13	17	30	
GREG	4 × 4			1 × 4		
	3 × 3	3 × 3	4 × 4	1 × 3		
	2 × 1	3 × 2	2 × 3	6 × 2		
	1 × 2	4 × 1	4 × 2	2 × 1	—	119
	29	19	30	21	20	
HANK			1 × 4	2 × 4		
	3 × 4	2 × 3	1 × 3	1 × 3		
	2 × 3	1 × 2	3 × 2	2 × 2		
	1 × 2	3 × 1	1 × 1	1 × 1	—	71
	20	11	14	16	10	
IRIS	4 × 2	4 × 3	5 × 4	3 × 4		
	7 × 1	7 × 2	6 × 2	8 × 2	—	131
	15	26	32	28	30	

Note: Study includes ST-driven recips, but not motor-driven recips.

In summary, let us be sure of one thing: Our only purpose here is to acquaint the reader with yet another method of identifying potentially vulnerable machinery. As is the case with so many numerical ranking methods, our complexity assessment approach will be helpful only if it is tempered by good judgment and solid experience of the implementor.

Chapter 12
Life-cycle cost analysis

It is almost self-evident that the entire subject of maximizing machinery uptime should start by asking what will be the cost of buying, installing, commissioning, operating, maintaining, and even ultimate disposing of the machine. While capital costs of new projects attract the most attention of management and vendors, we recognize that operating and maintenance (O&M) expenses are also recognized as significant. Unfortunately, evaluating the cost of running a plant on a common basis with the capital cost is difficult, so managers tend to give priority to the initial cost. Consequently, poor reliability and performance do not show up until the job is actually up and running.

Inexpensive systems are likely to have inferior materials, poor workmanship, and weaker designs. System designers frequently do not opt for redundant equipment because it is "too expensive," even though averted, lost production may pay for the initial cost many times over. Decisions based on a short-term outlook are ineffective for minimizing total long-term process and product expenses.

Today, shutting down a process costs tens of thousands of dollars for large plants. With predictive and preventive maintenance methods, processes are not shutdown as frequently – often for years. Sometimes, even batch processes are not idle for periods long enough to perform maintenance.

Life-cycle costing is a promising evaluation tool that makes it possible to quantify the long-term outlook. We are defining life-cycle costs and describe current calculation methods. Since LCC is not just an issue for financial management, the following also provides a model LCC policy for capital project management. Smoother startups, lower maintenance costs, and higher operating efficiencies are the net result. The company's bottom line will look better and the plant will become more competitive.

In petrochemical plants, for instance, maintenance and downtime costs can exceed the initial equipment cost. Life-cycle costing identifies and quantifies all associated project costs over the life of the project. It includes future costs of maintenance and repair, downtime, production losses, replacement, decommissioning, and incremental operating costs associated with the material choice as well as the initial costs.

Asset Management

There are various perspectives on asset management. Accountants consider depreciation and operational cash-flow important. Design engineers wrestle with performance and cost trade-offs. Quality inspectors want low reject rates, and maintenance professionals hope for few equipment problems. True asset management must combine these concerns in a multi-disciplined approach. The key to optimizing life-cycle costs is to combine all professional practices.

Life-cycle costing analysis is a tool that can assist, but it must be accompanied by other techniques and disciplines appropriate to the situation. Life-cycle costing has always been applied in an intuitive way in the form of cost-benefit deliberations. The main value of a formal LCC is that it quantifies life-cycle elements so that their relevance can be established and receive appropriate attention. Apply LCC early in the asset's life to achieve the greatest benefit. Start during concept formulation or, at the latest, during the design-and-specification stage. The U.S. Department of Defense (DoD, Washington) has found that decisions made in the early phases of developing a concept determine 70% of eventual life-cycle costs [1, 2]. Life-cycle costing may also be initiated in later project stages to audit O&M efficiency or to review the benefits of new modifications. In short, LCC is a valuable starting at any time. Department of Defense practitioners have found two valuable by-products of LCC:

1. Life-cycle costing requires a comprehensive review with a long list of questions and answers. As a result, the asset design is more detailed before bidding than when LCC is not used.
2. Budget forecasts are better, because more-realistic cost and time schedules are developed. Companies gain a more-comprehensive understanding of operating costs.

Unreliable equipment causes significant lost production and waste. However, reliability is a fuzzy concept to most project engineers and they do not know how to address it. Life-cycle costing provides a way to evaluate long-term costs for repairs, lost production, and the initial costs of design, procurement, and installation. Through the use of LCC, more-reliable equipment can be justified using a credible analysis that is acceptable to accountants and business planners.

Increasing the useful lifetime of any system costs money and causes an apparent trade against other benefits. Fig. 12-1 illustrates such a trade-off. It shows that life-cycle costs and benefits depend on good design integration and support. Hardware is only one factor in the overall picture.

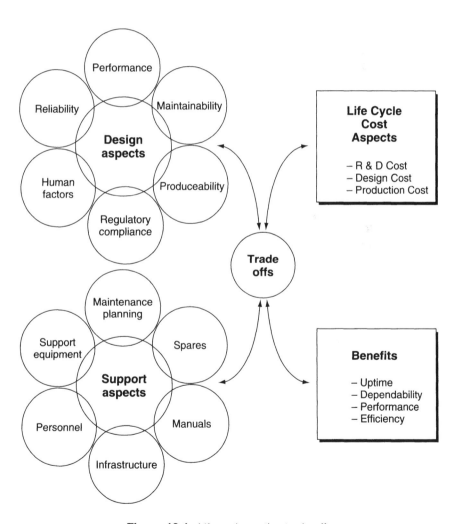

Figure 12-1. Life cycle costing trade-offs.

Twelve Steps in the LCC Process

Any application of LCC analysis is likely to involve certain fundamental concepts. The relative importance of each of these concepts, and hence their level of application, will vary according to the requirements of a particular LCC analysis. In general, a LCC analysis follows the 12 basic steps illustrated in Table 12-1. The details below expand upon the table. This analysis has to involve both the users and the producers of the physical assets. The cost estimates must be based on the experience of both organizations.

Step 1: Define the problem. This is an obvious starting point.

Step 2: Identify the feasible alternatives. Engineering must make preliminary designs of multiple configurations. This stage eliminates unworkable solutions. The concern here is with meeting performance parameters.

Step 3: Consider alternatives and the system requirements. This is the first look at operations and maintenance. Identify and categorize the life-cycle activities. If nothing else, this activity raises awareness that endurance is a parameter in the design process.

Step 4: Analyze the total lifetime of events for the physical asset. Include in these events all applicable future activities associated with research, development, production, construction, installation, commissioning, operation, maintenance, and disposal. In the analysis, identify all the applicable resources required during the lifetime of the asset. Some resources are used to construct the asset. Other resources are replacement parts and maintenance chemicals. Group the identified events,

---

Table 12-1
Major steps of a life-cycle cost analysis

1. Define the problem.
2. Identify feasible alternatives.
3. Consider the alternatives in terms of system requirements – operations and maintenance. Then identify and categorize life-cycle activities.
4. Develop the cost breakdown structure (CBS).
5. Develop the cost model.
6. Estimate the appropriate costs.
7. Account for inflation and learning curves as a function of time.
8. Discount all estimated costs to a common base period.
9. Identify the "high-cost" contributors, determine cause–effect relationships.
10. Perform a sensitivity analysis and calculate the final **LCC**.
11. Perform a risk analysis identifying trade-offs.
12. Recommend a preferred solution, select the most desirable alternative.

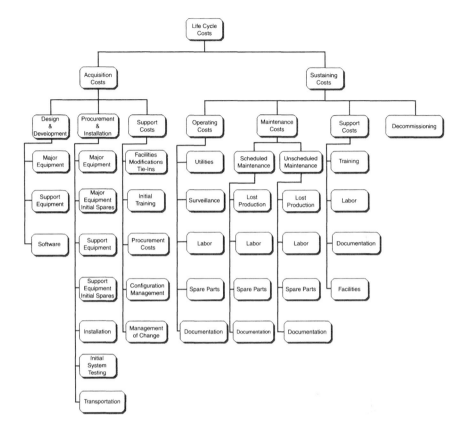

Figure 12-2. Cost breakdown structure (CBS).

activities, and resources into major LCC elements, and then break them down into sub-elements. This activity has been refined into what is known as the CBS concept (Fig. 12-2). It is a convenient way of dividing the life cycle into workable sized packages for cost estimating.

Step 5: Set up a model to define the cost factors and estimating relationships. These factors and relationships include items such as hourly labor rates, mandated profit margins, and fuel-consumption rates. The actual factors and relationships used in a LCC analysis vary according to the nature of the asset and the business operations of the user and vendors.

Step 6: Work up the cost of each of the life-cycle elements. The previously determined cost estimating factors and relationships are applied to cost models for each of the elements.

Step 7:   Account for inflation and learning curves. Set the accuracy required in the calculated life-cycle cost. Inflation will have strong effects on the life-cycle cost of today's physical assets. However, future changes in inflation rates are difficult to predict. This is a subject that requires the judgment of outside economists and accountants before being applied in a LCC analysis. Sometimes, it is easier to assume zero inflation and do the analysis rather than have no answers.

The effect of learning curves is probably a bit more predictable. It applies when several identical physical assets will be produced or constructed over time. A learning curve is the function used to describe the non-linear relationship between skill acquisition and time elapsed during a project or plant startup phase.

Step 8:   Discount all the estimated costs to a base period. Unlike inflation, discounting is not optional during LCC analyses, where two or more similar assets are being compared. The differences are important here because they will likely result in different levels of cash-flow requirements at different points of the life cycle. Discounting yields a common basis for financial comparison, by removing the effects of time differences. The process is based on finance mathematics and uses the concepts of sinking fund, present value, and capital recovery. Consult any cost estimating textbook for assistance.

Step 9:   Identify the high-cost contributors. There are facilities in which one or two costs overwhelm all the others. It is a shortcut to concentrate on such items, because they promise the highest payoff. The high cost is usually the result of an underlying cause. Search for the cause and eliminate it or mitigate it.

Step 10:   Calculate the final LCC, using an appropriate cost model. In many cases, this is likely to entail a straight summation of the cost breakdown elements. But, it can involve far more complex mathematics, according to the characteristics of the asset's life cycle and the management approach used for the LCC analysis.

In the overwhelming majority of cases, the model should include a sensitivity analysis. Sensitivity analysis consists of evaluating the results displayed by a model (mathematical or other) upon changing one or more input variables. In practice, this is seen as a very large spreadsheet activity. It is a lot of work, but has a big payoff. The Section "Repairing pumps" (p. 207) shows a simple example.

Step 11:   Perform a risk analysis. The LCC technique can be useful when applied to situations that consider alternative decisions on a cost basis. These are basically trade-offs. A few typical situations are:

- Balancing the relative levels of reliability and maintainability for a given asset against a desired level of availability.
- Deciding on the most cost-effective maintenance policy for sub-elements of a given asset. The usual choice is predictive, preventive, or emergency maintenance.
- Deciding which asset to procure when faced with two or more that will satisfy all specified requirements.
- Deciding whether to modify an asset or repair it without changing the current configuration.
- Deciding whether to retain or dispose of an existing asset.

Step 12: Recommend a solution. Life-cycle costing can be applied to assist in logical management of an asset, even without looking at alternatives. Examples of this approach are:

- Identifying the exact subsystems where design simplification and cost control will produce major cost reduction and longer life cycles.
- Establishing a more accurate budget for the actual project.
- Understanding the inner workings of an asset. This sets up a more effective management organization and better control procedures.

Examples

Here are some examples that show the application of the 12 steps. Of course, they are greatly simplified and each covers only a few steps. The idea is to show how beneficial this whole analysis can be. Unfamiliar financial terms are defined in Table 12-2.

Repairing Pumps

Pumps are important elements in chemical processing plants and make up some 10–20% of all components used. Frequently it is not obvious what repairs on process equipment really cost. Consider a population of centrifugal pumps in a refinery. Long-term records show that the mean time between repair (MTBR) of these pumps is 25 months. We want to find the equivalent capital cost of the repairs. The life of a pump is 15 years. This calculation discounts annual repair costs back to the date of purchase. Other data from the refinery are mean time to repair

Table 12-2
A few helpful definitions

Capital equipment cost	The cost of purchasing equipment and materials
Capital installation cost	The cost of installation, such as labor, and in some cases will include basic design options
Total capital cost	The sum of the equipment and installation costs, plus any engineering costs and tax, insurance, freight or other overhead charges. This does not include contingency cost, but actual costs
Annual maintenance cost	This is the costs of repairs, preventive maintenance, and condition monitoring for equipment
Annual operating cost	Cost of energy, catalysts, and waste disposal but excluding feedstocks unless there ia sn issue involving utilization or losses
Benchmark costs	The annual operating or maintaenance costs of a similar installation believed lowest known
Total-life maintenance costs	This is the total of all maintenance costs from startup trough the life of the project
Total-life operating costs	This is the total of all operating costs from startup through the life of the project
Present worth	This is the present value of a future expenditure, based on an assumed interest rate and number of years of useful life. The formula for this: $P = F/(1+i)^n$

(MTTR) = 5 d; C_G = \$7500 (see Table 12-1) current interest rate = 6.5%; annual repair costs, C_Y, can be calculated by:

$$C_{PV} = \text{PV}(\text{rate, years, } C_Y)$$

$$C_Y = \frac{8760 \times C_G}{[(MTBR \times 30.4 \times 24) + (MTTR \times 24)]}$$

$$C_Y = \frac{8760 \times 7500}{[(25 \times 30.4 \times 24) + (5 \times 24)]} = 3578$$

$$C_{PV} = \text{PV}(0.065, 15, 3578) = \$33,643$$

The present value of the repair costs, C_{PV}, can be determined by looking at the present value factor as a function of interest rate, years of life, and annual costs. This example employs the PV function in Microsoft Excel [3].

The sensitivity analysis now has a basis. Look for the benefits that could be derived from attempting to reduce repair costs. Evaluations will compare purchasing a more expensive and hence (hopefully) more-reliable pump or by making repairs more efficient and hence less costly.

Life-Cycle Costing of Pumps*

When a chemical plant is being designed, it is very often only the initial capital outlay, or first cost, for a pump that is considered. Operating costs associated with the use of the pump are often disregarded. Analyzing the costs incurred by existing pump systems provides a useful basis for assessing which type of pump or which measures are best able to minimize life-cycle cost. A service life (life cycle) of 10 years was selected as typical for pumps in the chemical industry. In the present study, an annual operating time of 8000 hr is assumed.

The life-cycle cost is computed on the basis of a simplified version of the NORSOK† standard. The method of calculation used is based upon the present-value method, in which all relevant costs are discounted to their present value (i.e., to the equivalent value in the first year). The result is the total present-value or life-cycle cost of the pump and represents the total cost over 10 years discounted back to the year in which the initial capital investment was made. A currently valid interest rate of 8% was assumed.

The costs arising from lost production during plant downtime and any costs incurred as a result of emergency measures taken during non-operational periods have not been included in the following analysis.

Scope of the Analysis

We are going to look at a comparison of different centrifugal pumps made of stainless steel and equipped with a variety of shaft seals:

- single mechanical seals (SMS)
- double mechanical seals (DMS)
- magnetically coupled pumps ('magnetic')
- canned motor pumps ('canned').

The pumps analyzed were in the size categories of 32–160 and 40–200 to 50–200.

Data Sources and Estimates of Future Costs

Table 12-3 shows the acquisition costs of the pumps. Repair costs for the period 1990–1998 were taken from the servicing and maintenance database and extrapolated linearly to cover a 10-year life cycle. All repair work was taken into account, including the replacement of mechanical

---

* Courtesy of Karl Ost of Degussa, Hüls, Germany.
† NORSOK is the Norwegian Offshore Petroleum Standards Organization.

Table 12-3
Capital outlay (Costs in DM)

Shaft seal	Capital outlay (Σ 1–5) DM	1 Purchase price of pump	2 Buffer fluid unit[a]	3 Engineering costs[b]	4 Installation costs[c]	5 Spare parts (interest payments)
Pump size: 32–160; Material: stainless steel (1.4408); Motor: EexT3						
SMS	10,700	6,500	–	3,300	450	450
DMS	21,450	8,700	7,500	3,800	900	550
Magnetic	12,600	8,300	–	3,300	450	550
Canned	16,000	11,700	–	3,300	450	550
Pump size: 50–200; Material: stainless steel (1.4408); Motor: EexT3						
SMS	13,700	9,300	–	3,300	450	650
DMS	25,000	12,000	7,500	3,800	900	800
Magnetic	15,550	11,000	–	3,300	450	800
Canned	21,350	16,800	–	3,300	450	800

[a] Buffer fluid system with automatic supply unit.
[b] Dimensioning and configuration, technical specifications, invitation to tender, procurement, documentation.
[c] Labor costs for installation excluding additional material costs.

seals, shaft and bearings, and the repair of damage arising from wear and corrosion. The highest costs are due to defective shaft seals and damaged bearings. To calculate the life-cycle costs for pumps with different shaft sealing systems, the average values of the following parameters were computed:

- mean time between failure (MTBF)
- cost per repair.

Pumps in the size category 40–200 to 50–200 were dealt with by assuming that repair costs for pumps in this category were similar. Energy costs were calculated for pumps in the 50–200 size class.

Costs for electrical, instrumentation, and control engineering were not included as these are due to change as a result of the new regulations covering equipment and systems intended for use in potentially explosive atmospheres (ATEX 100a).

Maintenance and Repair Costs

The cost and frequency of repair work were extracted from the service and maintenance database and the corresponding average value was computed (Table 12-4).

Table 12-4
Pump history and repair cost (in DM)

Shaft seal	MTBF (years)[2]	Repair costs/cycle[2]	Maintenance costs/year[3]	No. of pumps examined
Pump size: 32–160				
SMS	2	2500	120	15
DMS	2	2900[1]	480[1]	–
Magnetic	5	3850	80	9
Canned	4	3200	0	19
Pump size: 50–200				
SMS	2	2400	120	23
DMS	2	2900[1]	480[1]	–
Magnetic	5	2650	80	34
Canned	4	1900	0	24

[1] Average empirical values (feedback from operating staff).
[2] Average values computed from SAP* data.
[3] Empirical values as reported by service and operating personnel.

Energy Costs

Energy costs were calculated on the basis of the efficiencies of the pump mechanisms (assumed constant for each pump size analyzed) and the efficiencies of the electric motors, the magnetic couplings, and the canned motors (connected loads of the electric motors). The results are shown in Table 12-5.

Table 12-5
Pump energy cost (in DM)

Shaft seal	Pump size/ Power (kW)	Energy costs per year	Pump size/ Power (kW)	Energy costs per year
SMS	32–160/1.5	1450	50–200/7.0	6250
DMS	32–160/1.5	1530	50–200/7.0	6400
Magnetic	32–160/1.5	1760	50–200/7.0	7120
Canned	32–160/1.5	1760	50–200/7.0	7670

Note: If the pump motor power is >15 kW, the energy costs must be calculated separately. Assumed are 8000 operating hours per year. Electricity charges are 10 DM/100 kWh.

*Computerized Maintenance Management System (CMMS).

Results of the Life-Cycle Cost Analysis

Summary of the results are presented in Tables 12-6 and 12-7 as well as in Figures 12-3 and 12-4. For pump sizes 32–160 with a pumping capacity of $10\,m^3/h$, head capacity 28 m, motor speed $= 2900\,min^{-1}$, and power output $= 1.5\,kW$. For pump sizes 50–200 with a pumping capacity of $40\,m^3/h$, head capability 41 m, motor speed $= 2900\,min^{-1}$, and power output $= 7.0\,kW$.

Results

The foregoing analysis was based on 8000 operating hours per year. If the pump operates for fewer hours (e.g., in batch processes), the distribution of costs can alter significantly. A comparison of the relative magnitudes of the different types of costs for the two pump sizes indicates that energy costs become the dominant factor as pump motor power increases. The results show that pumps with an SMS have the lowest life-cycle costs.

Despite the higher purchase price and the higher energy costs associated with magnetically coupled pumps, the lower repair costs mean that these pumps are only about 0–4% more expensive overall and offer a higher degree of availability.

Table 12-6
Results of LCC analysis for pump sizes 32–160

Shaft seal	Life-cycle cost (LCC) DM	%
SMS	30,800	100
DMS*	46,100	150
Magnetic	30,700	100
Canned	34,000	110

*These values were calculated on the basis of experience.

Table 12-7
Results of LCC analysis for pump sizes 50–200

Shaft seal	Life-cycle costs (LCC) DM	%
SMS	68,200	100
DMS*	85,000	125
Magnetic	71,050	104
Canned	80,050	117

*These values were calculated on the basis of experience.

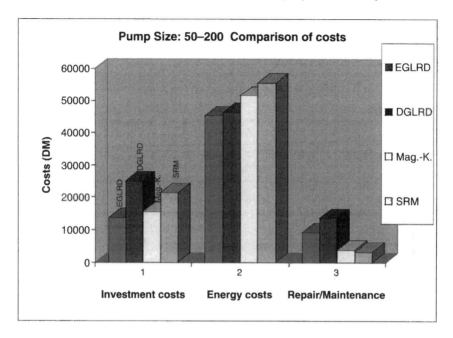

Figure 12-3. Result statistics I

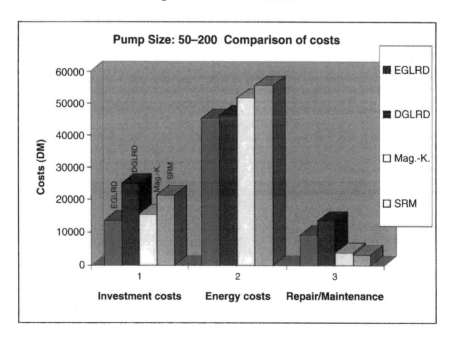

Figure 12-4. Result statistics II

The life-cycle costs of canned motor pumps are some 10–17% higher principally as a result of increased energy costs. The safety rating of canned motor pumps is however higher than that of magnetically coupled pumps because of their dual-enclosure design. A further benefit of canned motor pumps is that operating noise is significantly reduced even in the case of larger motors. An advantage of magnetically coupled pumps and those with canned motor is the long average period of 4–5 years between two repairs. This high level of availability means that, in certain cases, it is possible to do without standby pumps.

It was further found that the highest life-cycle costs arise for pumps equipped with a double mechanical shaft seal and automatic buffer-fluid pressurizing system. Today, double mechanical seals tend only to be used if no other seal is feasible (e.g., when pumping fluids with a high solid content). However, this statement is true only for pump motor powers <15 kW. As pump motor power increases, their life-cycle cost falls in comparison to magnetically coupled pumps and those with canned motor.

It is important to realize that the analysis can only indicate general trends. The pumps included in this analysis are used for conveying a wide variety of fluids at a broad range of temperatures and pressures, making direct comparison not always legitimate.

In many cases, the pumpage does not permit a choice of shaft seal to be made. Often pumping conditions dictate that only pumps with seals meeting very particular criteria may be used (DMS, magnetically coupled and canned motor). Figures 12-5 and 12-6 provide an overview of the foregoing work.

How to Select an Appropriate Pump with a Minimal Life-Cycle Cost

Calculate the annual energy consumption (operating hours, power demand of pump) and compute the life-cycle cost. Figure 12-7 shows the procedure.

Are the highest cost savings to be made by minimizing energy costs or by reducing the initial capital expenditure?

- Capital outlay >energy costs: optimize capital expenditure.
- Select the lowest cost pump offering the required degree of availability.
- Choose a pump type that is already in use in order to save on engineering and spare-part costs.
- Energy costs > capital outlay: optimize energy costs.
- Select an appropriate pump with the best efficiency.
- Accurately calculate the required pumping capacity Q and the head H. Remember, reserve capacities in Q and H are expensive!
- If pumping capacity and head vary, the use of a frequency converter should be considered.

Life-cycle cost (DM) of pumps of size 32–160 with various shaft sealing systems

Interest rate: 8.00%	Pump size: 32-160		Motor speed: 2900 rpm		Q: 10 m³/h		H: 28 m			
Year	1	2	3	4	5	6	7	8	9	10
Single mechanical seal										
Initial capital outlay	10,700									
Energy costs	1450	1450	1450	1450	1450	1450	1450	1450	1450	1450
Repair + maintenance costs	120	2620	120	2620	120	2620	120	2620	120	2620
Sum	12,270	4070	1570	4070	1570	4070	1570	4070	1570	4070
Sum (present value)	12,270	3769	1346	3231	1154	2770	989	2375	848	2036
Life-cycle cost	**30,788**									
Double mechanical seal										
Initial capital outlay	21,450									
Energy costs	1530	1530	1530	1530	1530	1530	1530	1530	1530	1530
Repair + maintenance costs	480	3380	480	3380	480	3380	480	3380	480	3380
Sum	23,460	4910	2010	4910	2010	4910	2010	4910	2010	4910
Sum (present value)	23,460	4546	1723	3898	1477	3342	1267	2865	1086	2456
Life-cycle cost	**46,120**									
Magnetically coupled										
Initial capital outlay	12,600									
Energy costs	1760	1760	1760	1760	1760	1760	1760	1760	1760	1760
Repair + maintenance costs	80	80	80	80	3930	80	80	80	80	3930
Sum	14,440	1840	1840	1840	5690	1840	1840	1840	1840	5690
Sum (present value)	14,440	1704	1578	1461	4182	1252	1160	1074	994	2846
Life-cycle cost	**30,691**									
Canned motor										
Initial capital outlay	16,000									
Energy costs	1760	1760	1760	1760	1760	1760	1760	1760	1760	1760
Repair + maintenance costs				3200				3200		1600
Sum	17,760	1760	1760	4960	1760	1760	1760	4960	1760	3360
Sum (present value)	17,760	1630	1509	3937	1294	1198	1109	2894	951	1681
Life-cycle cost	**33,963**									

Figure 12-5. Summary table 1.

Life-cycle cost (DM) of pumps of size 50–200 with various shaft sealing systems										
Interest rate: 8.00%	Pump size: 50–200		Motor speed: 2900 rpm		Q: 40 m³/h		H: 41 m			
Year	1	2	3	4	5	6	7	8	9	10
Single mechanical seal										
Initial capital outlay	13,700									
Energy costs	6250	6250	6250	6250	6250	6250	6250	6250	6250	6250
Repair + maintenance costs	120	2520	120	2520	120	2520	120	2520	120	2520
Sum	20,070	8770	6370	8770	6370	8770	6370	8770	6370	8770
Sum (present value)	20,070	8120	5461	6962	4682	5969	4014	5117	3442	4387
Life-cycle cost	**68,224**									
Double mechanical seal										
Initial capital outlay	25,000									
Energy costs	6400	6400	6400	6400	6400	6400	6400	6400	6400	6400
Repair + maintenance costs	480	3380	480	3380	480	3380	480	3380	480	3380
Sum	31,880	9780	6880	9780	6880	9780	6880	9780	6880	9780
Sum (present value)	31,880	9056	5898	7764	5057	6656	4336	5707	3717	4892
Life-cycle cost	**84,963**									
Magnetically coupled										
Initial capital outlay	15,550									
Energy costs	7120	7120	7120	7120	7120	7120	7120	7120	7120	7120
Repair + maintenance costs	80	80	80	80	2730	80	80	80	80	2730
Sum	22,750	7200	7200	7200	9850	7200	7200	7200	7200	9850
Sum (present value)	22,750	6667	6173	5716	7240	4900	4537	4201	3890	4927
Life-cycle cost	**71,001**									
Canned motor										
Initial capital outlay	21,350									
Energy costs	7670	7670	7670	7670	7670	7670	7670	7670	7670	7670
Repair + maintenance costs				1900				1900		950
Sum	29,020	7670	7670	9570	7670	7670	7670	9570	7670	8620
Sum (present value)	29,020	7102	6576	7597	5638	5220	4833	5584	4144	4312
Life-cycle cost	**80,026**									

Figure 12-6. Summary table 2.

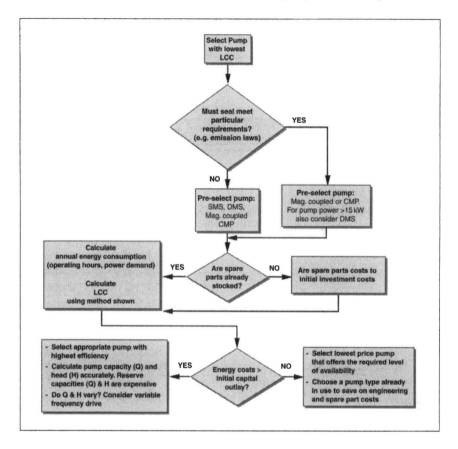

Figure 12-7. Decision chart.

Are spare parts for the pump under consideration already in store? If not, the spare parts costs for 2 years should be incorporated into the capital outlay used in the calculation of the LCC.

Piping

Our second example was selected for two reasons: (1) piping can be considered a surrogate for any major piece of equipment and (2) piping analysis is fully developed. In the past, either the piping and the equipment were installed knowing that they would be replaced at frequent intervals, or the materials were so exotic that they outlasted the useful life of the process. Repair frequency was estimated but not the total cost of ownership over the life, converted to present-value (PV) terms.

Years ago, it may have been more cost-effective to replace piping and equipment rather than spend additional funds for corrosion-resistant parts. Manual calculation methods did not allow time for analysis of alternate materials of construction. Previous experience or data from installing corrosion-testing coupons in the process vessels was used.

For a new process, experience was not available and there was no procedure to install coupons. Additionally, processes were rarely continuous and frequent downtimes for regular maintenance allowed windows of repair or replacement time for non-alloy materials of construction.

The SSINA Method

One form of LCC is a procedure adapted for the evaluation of piping materials. It was developed by the Specialty Steel Industry of North America (SSINA, Washington, D.C.) and it uses the standard accounting principle of discounted cash flow to change total unit costs incurred during a life cycle to present-day values. The SSINA method expresses LCC as the sum of five components (Fig. 12-8). These calculations, in PV (present value) dollars, accurately portray the true costs of using different materials. They are:

Initial materials acquisition costs (AC). AC is the total cost of materials, at today's values, used in the initial fabrication and installation of the unit, including discounts. AC totals the independent units being assessed, such as pipe systems, pump systems, distillation units, or heat exchangers. This includes plate and sheet metal, pipe, tubing, fittings, and all other miscellaneous parts.

AC may be calculated from a cost per dimension. Be careful when comparing the total material costs for a unit, to use the quantity of materials based on the best design, for each of the alternative materials. The design requirements, such as vessel-wall thickness and beam lengths,

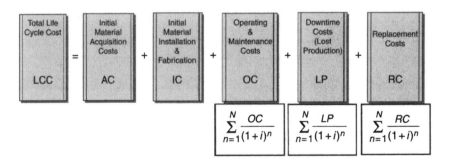

Figure 12-8. LCC calculated as sum of five costs.

may vary with different materials because of material properties. These variations affect the pricing of alternate materials. If material costs are included in the fabrication and installation costs in turnkey projects, these lump-sum costs are included in AC.

Initial materials installation and fabrication costs (IC). IC includes the total costs of cutting, forming, welding, assembling, and installing the independent unit. Methods of estimating, such as cost per foot of pipeline or cost per square foot of sheet metal, may be used based on experience from similar jobs or projects.

IC includes other installation costs, among them surface preparation and protection, such as painting, and the application of epoxy coatings and linings not previously included in AC. IC also includes special labor skills, such as certified engineers, materials consultants, and specialty-qualified operators to aid in the installation.

Operating and maintenance costs (OC). This includes maintenance costs at each maintenance interval and indirect operating costs that occur as a result of a material choice. Maintenance costs, at today's values, include labor costs plus specialized inspection equipment for X-ray, ultrasonic, or similar acoustic techniques, and external specialist skills such as consulting.

The frequency and costs of regularly spaced maintenance events are based on experience, supplier specifications, or similar industry norms. If no maintenance is planned, the maintenance interval time is the estimated life of the project. An overhaul or shutdown can be treated as a replacement material cost.

Include in OC any indirect operating costs, such as frequent painting of corroding steel or patching leaks from holes in process equipment. Another example is the requirement of spark testing a liner at regular intervals. Include these and similar items in annual material-related costs.

Lost production costs during downtime (LP). LP is the revenue lost as a direct or indirect result of the unit being out of service. Downtime of ancillary and related equipment may also affect lost production costs. Add lost production from these sources into the total. The number of lost production events in this category usually equals the number of maintenance events, because they often coincide in plants currently not using PdM.

Replacement materials costs (RC). RC includes replacing parts of the system too expensive to repair. Materials subject to high corrosion and wear applications may require replacing several times during the desired project life.

Estimate regularly spaced replacement intervals (15 years is a typical value) based on experience, operating conditions, manufacturer's specifications, or corrosion tables. NACE International (Houston, Tex.)

and other organizations have published corrosion rate tables for metallic and non-metallic materials of construction, which can be used to predict equipment life. A unit may have to be replaced once or several times during the commercial life of the host plant.

RC is not a maintenance cost. RC includes expenses like the following:

- Removal costs, per event, including the labor and material costs of removing, stripping, or demolishing the unit.
- Material and installation costs, per individual event, including the costs of replacing the material, the labor costs of reinstallation, transportation and delivery costs.
- Residual value of the material, per event, including the scrap value, which may be as high as 30% of the original material. This is a credit treated as a negative expense.
- Decommissioning costs, including cleaning and preparing for salvage, or otherwise meeting environmental, health, and safety regulations. The residual values and decommissioning costs are especially significant for large quantities of high-priced materials in short life-cycle applications. Decommissioning also can be significant in the production of hazardous materials to meet environmental regulations.

LCC is a Total

Computing the sum that equals LCC is easy. All future costs that will be incurred as a result of a material choice are discounted to PV. The PV of costs incurred at regular, but not necessarily annual, intervals plus operating costs are summed with the initial costs to become the total LCC. Total initial costs, the sum of material, fabrication, installation, and other installation costs, are not discounted as they are determined at PV.

A spreadsheet works well for making the calculation steps and presenting the comparisons. A computer program in spreadsheet format is available free from the SSINA. Using a spreadsheet allows a sensitivity analysis. Primary input variables can be changed, showing the effect each has on the total.

Evaluation of existing projects next, we present a method used to evaluate the LCC of an existing project and to compare it to benchmark data. We assume that data gathering was adequate. After a project has been installed and has operated for several years, review the performance data. Determine the initial capital cost and annual maintenance costs for the equipment. Also, determine operating costs such as those involving energy and feedstock utilization. Estimate a reasonable interest rate for the period such as the average prime rate over the period. Combine the total maintenance and operating costs for each year and calculate a present value back to the startup year. The life-cycle cost is the sum of the capital investment and the PV of the total maintenance and operating costs.

The next step is to determine the breakeven capital investment. This is the amount of capital that could have been spent, so that the annual maintenance and operating costs would have been equal to the benchmark costs, and the life-cycle cost would be the same as for the comparison case.

The ratio of breakeven capital to actual capital is used as a tool to evaluate whether the capital spending was optimal. The closer the ratio is to unity, the more optimized the capital, and the lower the life-cycle cost.

Benchmarking

There are three key assumptions that impact a basic life-cycle optimization strategy and affects the actual implementation of the LCC theory:

1. Maintenance costs will decrease with the increasing cost of initial capital, up to a benchmark point of maintenance costs.
2. Operating costs will decrease with increasing cost of initial capital up to a point, which is the benchmark operating cost.
3. The opportunity for minimizing the life-cycle cost is highest in the early phases of project development and lowest at the late stages of a project.

There are 13 activities or (BPs) (Table 12-8) in a project that can benefit from considering LCC. Optimization should be considered at each activity point. The impact lessens with the positioning of an activity in the list. Wherever you are in a project, start now to optimize. Do not cry over the lost opportunity of not being at the top of the list. LCC pays off even from the bottom, so implement the 12-step process for any activity.

From Theory to Practice

The goal is to spend the amount of capital that will produce the lowest possible life-cycle cost. These steps are an approach to this optimization.

1. Develop an annual maintenance and operating-cost curve as a function of the capital invested.
2. Plot the annualized cost of capital on the same graph.

Table 12-8
Project activities that benefit from LCC

---

1. Generation of specifications based on appropriate industry standards, as well as in-house standards
2. Avoiding reliance on "Vendor's Standard"
3. Specification and P&ID review by technical discipline experts
4. Determining and focusing on key factors affecting life-cycle costs
5. Obtaining quotations from vendors
6. Technical bid tabulation
7. Vendor selection and bid conditioning
8. Equipment design audit and review
9. Testing at vendor's facility
10. Installation practices
11. Commissioning practices
12. Operating practices
13. Maintenance practices

---

3. Add the two cost curves together to obtain the life-cycle cost graph.
4. Determine the amount of capital that corresponds to the lowest life-cycle cost.

See Figure 12-9. This graph can be constructed by adding the values representing the capital cost to the numbers representing the operating costs – including maintenance costs. The resultant minimum is indicating the optimum cost situation. Estimating annual maintenance costs on a future project is not straightforward. The following discussion describes a method for making this estimate.

Past projects have certain design and installation options that proved to have high maintenance or operating costs – step 9 of Table 12-8. Identify and evaluate these specific options, also referred to as "key reliability factors" that impact on costs. Assume that the key reliability factors can be identified for a project, and that historical data are available, on a common basis, for maintenance and operating costs versus capital cost. Add together all the options to develop an overall maintenance and operating cost curve.

For an example, we develop benchmark data here, demonstrating that the service factor of a gearbox affects the frequency of repairs and, thus, maintenance costs. This is a case history of two gearboxes and their maintenance costs for different service factors.* We show how to put this on a common basis.

---

* Service Factor is the ratio of design power capability to expected power required. This is the definition of the American Gear Manufacturers' Association (AGMA, Alexandria, Va.).

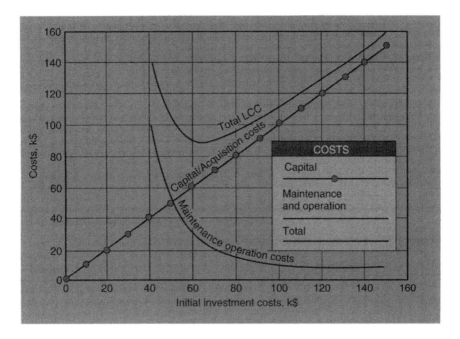

Figure 12-9. Minimizing total LLC (Top curve).

Gearbox LLC Evaluation

A mixer's gearbox experienced severe gear-tooth pitting after the first year of operation. Failures occurred each subsequent year, costing hundreds of thousands of dollars in production losses. By applying a failure analysis and design review, engineers determined the AGMA service factor was only 1.07. API (American Petroleum Institute, Washington) standards require a 1.70 service factor for equipment working with extrusion machinery. The engineers specified a new gearbox with a 2.0 AGMA service factor. The gears were wider and the gearbox bigger. As a consequence, the motor had to be moved several inches. The cost of the new gearbox and motor relocation together was $294,333. The original gearbox cost $160,000 (not including motor foundation work). The new gearbox ran five years before requiring any significant repairs.

Cost performance was compared on a 10-year basis. Although 10 years of history were not available, future years were assumed to require the average maintenance costs of the known years. The spreadsheets and maintenance cost graph are shown in Tables 12-9, 12-10, and 12-11. We will walk through them for greater understanding. The year column needs

Table 12-9
Life-cycle cost of case 1 (SF = 1.07)

Year	Cost category	Actual cost, $	Discount factor	PV of CF, $	Cumulative maint. PV, $	% of capital
0	Capital	160,000	1.0000	160,000	0	0
1	Maintenance	26,887	0.9259	24,895	24,895	15.56
2	Maintenance	64,867	0.8573	55,613	80,508	50.32
3	Maintenance	53,930	0.7938	42,811	123,320	77.07
4	Maintenance	48,561	0.7350	35,694	159,014	99.38
5	Maintenance	48,561	0.6806	33,050	192,064	120.14
6	Maintenance	48,561	0.6302	30,602	222,666	139.17
7	Maintenance	48,561	0.5835	28,335	251,001	156.88
8	Maintenance	48,561	0.5403	26,236	277,237	173.27
9	Maintenance	48,561	0.5002	24,293	301,530	188.44
10	Maintenance	48,561	0.4632	22,493	324,023	202.51
Net present cost				$484,023		

Table 12-10
Life-cycle cost of case 2 (SF = 2.00)

Year	Cost category	Actual cost, $	Discount factor	PV of CF, $	Cumulative maint. PV, $	% of capital
0	Capital	294,333	1.0000	294,333	0	0
1	Maintenance	0	0.9259	0	0	0
2	Maintenance	0	0.8573	0	0	0
3	Maintenance	0	0.7938	0	0	0
4	Maintenance	0	0.7350	0	0	0
5	Maintenance	50,000	0.6806	34,029	34,029	21.27
6	Maintenance	0	0.6302	0	34,029	21.27
7	Maintenance	8,333	0.5835	4,862	38,892	24.31
8	Maintenance	8,333	0.5403	4,502	43,394	27.12
9	Maintenance	8,333	0.5002	4,169	47,563	29.73
10	Maintenance	8,333	0.4632	3,860	41,423	32.14
Net present cost				$345,756		

no explanation. The cost category column sets a category of expense. In the year zero it is the capital. All other years are maintenance costs.

The actual cost column follows Table 12-10. For the Case 1 gearbox, there were three actual repairs before it was scrapped. The average of the three repairs is $48,561. This number is repeated to the tenth year. In Case 2, $8333 is the average of 6 years (50,000/6).

Table 12-11
Symbols and abbreviations

AC	= Initial material acquisition costs, $
C$_G$	= Average repair cost, $
C$_{PV}$	= Present value of costs, $
C$_Y$	= Annual repair costs, $
D	= Discount factor, dimensionless
F	= Future costs, $
i	= interest rate, dimensionless (decimal form)
IC	= Initial materials installation and fabrication, $
LP	= Downtime costs, $
MTBR	= Mean time between repairs, mo
MTTR	= Mean time to repair, d
n, N	= time intervals, yr
OC	= Operating and Maintenance cost, $
P	= Present worth, $
RC	= Replacement costs, $

Discount factor is calculated as $DF = 1/(1+i)^n$, with $i = 0.08$. C_{PV} of cash flow (CF) is the output of the PV function in Excel. Maintenance C_{PV} cumulative is tricky. The first line is zero, since 'capital is not maintenance.' Then start adding maintenance costs. For instance, Line $2 = 24,895$; Line $3 = 24,895 + 55,613 = 80,508$; Line $4 = 80,508 + 42,811 = 123,320$. The last column is the cumulative column divided by the initial capital cost and converted to percent. Finally, add up the C_{PV} of the CF column to get a net present cost.

Considering only maintenance costs and the initial expenditures, Case 1 (1.07 SF) has a 10-year cost of $484,000, while Case 2 (2.0 SF) costs $346,000. The difference would be much greater if production losses were included. However, the example as stated illustrates how lower life-cycle costs may be achieved despite higher initial cost.

In summary it can be easily seen that the lowest cost of initial capital may not produce the lowest life-cycle cost in the long run. Maintenance and operating costs can be quite significant and cannot be ignored in the economic analysis.

Making LCC Policy

Many reliability professionals are talking about LCC today. Frequently, that is where this subject remains, in the talking phase. To implement LCC practices, a company policy has to be implemented. It would help

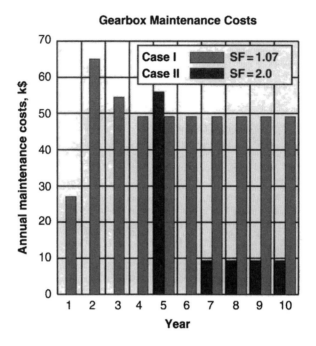

Figure 12-10. Two different gear boxes.

to educate plant personnel in LCC concepts by enforcing a few basic and simple administrative procedures.

This policy would apply to new or replacement projects, particularly where major equipment is purchased and installed. All phases of project development, from conception to startup, should be included in this policy of minimizing the life-cycle cost. However, the implementation of LCC will add additional steps and reviews in the project development process, as shown in Figure 12-11.

Audits of the LCC procedures should be conducted for projects 6 months after closing. This should be done during the regular project audits. Checklists assist project team members in conducting these audits. Their main purpose is to assure that the overall LCC policy is being adhered to in the project development process.

Set up 5-year audits as well. This is a good interval to re-evaluate life-cycle costs, and determine the break-even capital. Audit against the LCC assessment in the project work scope document and against benchmark data. Although the audit is conducted after 5 years, the maintenance and

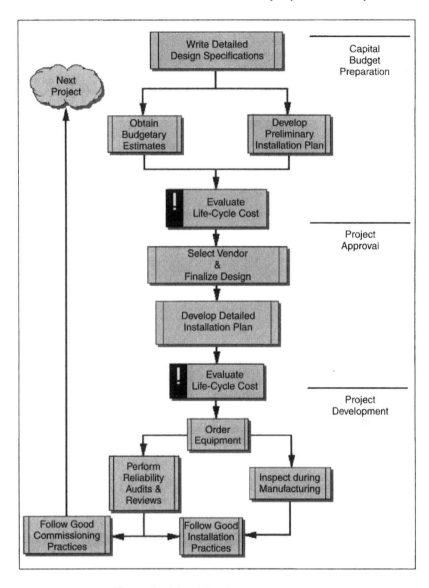

Figure 12-11. LCC policy implementation.

operating average annual costs will be extended to a 15-year (or other period specific to your installation) standard equipment life and the final evaluation made on that basis. Lessons learnt will be incorporated into a continuous improvement of the Life Cycle Policy.

References

1. US Department of Defense, *Acquisition of Major Defense Systems*, DoD Directive 5000.1, Washington, July 1971.
2. MIL-HDBK-276-1, *Life Cycle Cost Model for Defense Material Systems Data Collection Workbook*, p. 41, February 3, 1984.
3. Microsoft Corp., Excel Ver 5.0, Help Function PV. Redmond, Wash., 1997.

Bibliography

Fabrycky, W. J. and Blanchard, B. S., *Life-Cycle Cost and Economic Analysis*, Prentice-Hall, Inc., Upper Saddle River, N.J. 1991.
Paul, B. O., Life cycle costing, *Chemical Processing*. **57**(12), pp. 79–83, December 1994.
Bloch, H. P., *Improving Machinery Reliability*, 3rd edn, Gulf Publishing Co., Houston, Tx., 1998.
Galster, D. and Geitner, F. K., Using life-cycle costing tools, *Chemical Engineering*, pp. 80–86, February 2000.

Chapter 13
Starting with good specifications

Apply Mechanical Reliability Principles to Turbomachinery Design*: Use these Guidelines to Improve Availability

During the last decade, process machinery users have gradually become aware that the achievement of reliable, continuous operation of their machinery is perhaps the most important aspect affecting their decision to purchase a specific piece of equipment. As a result, there is more mention of the word "reliability" in equipment standards, and certain well-defined component lifetimes are now being specified. In earlier times parameters such as initial cost or higher efficiencies were the deciding factors in purchasing plant machinery. However, and as was brought out in our earlier chapters, the use of these "non-experience-based" decision-making methods led to the purchase of machinery which met all the project specifications at the time but resulted in unreliable, expensive to maintain plant equipment.

The design of a turbomachine has a direct influence on its life-cycle reliability and, therefore, its ability to operate with a maximum time interval between failures. A new mechanical reliability concept can shed light on what has so far been essentially a black art – based partly on science and partly on experimental verification – with little in the form of theory to guide the way. This concept involves creating a universal mechanical failure mode list that applies to all machinery and, from there, developing design rules and guidelines that eliminate these failure modes or isolate their effects. Applying these design principles leads to advanced turbomachinery that exhibits an extremely high resistance to mechanical

* By Abdulrahman Al-Khowaiter. Mr Al-Khowaiter is a rotating equipment consultant with Saudi Aramco Oil Company in Saudi Arabia.

breakdown. That said, this chapter illustrates the forward thinking and example-upgrading processes employed by a major oil producer and petrochemical manufacturer. While not necessarily applicable to a specific machine model, the principles highlighted here merit consideration.

Origins of Turbomachinery

The first practical turbomachine was developed in the late nineteenth century by professor Gustaf de-Laval of Sweden. His requirement in 1883 for high centrifugal acceleration on a newly designed centrifugal separator led to developing *a* supercritical rotor with speeds of 40,000 rpm and greater. In addition to these previously unheard of speeds, need for a high-speed prime mover to drive the new centrifuge led de-Laval to invent the first practical steam turbine [1].

de-Laval was the first to question the widely accepted, mathematically "proven" theory that there existed an absolute speed limit for every symmetrical body rotating around its axis of symmetry. It was thought at the time that this limiting frequency could not be exceeded without total destruction of the rotating body. However, de-Laval's deeper physical understanding of this phenomenon led him to believe that the observed limit was only a critical speed arising from convergence of the rotor's rotational speed with the natural frequency of the shaft's flexural vibration. His solution was to produce a supercritical rotor that became the first to successfully operate above these limits, through the use of a "flexible" shaft that allowed decoupling the lateral critical vibration from the unbalance forces acting on the rotor.

Soon after, in 1884, Charles Parson of Great Britain successfully developed the first large output, multiblade axial flow steam turbine that included a host of innovations [2]. This powerful turbine quickly brought an end to the domination of large reciprocating steam engines in power generation and mechanical drive services such as ship propulsion and opened up new, practically unlimited power capabilities. Both inventors ran into numerous mechanical reliability problems, and many difficult technological barriers had to be crossed to achieve successful operation. Through skill and determination, these early designers/researchers/manufacturers were able to single-handedly overcome at least eight major turbomachinery design problems:

1. High rotor vibration was due to supercritical speeds, limited mass balancing capability, and insufficient rotor-bearing damping technology.
2. High-speed shaft couplings were not available and had to be designed and developed explicitly for these early machines.

3. Due to the high speeds and close clearances, highly accurate fabrication and machining tolerances were necessary and required an advance in the state of the art.
4. The new turbomachinery exhibited an extreme sensitivity to misalignment and demanded new rigorous shaft-to-shaft and internal alignment standards.
5. Suitable shaft-to-casing pressure seals such as labyrinths were the subject of intense development.
6. Need for high-speed gearing to step down the advanced rotor speeds was also a design and reliability challenge.
7. Both pioneers were forced to develop novel turbine metallurgy to overcome the combined thermal and mechanical stresses acting on the blading.
8. Need to optimize thermodynamics by maintaining a high thermal efficiency was clearly recognized by both de-Laval and Parsons, and imposed mechanical design constraints.

Although the technological challenges listed were first overcome almost 120 years ago, these turbomachinery design characteristics remain applicable. Many of these same issues are responsible for present process plant machinery failures.

Interestingly, both de-Laval and Parsons ultimately resorted to brute force methods to achieve the reliability necessary for continuous operation. For example, in a bid to control the extraordinary rotor vibrations of their new turbomachines, drastic measures were the order of the day. de-Laval found a solution in the form of sleeve bearings with an unprecedented length to diameter ratio (L/D) of up to 8:1, while Charles Parson ingeniously designed and applied *triple-layer* squeeze film radial bearings (circa. 1890) to ensure sufficient damping in his new turbines (Fig. 13-1).

Machinery Reliability by Design

This section wants to highlight mechanical design weaknesses that commonly occur in turbomachinery and explain how their reliability can be improved during the design phase. Turbomachines share several unique characteristics and yet exhibit the mechanical characteristics of "normal" machinery. For example, like any machine, a centrifugal compressor includes a rotating shaft and stationary bearings. However, due to the requirement for rotordynamic stabilization at high rotational speeds, centrifugal compressor bearings must not only provide low frictional shaft support and positioning, but are also required to supply significant rotordynamic damping to minimize deflections and destructive vibration.

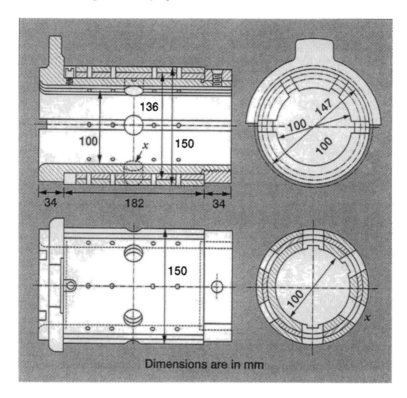

Figure 13-1. Triple-layer squeeze film radial bearing [1].

In addition, turbomachines are also unique in that while revolving at extremely high shaft velocities, they must also simultaneously retain the machine's internal pressure through the use of shaft-to-casing seals. Add to this mixture such extras as high temperatures, large power outputs, and extraordinary fatigue cycles and the recipe for mechanical disintegration is complete. Therefore, perhaps more than any other type of machinery, turbomachines are sorely in need of reliable mechanical design. Before going any further then, the meaning of "mechanical reliability" should be defined: "Mechanical reliability of a machine is a measure of its ability to operate continuously at design conditions, and without mechanical failure, for a specified time period."

From a long-term study of why some machinery are able to attain very high reliability levels while others suffer excessive breakdowns, clearly many failures were inborn or inherent in the unreliable machinery's design. On the other hand, the superior, highly dependable machines contained a minimum of inherent or "dormant" failure possibilities. As a consequence, the following general rule was formed: *The ability of any machine to achieve a satisfactory mean-time-between-failures (MTBF) is governed by its resistance to failure.* This statement is based on purely logical reasoning, since failure is the opposite of reliability.

After reaching this simple conclusion, the next step was to develop a universal mechanical failure list, the function of which is to describe the failure modes of all machinery. This list then becomes a qualitative design tool to apply when evaluating machinery at the design stage as shown in Table 13-1. By uncovering the elusive mechanical characteristics that are responsible for machine reliability, it should then be possible to evaluate a design while still in the blueprint stage of development. This was not possible in the past.

Table 13-1
Universal machinery failure mode list

---

1. **Lubrication breakdown.** Covers all machine failures caused by loss of lubrication due to:

   - Loss of lubricant, contamination.
   - Insufficient flow, film breakdown. Main components affected are bearings, seals, and sliding contact surfaces.

2. **Excessive vibration.**

   - Covers all vibration-included failures of machine components such as rotor shaft vibration, impeller vane vibration, turbine blade vibration, and others.

3. **Corrosion/adhesion.** Machinery failures caused by surface molecular action:

   - Corrosion results from a chemical reaction between the turbomachine component and its environment, such as process gas corrosion.
   - Adhesion occurs as a result of molecular attraction and is responsible for "sticking" and deposits adhering to machinery components.

4. **Erosion/wear.**

   - Includes abrasive wear, fretting, scuffing, and galling. This mode covers all failures caused by material loss through mechanical action.

5. **Foreign object damage.** Covers all failures attributed to ingesting an object (solid/liquid), which is either:

   - not an intrinsic part of the machine or
   - a part of the machine that has moved from its design position.

6. **Seal failure/static–dynamic.** Includes all failures resulting from:

   - Loss of sealing ability or containment at an interface. The sealing interface may be static or dynamic. Examples are static gaskets, O-rings and mechanical seals.

Table 13-1
Universal machinery failure mode list–cont'd

---

7. **Excessive thermal growth/degradation.**

   • Failures due to temperature changes exceeding normal design limits and material property degradation from excessive temperature variation.
   • Thermal cycling failures.

8. **Locking/holding mechanism failure.** Failures in assembled components, leading to loosening in joints:

   • Held by fasteners.
   • That do not rely on fasteners but utilize elastic properties to maintain the assembly.

9. **Failure due to improper internal geometry.** Includes failures that occur from:

   • Loss of design geometric alignment between components.
   • Rubbing and impacting between components that are designed to be noncontacting.

10. **Overstressed material.** Refers to all breakdowns attributed to material structural failure:

    • Fatigue failures, ductile fractures, surface deformation and other mechanical stress-induced component failures.

For example, in the late 1930s, military aircraft engine development intensified in anticipation of the impending World War II [3]. Most American and European piston engine manufacturers achieved the desired engine MTBF of approximately 250 hr through factory testing of prototype engines at supercharged conditions simulating almost double the actual required brake horsepower output. These severe, punishing tests naturally produced mechanical breakdowns at the weakest points in the prototype engine designs. From here, the designers would redesign the failed parts and then return to the test stand for further trials. This procedure would continue until the engines were finally able to consistently reach an MTBF of 250 hr minimum or whatever time-between-overhaul (TBO) limits imposed by Air Force guidelines.

Practically, all high-power reciprocating engine manufacturers of that era found through extensive testing that only silvered bearings could guarantee the required fatigue life at the design conditions of greater than 3000-psi bearing loading. Babbitt alloys were capable of only one-half the ratings of the silver-lined steel bearings [4].

This method, while successful, was essentially a "test-to-failure" reliability analysis method and is still commonly used. In general, many designers are not able to produce mechanically reliable designs on paper and are forced to obtain quantitative and qualitative reliability design improvements through trial and error. One major reason is that often machines are designed by multidisciplinary teams, with specialists in specific areas such as stress analysis, metallurgy, heat transfer, and manufacturing. The ability is lacking to tie the effects of each discipline's output and evaluate the interaction of each component while still on paper.

In some cases, however, an outstanding, highly gifted designer will break this rule and produce reliable, high-performance machinery straight from the drawing board. Our aim is to attempt to capture part of this intuitive mechanical design ability – which is clearly evident in the ingenious designs of de Laval and Parsons – and transform it into a readily understood, logical design procedure.

Mechanical Reliability Design Principles

After developing a failure mode list, design guidelines that the designer and user can apply are needed. The following guidelines maximize the component and overall machine's resistance to failure:

1. Design out failure modes. This is the most powerful design method possible. By eliminating the built-in failure modes of the individual components and the machine as a whole, overall mechanical reliability is raised.
2. Minimize the number of rotating and static parts. Machine reliability is equal to the product of its n-component reliabilities, with each component having a value less than one [5]:

$$R(machine) = R_1 \times R_2 \times R_3 \times R_4 \times \ldots R_n$$

   Therefore, as we saw before in this book, by reducing the actual number of components in this series, the magnitude of overall reliability will be increased.
3. Apply design safety factors to all failure modes. Practically, all designers are aware of, and apply, design safety factors to pure mechanically stressed components during the stress analysis phase (failure mode 10). However, few designers are aware that design safety factors must also be applied to the remaining failure modes such as lubrication failure, vibration, corrosion, wear, and others.

   For example, in turbomachinery all the design efforts and safety factors added to ensure an infinite mechanical fatigue life for the shaft may be wasted. Shaft thrust collar loosening from poor locking design will destroy this finely crafted rotor in seconds. Therefore, safety factors must be applied to all ten of the failure modes. This is because – as in a chain – the weaker links will always defeat the purpose of any stronger, heavier-duty links.

4. Add parallel redundancy to failure modes. If a failure mode on a component or major assembly cannot be designed out, use a redundant design to increase reliability against that specific mode. Redundancy improves reliability through the familiar law:

$$Parallel\ component\ reliability = 1 - (1 - R_a)(1 - R_b)$$

where R_a and R_b are two identical components acting in parallel. For double parallel redundancy, if $R = 0.90$ then:

$$New\ reliability = 1 - (1 - 0.90)(1 - 0.90) = 0.99$$

5. Apply the integral design principle. Integral design is integrating separate machine components into a single part through creative design. This eliminates the following:

- Fastener failures and shrink and press fits that loosen.
- Misalignment is reduced due to joint elimination.
- Sealing failures cannot occur at non-existent joints.
- Material overstress due to shrink-fitted parts.
- Vibration failures arising from component looseness, unbalance, and alignment errors.
- Crevice corrosion attack.
- Improper internal geometry failures, such as rubs from loose components.
- Fretting-induced failures.

6. Maximize the separation distance between moving parts and non-compliant static parts. This applies to all components except for bearings.
7. Minimize the number of wearing surfaces such as bearings and sliding contact points.
8. Eliminate static and dynamic sealing points where possible.
9. Add flexibility to the interface between moving and static parts such as at dynamic sealing locations.
10. Use direct instead of indirect force transmission. Reduce the number of linkages.

11. Minimize high mechanical stress points by adding the maximum possible radius or taper to all cross-sectional changes in mechanically loaded parts.
12. Reduce the number of mechanical components stressed in bending. Change to pure tension, compression, or shear.
13. Use mechanical drive shafts of maximum rigidity.
14. Use ductile as opposed to brittle materials.
15. Incorporate self-aligning seals and bearings.

Designing Out a Failure Mode

By analyzing design aspects of a typical built-up steam turbine shaft-disk assembly, the following mechanical failure modes can be found that are inherent in the design:

- *Locking/holding mechanism failure*: Loosening of the shrink-fitted disks from the shaft due to incorrect pre-warming or from slight overspeed. For a six-stage rotor, the reliability against loosening is equal to the product of each of the six individual stage reliabilities.
- *Vibration*: Rotor vibration from unbalance, misalignment, fluid excitation, and various other sources.
- *Erosion/wear*: Steam condensate erosion, wear at the disk bore through fretting and wear on the shaft at the labyrinth interfaces.
- *Excessive thermal growth/degradation*: Thermal shock from sudden startups and thermally induced shaft bowing.
- *Corrosion/adhesion*: Stress corrosion cracking can occur at the highly pre-stressed disk bore area, predominantly at keyway stress concentrations [6].
- *Material overstress*: Disk rupture from overspeed operation.
- *Foreign object damage*: Foreign objects in the steam path, such as liquid slugs and cracked internal fasteners.
- *Failure due to improper internal geometry*: Misalignment of the rotor in the casing.

Therefore, eight design failure modes are acting on any built-up steam turbine rotor. However, by specifying a monobloc shaft-disk construction, the user can eliminate two failure modes: the locking/holding mechanism mode and the corrosion/adhesion mode. As a result, properly constructed integral turbine rotors are only susceptible to six failure modes and, consequently, overall turbine reliability is improved compared to a built-up design. Process plant failure statistics over the past 20 years have confirmed the validity of this design conclusion.

Applying the Integral Design Principle

Integral Design General-Purpose Steam Turbine Bushings

Figure 13-2 shows a throttle valve assembly for a well-known, general-purpose steam turbine. These turbines are used worldwide in many process plants, and the author's company operates at least 200 of these machines in 600-psig steam services. A constant reliability problem on eight of the company's 1200-hp models was excessive steam leakage from the throttle valve stem bushings. As a result, repacking each turbine's steam glands every 4 months was necessary. In 1997, an in-house redesign was undertaken to improve reliability. The outer gland was converted to an integral bushing-lantern ring (Fig. 13-3). This resulted in the following design and operational improvements:

- The lantern ring is now an integral part of the valve stem bushing. There is no possibility of incorrect axial installation of the lantern

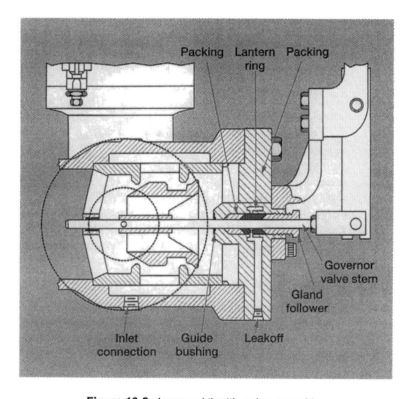

Figure 13-2. Improved throttle valve assembly.

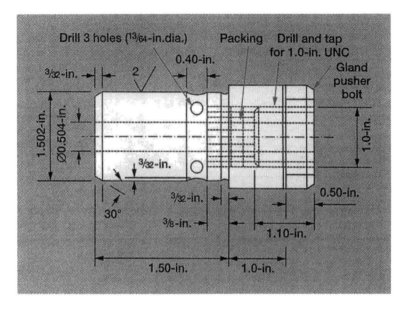

Figure 13-3. The outer gland was converted to integral bushing-lantern ring.

ring or packing rings, which can lead to partial blocking of the internal 60# steam leak-off line.

- Improved valve stem alignment and support by the accurately bored bushing, which adds a third guide point compared to two in the original design.
- The packing glands on all eight turbines have still not required repacking after 3 years of continuous operation. (This unusual life-span improvement was unexpected and is puzzling.) Valve stem wear has been reduced and steam leaks at this location have been eradicated.

Note: The inner, high-pressure bushing was retained as is. The new outer bushing was press-fitted into the machined valve cover with a 0.002-in. interference fit, using molybdenum disulfide grease.

Integral Design of Centrifugal Compressor Impellers

A two-piece welded or brazed impeller is superior to a three-piece design because the impeller vanes in the former are integrally machined with the back cover-plate. This reduces the number of welds by 50%, thus minimizing the possibility of weld zone cracking or braze metal bonding failure by an equal percentage.

Flexible support, tilt pad bearings are another example of applying the integral design concept to maximize reliability. In a conventional, horizontally split four-pad bearing, at least six separate components are necessary for the assembly. As can be seen from Figure 13-4, the flex-tilt design requires only two separate components. The following failure possibilities are eliminated by this design:

- Tilting pad pivot wear, which leads to rotor vibration.
- Tilting pad self-flutter vibration.
- Pad damping properties that vary with time.

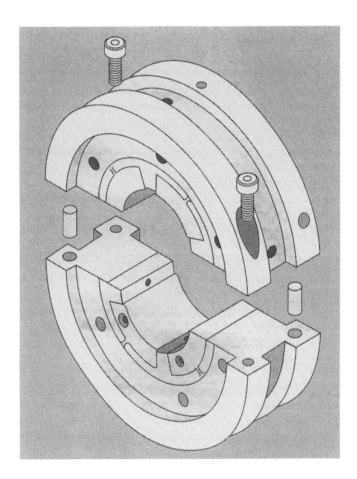

Figure 13-4. The flex-tilt design requires only two separate components.

Reducing Locking/Holding Mechanism Failures

For those assemblies where an integral design is not practical, the end result in most cases is that fasteners must be employed to join the separate members. For bolted, mechanically stressed joints, it is strongly recommended to incorporate the following joint design details [7]:

- Provide a thread engagement of 1.25 × bolt diameter or more.
- Use long bolts or studs to increase joint elasticity and, therefore, minimize the loss of initial tension from thermal cycling and vibration.
- The maximum possible alternating fatigue stress in the joint's fasteners should not exceed 10,000 psi, unless specially designed fatigue-resistant bolts and nuts are used [8]. A high bolt tension, or preload, tends to make the external fluctuating loads bypass the bolt itself, thus greatly increasing fatigue resistance.
- To reduce embedment, use hardened surfaces under nuts or apply thick, hardened washers.
- High-strength, rolled thread fasteners such as SAE 8 with a proof strength of 120,000 psi will increase bolted joint reliability.
- Minimize the number of surfaces or interfaces in the joint. For example, use a flanged nut (which is basically an integral nut-washer).
- All internal fasteners, whose failure in service can lead to foreign object damage inside the turbomachine, should be fully captive.

Coupling Bolting

From field experience, we have found that 1/4-in. diameter, high-strength coupling bolts are not as reliable as larger diameter bolts for turbomachinery drive shaft couplings. The reason is these smaller-diameter fasteners have a normal initial tightening torque limit of only 13 ft-lb to achieve 75% of their proof strength. The problem with such bolts is that during tightening, the combination of high torsional shear stress in the bolt body in addition to the shear stress at the threads greatly limits allowable torque [8]. Consequently, a relatively low tensile preload or joint clamping force results compared to larger diameter bolts, where a change occurs at 3/8-in. diameter size bolts and greater. From here on, torsional shear stress capability of these larger bolts begins to significantly exceed the thread shearing strength.

From the author's experience, at least 50% of high-speed compressors and steam turbines utilizing 1/4-in. bolts (SAE grade-8) included a relatively high number of loose bolts found during coupling inspection intervals of 2 years. Many bolts had essentially lost their original preload

Table 13-2
Transforming nut torque into tension

Torque absorption in a tightened bolt

	Percent of tightening torque	
	UNC	UNF
Bolt tension	15	10
Thread friction	39	42
Head friction	46	48
Total tightening torque	100	100
Loosening torque	70	80

tension (Table 13-2). However, for those couplings utilizing 5/16-in. diameter and greater fasteners, the incidence of loose bolts did not exceed 10% of all couplings inspected.

As a result, the company's centrifugal compressor standards now specify a requirement for minimum coupling bolt diameters of 5/16 in. There is no reason why these sturdier fasteners cannot be substituted for the existing turbomachinery industry minimum coupling bolt size of 1/4 in., since the 5/16-in. SAE grade-8 coupling bolts have almost double the torque capability (25 ft-lb) and, therefore, double the preload or clamping force (Table 13-3). In addition, from a human-factor's point of view, the larger diameter bolts are more resistant to overstress from mechanic error.

With gear couplings, users should be warned that the normal lubricated tooth sliding friction factor is in the range of 0.10. However, when lubrication becomes poor or non-existent, these couplings begin to lock up (torquelock) and sliding friction factors of 0.5 and greater occur. This implies that the gear coupling becomes more like a solid coupling and

Table 13-3
Torsional stiffness and strength comparison between bolts

Bolt torsional strength is proportional to D^3. Therefore, torsional strength of a 5/16-in. diameter bolt compared to a 1/4-in. bolt is:

$$(5/16)^3/(1/4)^3 = 0.0305/0.0156 = 1.95$$

Thus, a mere 1/16-in. increase in diameter gives double the torsional strength and, therefore, double the possible tensile clamping force.

loses its design flexibility. With high misalignment and excessive tooth friction, large shaft bending forces are now transferred directly to the coupling fasteners.

The bolts experience these bending forces as fluctuating tensile loads during each revolution. If the existing preload is less than this tensile force, then the bolts will now be fully stressed by alternating tensile fatigue stresses, *which they were not designed* to *handle.* The result is catastrophic bolt/coupling failure. See Ostroot [9] for one company's unfortunate experience with extensive plant damage caused by this situation.

Therefore, again, it is wise to use the largest coupling bolt diameter possible to safeguard against unplanned service conditions. In addition, this phenomenon of transferring misalignment-induced bending moments to the coupling bolts cannot occur in modern flexible metal dry couplings.

Thrust Collar Loosening

Loss of rotor axial fixation due to thrust collar loosening has caused many disastrous failures in the past. Incidence of such failures has been reduced in the past 25 years by incorporating thrust position monitoring and protective instrumentation such as non-contacting proximity probes. However, while thrust position monitoring can detect an impending rotor axial rub, it does not prevent the initial failure mode itself, such as thrust collar locknut loosening. As a result, a shutdown of the turbomachine is still required to repair the looseness. Therefore, reliability must be designed into the mechanical design of all components and reliance on instrumentation is insufficient.

The most dependable thrust collar will be an integral part of the shaft. However, removable thrust collars are necessary in many machines incorporating shaft seals such as centrifugal compressors. In these instances, the following common failure modes result from this removability:

- Thrust collar-induced rotor vibration due to fretting at the shaft–collar interface (vibration failure).
- Thrust collar locknut loosening leading to rotor axial rubbing (locking/holding mechanism failure).

To reduce fretting, present API standards require either thermal or hydraulic collar shrink fitting. This is an acceptable, reliable solution. For the remaining failure mode, locknut loosening, it would be prudent to specify a double locking arrangement. Figure 13-5 illustrates one turbomachinery manufacturer's solution to this problem. The split ring and double nut locking design is a highly secure thrust locking arrangement. This is an excellent example of applying the parallel redundancy rule to increase failure resistance.

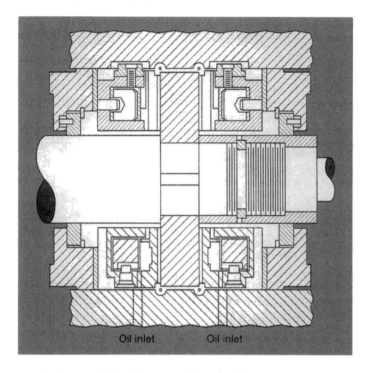

Oil inlet   Oil inlet

Figure 13-5. Highly secure thrust-locking arrangement.

Designing out Multiple Failure Modes

Problem: A number of horizontally split, centrifugal compressors in natural gas service (Table 13-4), utilizing bushing-type oil seals, were displaying a common failure symptom: excessive oil leakage from the compressor drains every 2–3 years. The normal repair procedure followed in the past was to remove the old shaft seals and install new OEM spare parts. Unfortunately, in most instances this was not enough to solve the problem, and leakages would remain at excessively high levels, above 75 gal/day per seal. Over the years, maintenance personnel discovered that a very shallow depression was actually being worn into the hardened Colmonoy®-coated shaft sleeves under the seal bushings. This depression was only 0.002–0.003 in. deep radially, but was sufficient to increase daily leakage by a factor of four or more. The worn area was difficult to notice since the seal repair was accomplished under field conditions, where unhindered visual access to this shaft zone was impossible without opening the casing itself.

Table 13-4
Natural gas compressor data

Compressor service	Sour natural gas
Installation date	1980
Quantity	four
Horsepower	4000–9000
Speed	7500–11,500 rpm
Suction pressure	50–250 psig
Discharge pressure	250–600 psig
Seals	Oil bushing type
Normal leakage	5–20 gal/day/seal
Differential seal pressure	5 psid (overhead tank)
Gas side bushing design	Steel ring with babbitt
Seal oil filtration	10 micron

The most likely reason for formation of this depression was that over many months of continuous operation, normal small debris in the seal oil such as pipe scale, silica, and varnish products would build-up against the stepped edges of the gas-side bushing. These foreign objects gradually adhered to this region and formed a hard deposit, which slowly ground into the shaft sleeve at this critical sealing area. The measured leakage rate, however, would remain relatively normal. More deposits would adhere to the seal bushings and decrease the clearance gap. After a certain time, the deposits would suddenly break off, and in one 24-hr period, the leakage rate would jump from an average of 15 gal/day–100 gal/day and more.

In addition to this shaft wear, in many instances the babbitted sleeve lands themselves were also heavily worn, which contributed to the excessive oil losses. This wear was occurring over time due to suspected transient shaft contact during startups and infrequent process upsets.

Solution: The solution undertaken was based on eliminating the following multiple failure modes:

- *Foreign object damage*: Hard deposit buildup wearing into the shaft.
- *Erosion/wear mode*: Shaft surface and bushing metal wear from light shaft rubs during startups and process fluctuations.
- *Thermal growth/degradation*: Bushing metal softening in the discharge end seal due to high temperatures increases the abradability.

First, a careful oil seal bushing design analysis led to the conclusion that the inner surface should have bearing qualities such as softness, embedability, and low friction. However, there is no need for the extreme softness of the standard OEM babbitt layer. We realized that a more durable aluminum alloy bearing surface would be far more reliable in this service due to its high load capability and wear resistance. Aluminum alloys also have greater temperature resistance (500 °F/260 °C versus 270 °F/132 °C for babbitt) and superior corrosion resistance. In addition, aluminum bearings have proven themselves in medium to large diesel engines on heavy-duty crankshaft service and have become the bearing of choice due to their extreme durability [10]. Therefore, the solution was to machine out the babbitt layer in the OEM bushings and install a shrink-fitted aluminum alloy insert in its place. The ideal alloy to use in this case would be 95% Al/5% tin. However, this was not available and annealed, low-alloy aluminum was used instead (Fig. 13-6).

To counteract the foreign object failure mode, the stepped bushing lands that trapped deposits were eliminated by adopting a straight, constant-diameter bushing bore. Again, a detailed design analysis of the original OEM stepped inner seal land led to the conclusion that the

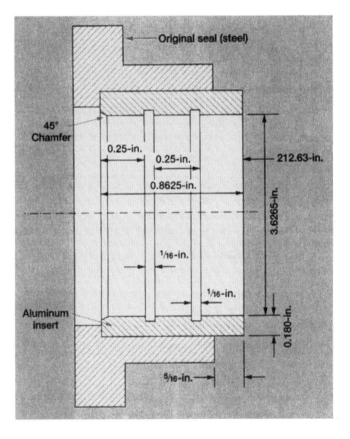

Figure 13-6. A more reliable aluminum alloy bearing surface.

function of the stepped lands was to allow "controlled" seal leakage when the tighter seal lands were rubbed by the shaft. In addition, it was surmised that the manufacturer might have been trying to reduce oil seal lockup effects on the shaft, which in some machines can cause vibration due to a shift in lateral critical speeds. However, since all of the compressors were rigid shaft machines, seal lockup would only lead to increased rotor rigidity, which is not a problem (below first critical). As can be seen from Figure 13-6, the new seal also has generous oil grooves that inhibit buildup of substantial hydrodynamic oil-film pressure.

Results: Installing the modified seals in May 1998 was a success from day one. The average shaft seal leakages were reduced from the original compressor manufacturer's variance of 5–20 gal/day/seal to a range of 2–10 gal/day maximum. Compressor shaft vibration levels were unaffected. After several months trial with the first compressor, a total of four compressors were subsequently modified with this design. They have been operating for 2 years without seal problems. The only solution available in the past was to either live with the high oil leakage rates or renew the shaft sleeves, which necessitated a complete rotor change out.

Improving Steam Turbine Trip Reliability

As a result of the extremely high energy density available in pressurized steam, and because of the nature of steam turbine design, there is no direct mechanical restriction to overspeeding a turbine rotor above its design rotation limit. Therefore, speed limitation must be artificially incorporated indirectly through multiple devices that finally block steam flow into the machine. On the other hand, prime movers such as electric motors are directly protected against overspeeding by the internal resistance of their electrical coils and the constant frequency of their alternating current power source (AC machines).

Unfortunately, as a consequence of the steam turbine's reliance on the series action of multiple devices to achieve overspeed protection, the resulting reliability of this action is inherently low [11]. The main reasons are that the devices involved in tripping the machine are considered and function as standby or fail-to-danger devices, which are naturally less reliable than active devices. In addition, the tripping action lacks parallel redundancy. The overspeed trip flowchart in Figure 13-7 portrays the series progression that must be completed to attain steam flow stoppage in a mechanical hydraulic trip system. For this flowchart, tripping action reliability, R_{trip}, is a function of:

$$R_{trip} = R_1 \times R_2 \times R_3 \times R_4$$

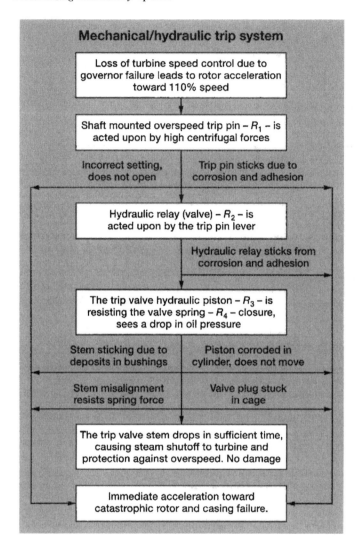

Figure 13-7. Many steps must be completed to stop steam flow.

Therefore, the crucial protection against overspeed failure is completely dependent on each device's reliability in this circuit. *This explains why steam turbine trip mechanisms are frequently tested and also sheds light on the reasons for the high frequency of dangerous runaway failures.* The following are general reliability rules recommended to improve this situation:

1. Add parallel redundancy to one or more trip circuit components. Modern digital electronic/hydraulic systems utilizing multiple speed

sensors add parallel redundancy to R_1. Notice that the shaft-mounted trip pin on older systems is a combined speed sensor and actuator.

2. Reduce the high dependence on components serving in standby mode, such as the stop valve, and increase utilization of active devices.
3. Minimize the number of devices in the series progression required for tripping.

The Hydraulic Governing/Trip Circuit

Hydraulic oil water contamination has led to trip component seizing and inability to protect the machine. Therefore, it is imperative that designers include real-world effects when analyzing critical systems. Since there will always be moisture in steam turbine hydraulic oil governing and trip systems, manufacturers should design the hydraulic trip components based on the conservative assumption of, say, 1% continuous water contamination in the hydraulic oil, i.e. 10,000-ppm water content. Therefore, regular carbon steel internal components are not acceptable and corrosion-resistant materials should be employed. This worst case analysis, by the way, is an example of applying safety factors to all machine failure modes, which in this case happens to be the foreign object mode.

Achieve Enhanced Stop Valve Reliability

Failures due to sticking: Turbine stop valve stem sticking due to steam deposits falls under the corrosion/adhesion failure mode. This problem continues to plague the industry mainly because most users are not aware that the typical stop valve stem-to-bushing assembly constitutes an almost ideal adhesive joint due to the following factors:

- An adhesive joint requires close clearances between the objects to be joined to achieve maximum adhesive strength [12]. This characteristic is available in all stop valves.
- An adhesive is necessary. This occurs frequently since steam piping from boilers carries an inorganic "cement" composed of silica and iron oxides as major constituents.
- In a bonded joint, it is preferred to have the mating surfaces as clean as possible before adhesive addition. In steam turbines, this requirement is always met, since normal steam leakage has a scouring effect on the stem and bushing surfaces, which effectively lays the ground for the future "adhesive" to bond properly.

- The stem-to-bushing interface comprises a joint that operates in shear, which unfortunately develops the maximum possible adhesive joint strength.
- A large bonding area increases resistance to adhesive joint failure. Most stems and bushings have a large contact area, as measured in square inches of actual surface area.

As a result, stop valve designers and users should strive to prevent adhesion by:

- Utilizing stem and bushing materials/coatings with a low affinity to the normally occurring inorganic deposits.
- Reducing the contact area or close-clearance zone between the stem and bushing.
- Separating the two functions, guidance/sealing, to minimize metal-to-metal proximity.

Mechanical Binding in Stop Valves

Valve stems binding in their bushings is a different phenomenon and falls under the improper internal geometry mode. However, in many cases these separate failure modes (adhesion and binding) could be acting together against the closing spring force. Therefore, to achieve dependable valve operation, all failure possibilities should be investigated and prevented by careful design detailing. Binding arises from misalignment of the stem's longitudinal axis with the valve and actuator guide bushings or from differential thermal growth. The following explanations clarify the sources of this binding:

- Incorrect stem alignment can occur during stop valve manufacture or assembly. Misalignment of the valve body to the actuator assembly is the result.
- Hot misalignment of the stem-to-bushing axis can result from differential thermal expansion of the valve body due to non-symmetry of the casting. This type of misalignment cannot be observed when cold.
- Elastic deformation of the valve body from internal steam pressure can also produce internal misalignment.
- Differential thermal expansion between the steam sealing bushings and their surrounding valve casing walls will cause inward growth of the bushing walls toward the stem. This results from the mechanical constraint of the high-temperature bushing by the relatively colder valve body walls. This highlights the importance of full thermal insulation of the valve, enclosing all high-temperature components.

Reducing Mechanical Binding through Improved Design

- Add 3/32-in. internal bore edge radii to all stop valve bushings. These radii will minimize hang-up of a misaligned valve stem on the normally sharp bushing edges.
- For large, critical turbines, valve stem electroplating at the bushing contact areas with a low sliding friction metal can greatly reduce the stem-to-bushing frictional resistance force (Table 13-5 and Fig. 13-8). For example, with silver rubbing against a steel surface, the static friction factor is only one-third that of steel against steel [13, 14]. Silver has also been proven to be an excellent high-temperature solid lubricant.
- Another, more conventional option for reducing the sliding friction of steel upon steel is to specify a high stem and bushing surface hardness, such as Rockwell C-45 or greater.
- Test all stop valves for internal misalignment due to pressurization by shop hydrostatic tests of the valve at water pressures equivalent to the service pressure. The valve stem, with its actuator removed, should move easily by hand during this test.
- A larger diameter stop valve stem will exhibit increased reliability compared to a smaller diameter stem. The greater stem rigidity reduces flexing and bowing, which lead to binding during operation.
- Beware of steam piping flange-to-valve body flange strain caused by bolting together non-parallel flange faces since this produces internal valve strain.

Table 13-5
Coatings used for various functions

Coating function	Coating used
Reduce wear	Titanium carbide, nitride
Reduce friction	Teflon, MoS_2
Increase friction	Titanium, bonded abrasives
Improve lubrication	Copper, lead
Increase temperature or load capacity	Electroless nickel
Prevent adhesion	Silver/gold plate
Imbed particles	Indium, lead
Reduce corrosive wear	Chromium plate or diffusion
Retain fluid lubricants	Phosphating, nylon
Rebuild surface	Steel hard surfacing
Reduce surface roughness	Silver plate
Prevent drop erosion	Polyurethane, neoprene
Prevent particle erosion	Cobalt alloy, molybdenum

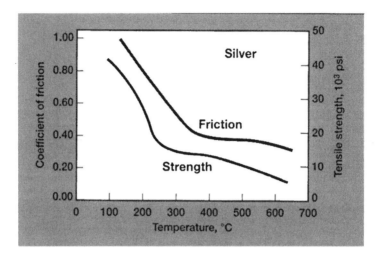

Figure 13-8. A low sliding friction metal can reduce resistance friction force.

Specifying Highly Reliable Turbomachinery

The following specifications for centrifugal compressors and special-purpose steam turbines are based on incorporating the mechanical design principles given in this section. Many of the design improvements highlighted here add little to the final purchase cost, but greatly improve life-cycle reliability of the turbomachine. Sometimes, however, a conflict will arise between the necessity for efficiency and the requirement for mechanical reliability. At this point, the user should carefully assess the situation. For example, some centrifugal compressors can be made more thermodynamically efficient when a flexible rotor is used compared to a rigid rotor, whose larger diameter shaft leads to an increased cross-sectional clearance area at the interstage seals and reduces impeller inlet area.

However, remember that many variables are involved. A good designer/user can still retain highly reliable design characteristics while maintaining the required efficiency. For instance, to increase efficiency of larger diameter, stiff shaft rotors, some designers resort to shaft scalloping, which increases the inlet flow area while still maintaining a high stiffness. In addition, the more rigid shaft can incorporate tighter interstage seal clearances due to the reduced shaft internal deflection compared to a flexible rotor. This again minimizes thermodynamic differences between the two rotors.

Centrifugal Compressor Specifications for High Reliability

Specify an API-617 machine with the following additions:

1. A stiff shaft machine (i.e., operating below the first lateral critical speed) is highly recommended where possible (Table 13-6). The recommended maximum allowable design speed is 15,000 rpm. Centerline support of the casing is required.
2. *Impellers*: welded, brazed or machined from solid. A two-piece, as opposed to three-piece, welded/brazed construction is preferred. Minimum allowable impeller and shaft keyway internal edge radius is 3/32 in.
3. *Couplings*: flexible metal, dry-type couplings only with 12–24-in. spacers, depending on machine size and temperature. Aircraft-quality coupling bolts with a 5/16 in. minimum diameter should be used. Couplings must be designed for infinite fatigue life at the following simultaneous conditions: 1.25 × design torque, 0.060-in. axial deflection and 1/2° misalignment. The couplings must be self-retaining upon flexing element failure. A fully-enclosed coupling guard is necessary.
4. *Coupling hubs*: Tapered, hydraulic shrink fit or double-keyed with multi-jackbolt tensioner (MJT) design coupling locknuts. The minimum shaft and keyway inside edge radius is 3/32 in.
5. Self-aligning tilting pad bearings are mandatory. Remember that large shaft journal diameters result in greater bearing – shaft damping.

Table 13-6
Rigid shaft advantages

- Less shaft deflection, thus less possibility of rubbing at internal stationary seals – allowing use of tighter interstage seal clearances.
- Stronger shaft, greater fatigue resistance due to larger diameter as compared to a flexible-shaft machine.
- No possible excitation of non-existing lower lateral critical frequencies.
- Simpler rotor balancing, no need for full-speed rotor balancing. Reduced sensitivity to unbalance from deposits and other sources.
- Reduced rotor thermal bow compared to equivalent flexible shaft.
- Negative damping produced by a rotor is proportional to the lateral pk-pk shaft deflection, the more positive damping is required at the oil film bearings to dissipate this flexural energy. Therefore, the more rigid shaft will require less damping at the bearings.

---

6. *Seals*: Non-contacting dry gas seals are mandatory. The minimum allowable O-ring cross-sectional diameter is 1/8 in. nominal. The mechanical seal's rotating sleeve should incorporate two O-rings in series to maximize sealing reliability (redundant sealing design).
7. Thrust collars should be removable, with 0.75-in. thickness disks as a minimum and be double-locked to the shaft. The thrust disk should be provided with 12 holes drilled and tapped radially to allow two-plane field trim balancing with setscrews.
8. If the compressor operates in services with significant rotor fouling, a low surface friction coating should be applied to the rotor to inhibit adhesion and deposit buildup.
9. All internal casing fasteners should be of 400-series, heat-treated stainless steel. These fasteners should be fully locked and captured.
10. *Lube oil piping*: All inlet and drain piping should be of 300-series stainless steel. Use stainless steel conical perforated metal screens before each bearing. The minimum allowable lube oil orifice diameter is 3/16 in. A stainless steel lined steel tank is a minimum requirement.
11. No shaft-driven lube oil pumps above 4000 rpm.
12. *Lube oil heat exchanger*: Either air-to-oil fin-fan coolers or U-tube design shell and tube with brazed or welded tubing. U-tube designs are superior due to the single tube-sheet leakage path.

Special Purpose Steam Turbine Reliability Specifications

Specify an API-612 machine with the following additions:

1. Integral rotor design only, including an integral thrust collar with a 1.0-in. thickness disk as a minimum. Rotors to be of stiff shaft design for backpressure-type turbines. The maximum recommended design speed is 15,000 rpm.
2. All backpressure turbine casings should be of two-piece axially split design only. The casing and bearing housing materials shall be of steel or steel alloys only. The packing gland housings shall be made integral with the turbine main casing. Rubbing contact-type shaft seals are not acceptable.
3. Turbine blades should be forged or milled only.
4. Casing axial expansion by minimum 1/2-in., thickness wobble plate.
5. *Couplings*: Flexible-metal dry couplings with 5/16-in. bolt diameter as a minimum. A full enclosure-type guard is necessary. The couplings must be self-retaining on flexing elements failure. Couplings

shall be designed for infinite fatigue life with the following simultaneous conditions: 1.25 × design torque, 0.060-in. axial deflection and 1/2° angular misalignment. An 18-24-in. spacer is required, according to the turbine size.

6. Self-aligning tilting pad radial bearings are mandatory unless a rotordynamic analysis recommends a different design.

7. The shaft should incorporate stainless steel steam deflector disks whose outer diameter is 1.8 × shaft OD. Disks to be shrink-fitted to the shaft.

8. *Coupling hubs*: Either integral, hydraulic fit or dual key tapered hub with a MJT locknut.

9. All casing and valve chest internal fasteners to be of 400-series, hardened stainless steel, fully countersunk, locked, and captured.

10. No shaft-driven lube oil pumps above 4000 rpm.

11. *Lube oil piping*: All piping of 300-series stainless steel. Each bearing should be protected by a stainless steel conical mesh screen, with a minimum lube oil orifice diameter of 3/16 in. A stainless steel lined oil reservoir is a minimum requirement. The heat exchanger shall be air-cooled or of U-tube shell and tube design, with brazed or welded tubing.

12. Speed governor and trip circuit hydraulic piping or tubing must be of 300-series stainless steel only.

13. An electronic/hydraulic or electronic/pneumatic speed governor is mandatory. Triple-redundant speed sensors and solenoid valves are required. Solenoid coils shall have H-rated insulation.

14. All hydraulic components in the trip circuit such as valves or actuators shall have corrosion-resistant stainless steel internals. The number of series-action devices in the trip circuit should be minimized.

15. All stop valve bushings should utilize a 3/32-in. radius on their inner bore edges. The valve shall incorporate design features that resist binding and adhesion of the stem and bushings.

16. For steam turbines >25,000 hp, it is advisable to install a second trip valve in series with the main trip valve. The second valve is to be of simple construction and mechanically latched by an instrument air-operated actuator. The actuator shall be fail-safe and is activated by an air-dump solenoid valve receiving overspeed signals from shaft speed indicators.

Incorporating these simple design concepts leads to rugged turbomachines that exhibit an extremely high resistance to failure. These design principles originate from a detailed study into hundreds of process machinery. Thus, we have extracted the underlying design choices that lead reliable machines to operate trouble free and discovered those elements of design that cause unreliable machines to perform poorly.

Building Reliability into Your Reciprocating Compressor Specifications*

Present-day API-618 Standards and other International Standards recognize this fact and impose design guidelines which force manufacturers into producing higher quality machinery. In the following, we outline a method of achieving extremely high reliability by concentrating on the weaknesses of reciprocating compressors. We want to introduce design rules which can lead to a doubling or tripling of existing industry-wide MTBF.

For example, if a reciprocating compressor operates continuously except for piston ring and packing failures once per year on average, then the machine is considered to have achieved an MTBF of 1 year. A second compressor which can achieve 2 years of trouble-free operation before requiring dynamic seal replacement has double the mechanical reliability of the first, i.e. the MTBF is 2 years.

However, as we already saw, the mechanical reliability of a machine depends upon a range of design factors, any one of which can severely limit the actual MTBF. For example, if the frame lubrication system of a compressor fails, the bearings will be destroyed in a matter of seconds, as there is no possibility of further operation without incurring major mechanical damage. The same is true if a piston rod fracture occurs suddenly; there is an immediate impact on reliability.

As a general rule, the ability of any machine to operate continuously without mechanical breakdown is dependent upon its resistance to the following general machinery failure modes:

1. Lubrication breakdown (loss, contamination, etc.)
2. Excessive vibration
3. Corrosion/adhesion (sticking)
4. Erosion/wear
5. Seal failure (static/dynamic)
6. Foreign object damage (FOD)
7. Overheating
8. Locking or holding mechanism failure
9. Failure due to improper internal geometry (rubs, misalignment)
10. Overstressed material (fatigue, rupture)

The resistance to these failure modes can be increased through proper design by incorporating inherent features which serve to extend the MTBF of the various components of the machine.

---

* By Abdulrahman Al-Khowaiter. Mr Al-Khowaiter is a rotating equipment consultant with Saudi Aramco Oil Company in Saudi Arabia.

Improving Mechanical Design Reliability

1. *Designing out a failure mode*: This is the most powerful technique available to increase reliability. By eliminating one or more failure modes through careful design, all the failure events tied to these modes are cancelled. For example, in reciprocating compressors, the corrosion failure mode is exemplified by valve spring/plate failures and piston rod fractures (Fig. 13-9) resulting from corrosive gas services. In these cases, corrosion degrades reliability by reducing the designed infinite mechanical fatigue life of a part resulting in a limited life component. By choosing proper corrosion-resistant materials for these highly stressed components, the possibility of corrosion failure is eliminated, which leaves only nine remaining failure modes to act upon the compressor. In the majority of corrosion-induced failures, specifying PEEK* valves, with INCONEL X-750 springs, and piston rods of 17-7 PH stainless steel will eliminate this failure mode.
2. Minimizing the number of rotating and static parts, and applying integral design philosophy as opposed to multiple elements.
3. Applying the concept of Design Safety Factors to all failure modes of the machine and not merely to mechanical stress calculations.
4. Reducing the interaction between failure modes, where one failure mode initiates another.

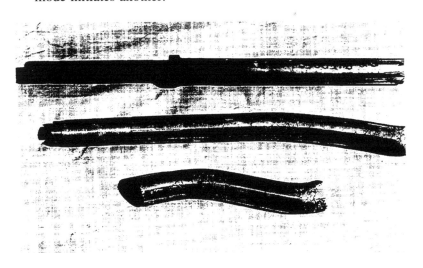

Figure 13-9. Fractured piston rods.

---

* Polyetheretherketone, a high-performance polymer.

5. Increasing the separation distance between parts with relative motion (excluding bearings).
6. Adding flexibility at the interface between dynamic and static parts, and incorporating self-aligning capabilities.
7. Reducing the number of bearings and sliding contact surfaces.
8. Adding parallel redundancy to the failure modes experienced.

Reasons for Lack of Reliability Rooted in Design

One of the major causes of low observed reliability in reciprocating compressors is the tendency for designers to produce machinery designed for an ideal world. Needless to say, actual or field conditions are often far from ideal. The following are several idealizations which manufacturers base their calculations on, with recommended solutions:

On paper	Real world	Solutions
Stress and strength calculations for crankshafts and piston rods are based upon a 100% geometric alignment. For example, when sizing the crankshaft diameter, the designer assumes perfect alignment of the main bearing saddles.	There is no such thing as perfect main bearing alignment; normal operating differential temperatures of the crankcase, frame loading deflections, and foundation distortion combine to drive the main bearing centerline out of alignment, even with accurate cold alignment. This misalignment introduces bending stresses which cause a significant reduction in fatigue life (Fig. 13-10).	The designer should assume an actual misalignment condition is always occurring, assign an average value, and include this real-world effect in the crankshaft and piston rod fatigue strength calculations.
The frame lubricant is a clean lube oil of specific viscosity, in as new condition.	Oxidized lube oil with moisture content up to 1000 ppm, salts, a PH ranging from 4 to 7, and varying viscosity.	Design bearings and other lubricated components to operate properly with used, partially contaminated oil.
Frame and cylinder vibration effects are ignored with regard to the inertial loading arising from vibration, and its impact on all static and mechanical components. Example: The loss of the separating boundary oil film on sliding elements such as crankpin bearings, due to relative motion.	Excessive vibration (above 0.35 in/s RMS) occurs on many compressors during normal operation. This reduces the life of wearing components and leads to fatigue failures.	Design all compressor components to continuously withstand 0.50 in/s RMS vibration amplitudes, with major frequency components occurring at 1X, 2X, 4X crankshaft rpm. Alternative: Design compressors with very low inherent vibration, such as the Super-Balanced Crankshaft design of Nuovo-Pignone, an Italian compressor manufacturer.

On paper	Real world	Solutions
Process loading: All cylinders are operating within design process parameters of suction and discharge gas pressure.	Cylinders operate at off-design conditions, leading to torsional vibration and overstress of crankpin bearings from cylinder overload. This leads to early bearing failures and fractured crankshafts.	Include different cylinder unbalance scenarios in the worst case fatigue stress analysis of the crankshaft, piston rod, and bearings. For example, design for a 20% drop in rated suction pressure, with constant rated discharge pressure.
An army of operators and maintenance technicians are stationed adjacent to the compressor, catering to every malfunction or component failure, and ready to tighten loose bolts, leaking gaskets, and other nuisances.	Compressors are visited by a relatively unskilled operator once in 24 hr, and a thorough evaluation of compressor condition is not made.	Design robust compressors that do not require constant bolt tightening, gasket and leak repair, and other "nursing."

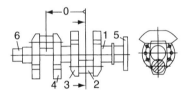

Figure 13-10. Example for a crankshaft fatigue failure location. A throw distance dimension 1 main bearing journal 2 crankpin for the connecting rod 3 web flange 4 counter weights 5 coupling flange 6 auxiliary drive end.

Failure Modes of Reciprocating Compressors

Improving reliability starts with understanding un-reliability

The improvement of a machine's design reliability begins with a careful analysis into the effect of each failure mode on the various elements. However, it is important to recognize that failure modes are highly interactive; meaning that the start of one failure usually leads to the initiation of a second and third failure mode. Thus, separating the primary causes from secondary faults is critical to the success of a scientific analysis into the machines reliability. For reciprocating compressors, the following association between failure modes and actual component damage is the first step toward evaluating reliability.

Failure mode	Primary damage attributed to
Lubrication breakdown This includes loss of oil, contamination, insufficient flowrate, and film breakdown.	Lubrication failures cause the following primary damages: Wiping of crankshaft bearings, seizing of crosshead slippers, piston scuffing, and seizing in cylinders. Excessive wear of packing, piston rings, rider bands, and piston rods.
Excessive vibration This mode covers all vibration induced failures due to frame vibration, torsional vibration, and gas pulsation-induced vibration.	*Frame and cylinder vibration* • Accelerated bearing wear due to fretting on sliding and rolling element bearings. • Fatigue and failure of static and mechanical parts, such as frame structural cracking. • Loosening of bolted components. • Foundation damage. *Torsional vibration* • Excessive shearing stresses on the crankshaft, leading to fractures at stress concentration points. • Coupling and flywheel bolt loosening/shear. *Gas pulsation* • Fatigue failure of inlet/discharge valves • Piping fatigue failures • Frame vibration • Foundation damage
Corrosion/adhesion This mode covers failures caused by surface molecular action. Corrosion in reciprocating compressors generally results from a chemical reaction between the process gas and components in contact with the gas such as valves, cylinder liners, pistons, and piston rods. Adhesion occurs as a result of molecular attraction, such as the tendency for lapped valve plates to adhere together when a layer of oil is trapped between the plates. Adhesion is also responsible for the adherence of deposits to machinery components.	• Corrosion/deposits on cylinder jackets causing overheating of cylinder liner. • Valve component corrosion induced failures of the springs, plates. • Piston rod failures due to corrosive attack. • Contamination of frame lube oil due to corrosion in lube oil tanks and piping. • Sticking valves causing hammering. • Sticking unloader mechanisms. • Corrosion of bearing materials (such as babbitt).

Failure mode	Primary damage attributed to

Erosion/wear
This includes abrasive wear, fretting, scuffing, and galling. This mode covers all failures which are caused by material loss.

- Cylinder excessive wear causing loss of efficiency and overheating due to gas recirculation in cylinder.
- Packing excessive leakage due to worn clearances.
- Piston rod misalignment due to rider band wear. Piston rod wear.
- Valve leakage due to wear of sealing surfaces.
- Crankshaft bearings/journals excessive clearance.
- Pistons: ring and rider band groove wear.

Seal failure
This includes all failures attributed to the loss of sealing ability or containment at an interface.

Static seals: Gasket leaks at valve cover, cylinder head, and valve seat.

Dynamic seals: Piston ring failure, pressure packing failure, valve leakage.

Foreign object damage
This mode includes all failures caused by the ingestion of an object which is either:

- not an intrinsic part of the machine or
- a part of the machine that has moved from its designed position.

- Solid particles in the gas damaging valves, eroding cylinder liners, and piston rings. This includes pipe scale, dust particles, and corrosion products.
- Liquid carryover/slugging.
- Frame and cylinder lubrication failure due to plugging of lubrication lines.
- Bearing wear due to solid particles in lube oil.
- Valve parts entering cylinder causing impact on piston and seizure.
- Water contamination of frame lube oil from the oil cooler.

Overheating
Failures due to temperatures exceeding normal design limits.

- Bearing babbitt melting in crankshaft and crosshead guide bearings.
- Valve element failure due to high temperatures.
- Piston and packing ring material degradation due to excessive temperature.
- Cylinder overheating lowers lube oil viscosity leading to high wear at piston–cylinder interface.

Failure mode	Primary damage attributed to
Locking/holding mechanism failure • Joints held by fasteners • Joints which do not rely on fasteners but utilize elastic properties such as shrink fits to maintain the assembly.	*Looseness in static parts*: Valve cover studs, distance piece bolts, cylinder head studs, and foundation bolts. This looseness leads to misalignment, valve leakage, and frame vibration. *Looseness in mechanical parts*: Flywheel, crosshead jam nut, piston nut, and crankrod bolts. *Result*: Impact and fracture.
Failure due to improper internal geometry • Rubbing and impacting between components that were designed to be non-contacting. • Loss of design geometric alignment between components.	• Crankshaft failures due to foundation unevenness/heat growth leading to main bearing misalignment and bending fatigue failure. • Piston rod fractures at the crosshead connection due to misalignment induced bending stresses. • Piston to cylinder head impacting due to an insufficient axial clearance gap. • Pressure packing excessive wear due to misalignment with the piston rod axis.
Overstressed material This mode covers fatigue failures, ductile fractures, surface deformation and other mechanical stress-induced component failures.	• Piston rod fatigue failure • Piston cracking and fracture • Valve spring/plate fracture • Bolt and stud fractures • Sleeve bearing babbitt failure due to surface fatigue • Rolling element bearing fatigue failures • Crankshaft fatigue failures • Auxiliary piping fatigue failures at joints

Mechanical Design Specifications

The following are specific design guidelines that enhance reliability by either eliminating a failure mode or increasing the compressor's resistance to failure modes.

Lubrication Failures

Designing out cylinder lubrication failures: Eliminate the cylinder lubrication system entirely by specifying a dry lubricated compressor design. The following are the two available options.

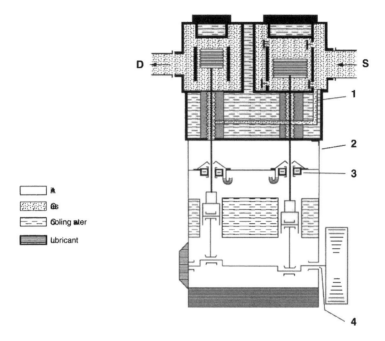

Key:
- □ Air
- ▨ Gas
- ⊟ Cooling water
- ▤ Lubricant

Figure 13-11. Compressor with non-contacting piston-cylinder design (Sulzer-Burckhardt). D Distance S Suction 1 gland leakage return to suction 2 open distance piece vented to atmosphere 3 guide bearing with oil scraper 4 oil seal-not gas tight.

Labyrinth Design. These compressors (Figs. 13-11 and 13-12) remove the need for cylinder and packing lubrication by utilizing a non-contacting labyrinth piston design (Sulzer-Burckhardt). Process plant experience has shown that this design consistently achieves double the reliability of conventional API-618 machines. Notice that two failure modes are greatly reduced: the lubrication failure mode (cylinder and packing) and the erosion/wear failure mode. No rubbing occurs on the piston, only at the packing gland which utilizes dry running carbon rings.

Conventional Dry Lubricated Compressor. Utilizing filled PTFE Rider bands and piston rings and carbon or PTFE packing, cylinder lubrication is eliminated as a failure mode, with the following limitations to ensure long-term reliability:

- Relatively clean process gas (or highly filtered suction).
- Discharge pressures ≤ 1500 psig.
- Preferably, discharge temperatures ≤ 250°F

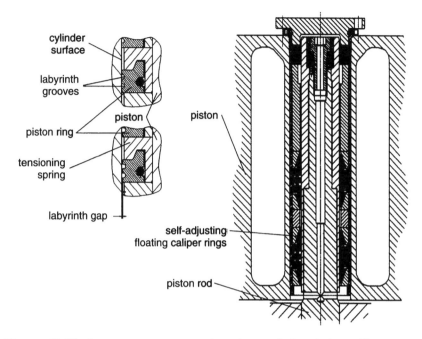

Figure 13-12. Compressor non-contacting piston-cylinder designs (Neumann & Esser).

- Internal metal sprayed hard coating of cylinder liner, hardness: RC ≥50 (using Linde gun or HVOF* process)
- 600 ft/min maximum average piston velocity.

Increasing Resistance to Failure of Conventional Lubrication Systems

- Specify divider block cylinder lubrication systems.
- *Interaction between cylinder lube and process gas*: Request the compressor manufacturer to study the effects of interaction between the compressed gas and cylinder lubrication oil, such as loss of viscosity and lubricity.
- Specify an all stainless steel lube oil circuit (frame and cylinder lube system).
- Specify direct shaft driven lube oil pumps. Chain driven pumps and splash lubrication are not acceptable.
- Specify an auxiliary electric motor driven frame lube oil pump.
- The capacity of the cylinder lubrication oil reservoir should be sized for 1 week of consumption at the design oil flow rate.

---

* High – velocity oxygen fuel process.

Excessive Vibration

Frame and Cylinder Vibration
- Specify a maximum allowable casing vibration of 0.35-in./s RMS at any point on the compressor. This forces the manufacturer to produce a rigidly constructed, well-balanced design.
- Specify epoxy grouted foundations with a minimum of 4.0-in. thickness epoxy mat. This adds damping and reduces vibration impact loading transmission to the foundation.

Torsional Vibration. Specify elastomeric/spring couplings between the compressor and the driver, except for low-speed synchronous motor direct coupled applications. For lower horsepower applications, banded V-belt drives produce very high torsional damping due to normal slippage of the belts in the sheave grooves.

Gas Pulsation. This type of vibration is well covered by API-618.

Corrosion/Adhesion

Designing Out Corrosion. In general, choose chemically inert materials to eliminate the possibility of corrosion. For example, use Teflon, PEEK, or other high performance polymers as valve materials, high-strength stainless steels for piston rod material.

Resisting Corrosion/Adhesion
- Deposits on discharge valve surfaces (such as carbon buildup) can be reduced by specifying lower discharge temperatures and mini-lube lubrication.
- *Deposits in cylinder cooling jacket*: Specify a minimum water flow velocity of 4 ft/s through each individual water jacket.
- *Intercoolers*: Apply high temperature, anti-corrosion coatings on shell internal walls.
- Specify Plug type unloaders as opposed to finger type, to reduce sticking (Fig. 13-13).

Erosion/Wear

Designing Out Wear. Use non-contacting designs such as the Free Floating Piston™ design (Thomassen Compression Systems) which relies upon

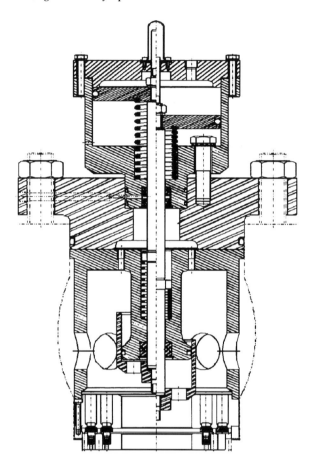

Figure 13-13. Plug unloader (Cook Manley).

a floating cushion of gas to continuously separate the piston from the cylinder walls (Fig. 13-14). In contrast with conventional technology, the Free Piston™ compressor has rider rings which feature flow nozzles and an aerostatic bearing profile on the underside. During the compression process, gas at discharge pressure enters the piston through small valves in the piston faces. This gas buffer volume inside the piston provides a continuous gas supply to the flow nozzles, resulting in an upward force on the piston, so that the piston floats on process gas.

Resisting Wear. Increase surface hardness values for components under constant wear. Specify a tungsten–carbide coating on piston rods, with

Figure 13-14. Free-floating piston (Thomassen).

a hardness ≥RC-60. Specify a pressure packing cup hardness ≥RC-50. For dry lubricated compressors, specify a cylinder liner surface hardness ≥RC-50.

- Wear is a function of rubbing velocity, therefore, reducing the velocity will minimize wear. Specify a maximum average piston velocity of 800 ft/min in lubricated compressors, 600 ft/min in conventional dry lubricated compressors. For Labyrinth compressors, follow the manufacturer guidelines.
- *Aluminum pistons*: Rider bands and piston rings cause excessive wear of machined grooves in the piston leading to early failure of the piston and its sealing capability. Specify that aluminum pistons shall have a hard ferritic insert (cast iron or hardened steel) in the

piston ring/rider band groove area. As a minimum alternative, specify anodized aluminum pistons.

- *Piston ring and rider band wear*: Specify thermoplastic low surface friction materials only, such as Teflon, Vespel®, or PEEK, reinforced with fibers to give stiffness and toughness to resist extrusion.
- Specify wear bands for all pistons (no metal-to-metal contact).
- *Bearings*: Specify a minimum 50,000-h design operating life for all sleeve and rolling element bearings.

Seal Failures (Static/Dynamic)

Static Seals. Specify that O-rings be used for cylinder head covers and valve covers. This produces a metal-to-metal joint which eliminates gasket relaxation and provides superior leakage resistance.

- Minimum allowable O-ring cross-sectional diameter = 5/32 in. (4 mm). This size limit assures a minimum tensile strength of O-rings and increases the reliability of sealing due to a greater O-ring compression dimensional tolerance.
- Specify that all O-ring and gasket materials have a minimum design temperature safety factor of 1.3.

Dynamic Seals

- Specify poppet valves as much as possible, as their design achieves a higher sealing reliability compared to plate, channel, or ring valves. The main reason for this is due to the molded thermoplastic poppets which have a tapered sealing contact area. A tapered seat for each poppet allows three-dimensional sealing as compared to only two-dimensional sealing with all other valve designs (Fig. 13-15).
- Specify high conductivity metallic backup rings to reduce pressure packing deformation and frictional heat buildup.

Foreign Object Damage

The reciprocating compressor is highly sensitive to gas cleanliness due to its inherently close clearances and sliding contact surfaces. As a result, reciprocating compressor maintenance incidence rates in general far exceed those of centrifugal machines. Therefore, cleaner suction gas will enhance the reliability of reciprocating compressors. By far, the most common FOD occurring is that due to liquid slug ingestion and solid particles in the process gas. To eliminate these failures, high quality

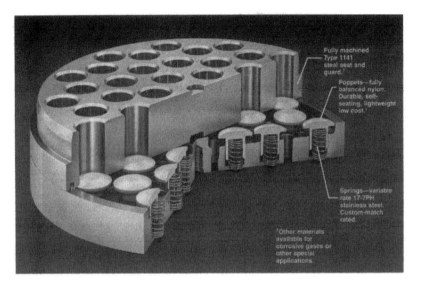

Figure 13-15. Poppet Valve (Cooper Energy Services Group, Mount Vernon, Ohio).

and efficiency coalescing suction filters should be specified for all compressors. The experience of a major oil company* over a 4-year period (1991–1994) has shown that installing modern high efficiency coalescing filters leads to a 75% reduction in repair incidence. This was with compressors in refinery service handling relatively dirty gases.

- Specify coalescing filters with 5-micron filtration capability at 99.99% efficiency. A self-cleaning capability and cartridge design is preferred.
- *Interstage cylinder suction*: Install permanent conical screens (#40 mesh) to protect against pipe scale and maintenance-related FOD.

Other Foreign Objects

- *Valve parts falling into cylinder*: Subject to verification of prior experience, specify non-metallic valves only. These cause minimal piston/cylinder damage upon failure.
- *Plugging of frame lube oil circuit*: Specify a minimum allowable orifice or opening of 3/16-in. diameter.
- *For large compressors* ($\geq 2000\,hp$): Specify air-cooled lube oil heat exchangers to eliminate the possibility of water mixing with oil.

---

* Chevron.

- *Suction knockout drums (addition of parallel redundancy)*: In addition to standard automatic liquid level float type drainers, install timer-actuated solenoid drain valves. Since these are time interval activated, they do not have the weakness of relying upon a float or other mechanical device. Install a mesh screen filter before the solenoid valve.

Overheating

Lower discharge temperatures are directly related to increased MTBF. According to a 1996 Industry Survey involving 60 users, compressor cylinders with gas discharge temperatures of 245 °F or less tended to experience ring and packing life as long as 3 years (25,000 hr). This is more than double the average dynamic seal lifetime.

- *Cylinder overheating*: Specify forced cooling systems only and incorporate a three-way temperature controlling valve (TCV).
- *Intercooler poor heat transfer*: Specify that intercoolers be sized to cool the process gas with 10% of tubes plugged. Minimum tube-side water velocity = 7 ft/s. The higher flow velocity specification is to discourage fouling.
- *Crankpin and main bearings*: High temperatures cause melting of babbitt and accelerated bearing fatigue. Specify aluminum–tin alloy sleeve bearings only. For babbitt bearings, the design temperature limit is 250 °F, while aluminum/tin alloys have a 300 °F upper limit.

Locking or Holding Mechanism Failure

Reciprocating compressors utilize bolted joints to unite the various component assemblies into one structure. Therefore, there is a high dependence upon bolted joint integrity to maintain reliable operation and any loosening of these joints makes a mechanical failure inevitable. Estimates of compressor failures due to this failure mode alone vary from 10 to 20% of all recorded failures. To improve bolted joint reliability, and achieve a significant reduction in failure incidence, the design of all joints must be analyzed carefully and a maximum incorporation of reliable design details should be done.

Designing Out Looseness. By applying the integral design approach, a design study of the machine should be made to minimize the number of bolted joints and change to one-piece components. An example of this is to specify one-piece valve cage-cover designs as opposed to the standard two-piece construction (Fig. 13-16).

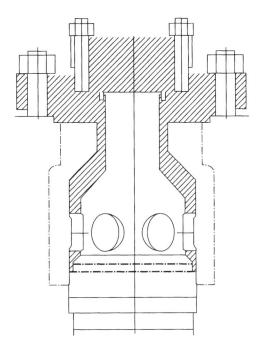

Figure 13-16. One piece valve cage cover (Cook Manley).

Resisting Loosening Failures. The vast majority of failures in reciprocating compressor joints are due to a loss of bolt/stud tension arising from joint relaxation and vibration loosening. These factors are caused by the following:

- A low design elasticity of the bolt/stud connection.
- Embedment of joint surfaces and nut surface.
- Insufficient thread engagement.
- Oversized holes.
- Temperature cycling of the joint.
- Excessive joint surfaces.
- Vibration overcoming the friction forces between the nut and stud threads, leading to nut rotation.

To Add Reliability, Increase Bolt Elasticity and Incorporate Nut Locking

1. For all fasteners, specify a minimum bolt/stud elasticity of:

$$\frac{\text{Diameter of stud/bolt (inches)}}{100}$$

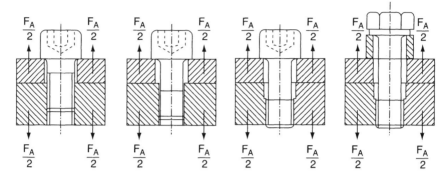

Figure 13-17. Bolted connections – increasing resistance to loosening from left to right.

This can be achieved by using longer studs tapped into deeper holes and by adding spacers under the nut (Fig. 13-17).

2. For all studs 3/4 in. and greater, specify that the shank diameter shall be equal to the thread root diameter. This reduces stress concentration and increases stud elasticity.

3. Specify MJTs (see Section 17, page 447) for critical joints such as the crosshead locknut, piston nut, anchor bolt nuts, and the flywheel shaft locknut (Fig. 13-18).

4. All other nuts ≥ 1.0 in. should be castellated with cotter pins, and rest on a minimum 1/4-in. thick hardened steel washer. This minimizes embedment and hole size effects. An alternative to castellated nuts is to use self-locking flexible nuts such as Flexnuts™ (Fig. 13-19).

5. *Anchor bolts*: Use the largest diameter possible to increase hold-down forces on the frame. Specify a minimum length of 48 in. to increase bolt elasticity and reduce tensile stresses in the concrete foundation.

Failure Due to Improper Internal Geometry

Changes in design internal geometry lead to misalignment between parts and off-design contact. To inhibit this failure mode, the designer should develop compressors which incorporate self-aligning component features to eliminate misalignment-induced failures, and add protective features that prevent mechanical damage arising from off-design contact.

- A distance piece with insufficient rigidity allows excessive cylinder deflection and misalignment. When evaluating different manufacturer designs, prefer stiff, heavily ribbed constructions for the distance piece.

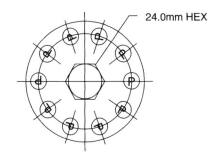

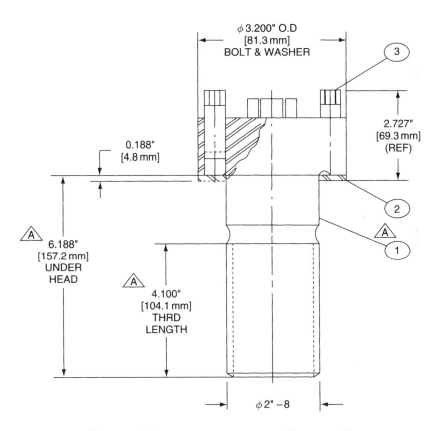

Figure 13-18. Multi-Jackbolt tensioner (Supernuts™).

- *Piston contacting cylinder head*: Specify a minimum allowable design operating clearance (axial) between the piston and the cylinder head of 1/8 in. Setting this minimum requirement reduces the possibility of impact by increasing the design separation gap.

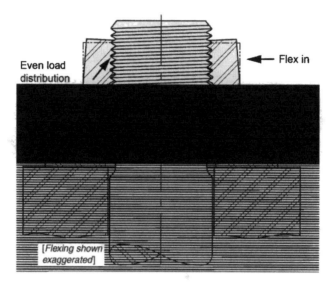

Figure 13-19. Flexnut (Flexnut™).
Note: They are designed to flex out at the bottom and flex in toward the top of the nut. This distributes the bolt load along many threads, adds elasticity, and prevents stress concentrations in the first few threads, thus reducing the possibility of stud breakage.

- Specify rail type foundation supports for all compressors with three or more stages. Minimum rail height above grout is 3 in. (~75 mm). This reduces differential temperature-induced distortion of the bearing saddles by equalizing temperatures under the crankcase.

Future Design Improvements to reduce this Failure Mode (These are not Presently Available):

- A continuously self-aligning crosshead guide to cylinder axis feature, to eliminate bending stresses on the piston rod, packing wear, and excess loading on crosshead shoes. One possibility is to use a spherical seating for the crosshead guide. Presently, two compressor manufacturers* (Fig. 13-20) have design options that allow piston rod self-alignment, but these mechanisms require shutdown and manual adjustment, they are not truly automatic aligning devices that will align "on the run."
- Self-aligning main bearing saddles.

---

* Sulzer-Burckhardt and Dresser-Rand.

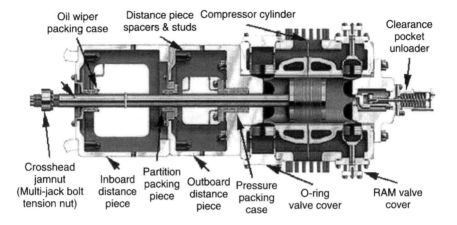

Oil wiper packing case   Distance piece spacers & studs   Compressor cylinder   Clearance pocket unloader

Crosshead jamnut (Multi-jack bolt tension nut)   Inboard distance piece   Partition packing piece   Outboard distance piece   Pressure packing case   O-ring valve cover   RAM valve cover

Figure 13-20. Piston and piston rod (Dresser Rand).

Overstressed Material

Resisting this failure mode is achieved by incorporating greater mechanical stress safety factors, reducing all stress concentrations to a minimum and increasing the use of ductile materials whose properties offer greater resistance to cracking than brittle materials.

- *Crankshaft fractures*: Specify a minimum crankshaft fatigue life of 150,000 hr at maximum design torque/speed with a continuous main bearing parallel offset of 0.004 in. in any single bearing.
- *All crankshaft, coupling, and flywheel keyways*: Reduce keyway stress concentrations by specifying a minimum keyway inside corner radius: radius = D/24 in. where D = shaft diameter at keyway. Specify "sled runner" type shaft keyways only. The resulting stress concentration factor = 2.3. This is very low in comparison to the industry standard of 3.0 or more. Result: A 23% reduction in peak bending and torsional stresses at this shaft location (Fig. 13-21).
- Apply gusseting ribs to all auxiliary piping (4.0 in. and below) welded to larger diameter piping or vessels.
- Specify a minimum coupling/flywheel bolt diameter of 1/2 in.
- For Piston rod threaded crosshead end, Maximum allowable tensile loading not to exceed 7000 psi at thread root. Rolled threads mandatory. The thread profile should be full radius thread root. Apart from fatigue life enhancement, the 7000-psi rating limits the minimum allowable piston rod diameter which increases stiffness and leads to reduced deflection and packing wear.

STRESS CONCENTRATION=1.03

STRESS CONCENTRATION=1.75

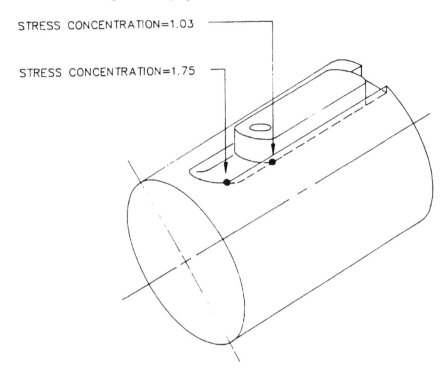

Figure 13-21. Shaft keyway with low stress concentration.

- For Piston rod non-threaded crosshead end design, Maximum allowable tensile loading is 9000 psi.
- Minimum allowable piston rod material yield strength is 100,000 psi (non-sour services).
- For Piston cracking failures, specify a minimum inside corner radius in the piston ring and Rider band land of 1/32 in. Minimum inside radius in the machined piston nut and collar counterbore is 1/8 in.

Monitoring

Although this chapter focuses on the mechanical design reliability of compressors, the role of proper monitoring and protective shutdown instrumentation should be clarified. It is important to understand that monitoring and protection devices do not usually prevent machinery failures. They act to reduce the magnitude of damage and repair time by stopping the progression of primary failures into secondary and catastrophic failures.

For example, some system failures – such as a loss of cooling water to the cylinder jackets which then leads to cylinder overheating – can be prevented by continuous monitoring. But it should be noted that many other mechanical failures cannot be prevented by monitoring. If a compressor with a single frame lube oil pump experiences pump damage, then by utilizing pressure switches in the frame lube oil circuit, the compressor can be automatically shutdown before catastrophic damage from low oil flow occurs. However, the machine has still failed because mechanical repair of the lube oil pump is required and the compressor is now shutdown.

If the lube oil pump had been spared (parallel redundancy), the compressor would have continued operation with no stoppage. In such situations, instrumentation has achieved an improvement in the MTTR of the compressor by reducing the damage but has little effect on the MTTF. This is because most instrumentation delivers after-the-fact information for a failure mode, i.e. the instrument has detected a failure, but has not prevented it.

Therefore, achieving machinery overall uptime starts with mechanical reliability. By themselves, instrumentation and protective devices are not sufficient; the underlying mechanical design of a machine must be reliable, otherwise constant shutdowns and failures will occur.

Minimum Recommended Instrumentation

Non-critical machine, Size < 2000 HP	Additional for critical machines ≥ 2000 HP
• Lube oil flow or pressure switches • Excessive discharge gas pressure switch, and low suction pressure switch • Compressor discharge excessive temperature switch • Two permanently mounted frame vibration accelerometers or velocity meters* with monitor	• Rod drop monitoring; by either proximity probe or calibrated microswitches (alarm/shutdown) • Valve temperature monitoring (alarm only) • Main bearing temperature monitoring. • Pressure packing temperature monitoring (alarm only) • Cooling water temperature alarm

Future Monitoring Developments. For large, critical compressors, a microprocessor-controlled cylinder balancing system utilizing the compressor discharge gas (recycle) as a method of maintaining suction pressures at each stage to within 10% of design pressure should be used. Activation occurs whenever suction pressure drops below 90% of normal. This is to reduce cylinder unbalance and overload-induced failures.

---

* Velometer™

Concluding Comments

In essence, an attempt was made here to provide a comprehensive approach to all design aspects which have an effect on the mechanical reliability – uptime – of reciprocating compressors. The failure modes and design solutions described apply equally well to reciprocating compressors of all sizes and in a multitude of gas service.

References

1. Jung, I., *Gustaf de Laval – The Flexible Shaft and the Gas Turbine*, de Laval Memorial Lecture, Stal-Laval Turbin AB, 1973.
2. Stodola, A., *Steam and Gas Turbines*, Vol. 2, Peter Smith, 1945, p. 496.
3. Gunston, B., *The Development of Piston Aero Engines*, Patrick Stephens Limited, 1993.
4. Taylor, C. F., *The Internal Combustion Engine in Theory and Practice*, The MIT Press, 1985, p. 516.
5. Villemeur, A., *Reliability, Availability, Maintainability, and Safety Assessment*, John Wiley Sons, 1992, p. 5.
6. Clark, E. E., Rotating equipment loss prevention – an insurers viewpoint, *Proceedings of the Twenty-Fifth Turbomachinery Symposium*, Turbomachinery Laboratory, Texas A&M University, College Station, Texas, 1996, pp. 103–121.
7. Bickford, J. H., *An Introduction to the Design and Behavior of Bolted Joints*, Marcel Dekker, 1997.
8. Heywood, R. B., *Designing Against Fatigue*, Chapman and Hall, 1962.
9. Ostroot, G., *Lessons Learned from a Coupling Failure and Lube Oil Fire*, Loss Prevention Symposium, AICHE, New York, N.Y., 1967.
10. Stinson, K. W., *Diesel Engineering Handbook*, Business Journals Inc., 1983.
11. Nelson, W. E., and Monroe, P. C., Understanding and preventing steam turbine overspeeds, *Proceedings of the Twenty-Sixth Turbomachinery Symposium*, Turbomachinery Laboratory, Texas A&M University, College Station, Texas, 1997, pp. 129–141.
12. Skeist, I., *Handbook of Adhesives*, Van Nostrand Reinhold, 1977.
13. Rigney, D. A., *Fundamentals of Friction and Wear of Materials*, American Society for Metals, 1981, p. 343.
14. Clauss, F. J., *Solid Lubricants and Self Lubricating Solids*, Academic Press, Inc., 1972.

Other Sources

Standards

1. American Petroleum Institute and ANSI Standard 617, 7th edn, July 2002, Centrifugal Compressors for Petroleum, Chemical, and Gas Service Industries.
2. American Petroleum Institute Standard 618, "Reciprocating Compressors for Petroleum, Chemical, and Gas Industry Services, 4th edn, June 1995.

3. Saudi Aramco Standard 31-SAMSS-003, Reciprocating Compressors for Process Air or Gas Service, 1997.
4. Chevron Specification CMP-MS-1626-H, Reciprocating Compressors for Continuous Duty Services, 1988.

Further Reading

1. *Reciprocating Compressors: Operation and Maintenance*, Heinz P. Bloch, John J. Hoefner, Gulf Publishing, 1996.
2. *Metals Engineering Design*, Oscar J. Horger, McGraw Hill, 1964.
3. *Industrial Reciprocating And Rotary Compressors: Design and Operational Problems*, Institution of Mechanical Engineers, London, 1970.
4. *An Introduction To The Design And Behavior Of Bolted Joints*, John H. Bickford, Marcel Deckker Inc., 1997.
5. *Approximate Methods In Engineering Design*, T. T. Furman, Academic Press, 1981.
6. *Designing Against Fatigue*, R. B. Heywood, Chapman and Hall, 1962.
7. *Marine Diesel Engines*, C. C. Pounder, Butterworth, 1972.
8. *Plain Bearing Design Handbook*, R.J. Welsh, Butterworth, 1983.
9. *Handbook of Adhesives*, Irving Skiest, Van Nostrand Reinhold, 1981.
10. Haviland, G. S., Designing with threaded fasteners, *Mechanical Engineering*, ASME, October 1983, pp. 17–29.

Conference Proceedings and Papers

1. Using high efficiency coalescing filters to improve reliability, N. C. Ruegsegger, *1994 Saudi Aramco Rotating Equipment Exchange*.
2. Update on thermoplastics in reciprocating compressors, Tom Watkins, Brian Perry, Ray Dodd, *1992 Chevron Machinery Conference*.
3. Increase reliability of reciprocating hydrogen compressors, S. M. Leonard (Dresser-Rand), *Hydrocarbon Processing*, January 1996.
4. The re-design of machinery for increased reliability, Abdulrahman Al-khowaiter, *1998 Process Plant Reliability Conference*.
5. Monitoring reciprocating compressors, Steven Schultheis, Michael Hanifan (Bently Nevada), *P/PM Technology*, June 1998.
6. Field survey of aramco reciprocating compressor problems, Anon., *Saudi Aramco Internal Document*, 1977.
7. Offshore gas compression: an operator's experience, DST Raubenheimer (Shell International), *Fluid Machinery for the Oil, Petrochemical, and Related Industries*, I.Mech E Conf. Publications, March 1981.
8. Availability and reliability in reciprocating compressors evaluated according to modern, criteria, E. Giacomelli, M. Agostini, M. Cappeli (Nuovo Pignone), *Quaderni Pignone*, **42**, 1986.
9. Crankshaft protection: guidelines for operators of slow speed integral engine/compressors, A. J. Smalley, *The Gas Machinery Research Council*, January 1997.
10. Compressor Anchor Bolt Design, P. J. Pantermuehl, A. J. Smalley, *The Gas Machinery Research Council*, December 1997.

Chapter 14

Owner–contractor interfaces and equipment availability

Introduction

Machinery uptime is related to availability but not synonymous with it. Availability or service factor may be simply stated as:

$$A = \frac{MTBF}{MTBF + MTTR}$$

where A = availability, MTBF = mean time between failures, MTTR = mean time to repair.

Availability in our context is recognizing the fact that many plant items can be repaired when they fail or their probability of failure can be reduced. This is illustrated in Figure 14-1. It shows that the securing of a required function of plant equipment is a combination of its built-in reliability and its maintainability.

Reliability, shown as a component of quality in Figure 14-2, exists as a requirement of plant equipment from its beginning to the end of its working life. The creation of reliability lies essentially in the sphere of design.

We have seen that acceptable built-in equipment reliability is achieved by the evolution of design procedures, codes of practice such as API standards, or methods that have shown good results in the past. As a consequence, the hydrocarbon processing industry, for example, usually refers to equipment as "reliable" or "unreliable" in a qualitative sense.

In certain services, where high reliability is required, the industry has, often intuitively, made use of the redundancy principle. Examples are installed spare pumps, compressors and heat exchangers. However, in the

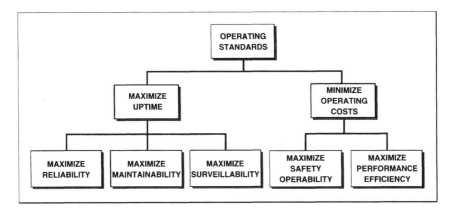

Figure 14-1. Machinery availability components.

recent past, many companies have provided non-spared installations in traditionally redundant services. It is here where built-in reliability is of utmost importance.

Maintainability is defined as the ability of an item of plant equipment, under stated conditions of use, to be retained in, or restored to, a state in which it can perform its required functions, when maintenance is performed under stated conditions and using prescribed procedures and resources [1]. This means that, if maintenance has to be performed at all, it should at least be possible. To someone maintaining plant equipment in terms of inspection, overhaul, and repair, maintainability is defined as shown in Figure 14-3. Maintenance strategies for optimization of resources are outlined in Figure 14-4. The most important strategy in our context is maintenance prevention. As we shall see, maintenance prevention has as its goal the detection and elimination of potential maintenance causing problems before they are built into the plant facility.

Despite a high commitment to the general quality idea, capital project organizations in the petrochemical industry often fail to stay focused on what ought to be their main goal, namely availability and its two components reliability and maintainability. The reasons are schedule compression or delays, frustrations, and managerial errors. To escape these problems, project professionals frequently turn to sophisticated techniques that range from critical–path diagrams to integrated project–management software. In the meantime, very little energy is directed toward the interfaces and relationships with contractors, such as engineering and construction companies, equipment vendors, and other suppliers to the project. The quality of these relationships may well have the most important influence on the ultimate service factor of a new plant.

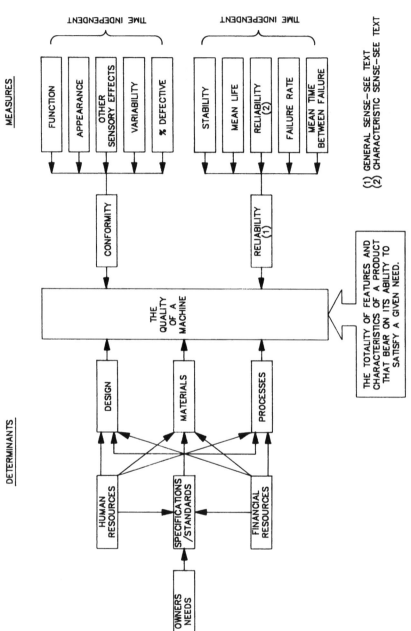

Figure 14-2. Machinery reliability and quality.

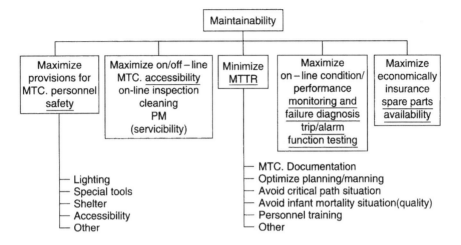

Figure 14-3. Maintainability components.

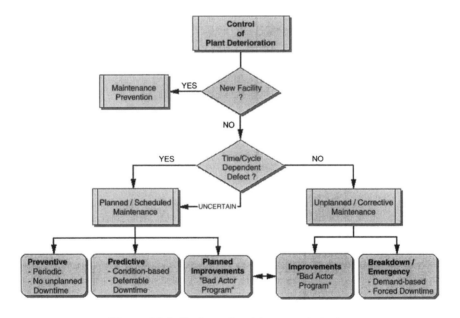

Figure 14-4. Equipment maintenance strategies.

Project Phases and Plant Availability

Capital projects in the petrochemical industries have three distinct phases as shown in Table 14-1. In our context, the first two phases are of special interest.

Table 14-1
HPI capital project phases

Project development
Project screening
Budget item
Class estimate (A)
Process design specification
   In-house
   By contractor
Class estimate (B)
Appropriation

Project execution
Contracting
Mechanical design
   In-house
   By contractor
Site preparation
Construction
Mechanical completion

Start-up and operation
Operation manuals
Operator training
Unit check-out
Start-up activities
Operation
Project close-out

Project Development or Definition

This first phase usually consists of preliminary process and economic studies. These studies serve the purpose of determining if the project appears to be viable from both a process and an economic point of view. Such investigations are usually conducted by the owner's technical and economic staff without contractor involvement. If, after these preliminary studies are completed, the project is still considered viable, a stage of very detailed process and economic studies gets underway. During this step of project development, several objectives must be accomplished:

- *Prepare an estimate by addressing project strategy and schedule*: Questions must be asked such as: Is it advantageous to go out for fixed-price bid or a reimbursable cost arrangement for the detailed design and construction contract? What is the labor situation likely to be? Should the project be split into individual bid packages or let as one large package? How much "in-house" manning is available?

- *Timing*: Overall timing must be looked at in detail because financial requirements and project returns have a strong bearing on project economics.
- *Definition of reliability and maintainability*: Equipment and overall plant availability consistent with life-cycle cost considerations are determined as part of an operating philosophy statement.

An important stage in the project development and definition phase centers around the preparation of the process design specification. This stage may involve the following activities:

- Conduct enough design work to produce a more realistic cost estimate.
- Start interfacing with outside design firms, if the project scope is beyond in-house capabilities.
- Enter into contract negotiations with the outside firm or firms selected to do the principal design work and/or the principal construction work.
- Prepare mechanical flow sheets.
- Determine the major, critical long-delivery equipment.
- Obtain quotations on long-delivery equipment.
- Place orders for major long-delivery equipment, subject to cancellation, as the project is still in its design phase. Normally, obtaining of quotations and placing of orders is done by the contractor. However, there have been many occasions, where critical plant equipment, such as non-spared machinery, has been pre-ordered by the owner organization and then handed over to a contractor for installation.
- Develop a final realistic estimate.
- Refine the project economics based on the cost estimate.

It is during these last stages of the project definition phase where the foundation for successful interfaces with contractors during project execution is laid.

Project Execution

Upon appropriation, which concludes the project development and definition phase, the project is executed. The design is now completed in detail. During this phase, a detailed engineering model of the facility is usually constructed. On a 3/8-inch model all equipment such as compressors, exchangers, furnaces, towers, vessels, valves, and other important components are shown to scale. The scale model is an essential tool to assure plant maintainability and avoid future maintenance load.

Table 14-2
HPI project coordination
procedure

1. Introduction
2. Scope of contract
3. Parties concerned
4. Communication between parties
5. Documentation filing and registration
6. Reprographic service
7. Estimating and cost control
8. Accounting
9. Planning and progress
10. Reporting
11. Design engineering
12. Procurement
13. Spares
14. Safety reviews
15. Quality assurance
16. Job close-out procedure

During the project execution phase the necessary steps toward design completion are carried out: plot plans, foundations, sewer systems, pipe spools, and other details are finalized on paper. After site preparation, plant construction is begun. Frequently 35–50% of the mechanical design work is completed at this time. During this phase a very intensive interface between owner representation and the contractor organization takes place.

From an organizational point of view, owner–contractor interfaces during the project execution, and frequently during the latter part of the definition phase, are systematized by a project coordination procedure. As an example, Table 14-2 lists the main points of this agreement between owner and contractor.

The final stage of the project execution phase is mechanical completion. Mechanical completion usually terminates the contractor's involvement and leads to start-up and operation of the new facility by the owner.

Start-up and operation, which we shall deal with later.

Operating Standards are Needed to Define Availability Goals

The purpose of all interfaces between owner and contractor should be the attainment of project service factor goals. These goals must be set by appropriate reliability and maintainability input during the project development and definition phase. A prerequisite for this input are operating standards.

Operating standards are a set of rules based on an operating philosophy incorporated into the project during its planning and definition phase. As the basis for all agreements between the parties involved, it usually begins with a vision statement such as: "We want to be quality leaders," "We want to be the best," "We want to be an excellent manufacturing operation." All these catch words could be meaningless, if they are not followed by detailed guidelines reflecting reliability and maintainability goals. An enlightened senior management would, for example, issue the following introductory statements:

Our prime objectives: We want to build and operate a plant to achieve excellence in everything we do. This means that we shall start up efficiently and on time. The facility will then be operated reliably and predictably. It will perform following strict, pre-established standards and procedures. Our plant will seldom experience upsets or shut-downs from avoidable causes.

Our equipment reliability and safety goals: We shall require a high service factor. All equipment will be selected to enable our plant to run for at least 3 years between shut-downs. We will make extensive use of PdM tools such as machinery monitoring, corrosion and leak detection, state-of-the-art instrumentation to measure force, temperatures, and electrical values (Table 14-3).

Table 14-3
Reliability and safety guidelines for a capital project

1. All plant notices and labels to conform to an accepted standardized design.
2. Gauge glasses minimized to the number required for safe operation.
3. Plant lighting to be "better than normal."
4. Instruments with a high reliability and service factor to be purchased for all important services.
5. Onstream testing, with low risk, of all safety trips and alarms to be incorporated.
6. All regular and irregular sampling points to be well designed (to improve safety and losses) labeled, well lit, and accessible.
7. All vessels and large pipes will be designed with enough vents, drains, and access points to enable blowers and extractors to be used for personnel entry without breathing apparatus.
8. Installed spare equipment to be minimized – each spare to be justified.
9. All small bore piping to be engineered with safety as the important criteria.
10. All critical machinery will be subjected to FMEA.

Table 14-4
Maintainability and serviceability guidelines

---

1. Interchangeability of parts and standardization of equipment to be optimized.
2. Equipment vendors to provide written maintenance procedures for all major equipment.
3. Identification, labeling, and access to all maintenance blanking points.
4. Standardization of valves.
5. Rising spindle valves versus non-rising spindles. See status at a glance.
6. Identification of non-return valve direction.
7. Spriral wound gaskets a must.
8. Power-aided operation of all large valves.
9. In place testing of safety valves.
10. Permanently installed connections for chemical cleaning or back-washing of all exchangers in fouling service.
11. Permanently installed positive blinding device at all regularly used blanking points.
12. Selection of materials and coatings to minimize need for regular painting.
13. All ground level areas inside process areas to be easy to maintain. Tile, concrete, tarmac, and grass are permissible.
14. Process area drains to be designed to eliminate "sanding down" for hot work.
15. All regular maintenance areas to be identified and made accessible from permanent surfaces. Working surfaces to be strong enough to take the largest piece of maintenance mobile equipment.
16. Low point and high point vents and drains easy to cap or blind. Screw in plugs not acceptable.
17. Exchangers to be designed to facilitate use of mechanical bundle pullers.
18. All process pumps to be back pull-out design.

Our work environment: We want to minimize all routine and low skill work when operating our plant. Our goal is to have our designers consider this in the detail design of the plant. Here is our list of requirements (Table 14-4). Each of these points will have, of course, an economic limit and in most cases there will have to be good judgment applied. We shall appoint sponsors and change agents* to help in the decision processes.

Our turnover and start-up procedures: The facility will be completed and turned over for commissioning during a 4–6-month time frame. The turnover will be sequential. There will be problems of lack of skilled manpower resources and of interference with ongoing construction work. These problems can be minimized by planning and designing during the detailed engineering phase. We shall define all "systems" as early as possible. Our goal is to engineer by systems in order to make construction and commissioning easier.

---

* Change agents should be internal consultants promoting change in relationships between people that have to interface to work successfully.

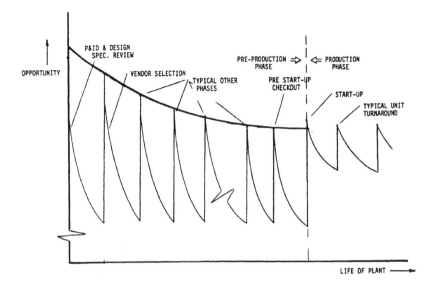

Figure 14-5. Opportunity for availability improvement vs. life of plant.

It can be costly not to begin project activities with well thought-out operating philosophies. Remember, "For every one dollar it costs to fix a problem. At the conceptual stage it will cost

$10 at the flowsheet stage
$100 at the detailed design stage
$1000 after the plant is built and
over $10,000 to clean up the mess after an accident"[1].

Conversely, the opportunities for reliability and maintainability input will diminish with the advancement of the project as shown in Figure 14-5.

Once operating standards have been issued and communicated they need to be interpreted and constantly reinforced throughout the duration of the project. This is done by daily interfacing with contract personnel, such as designers, project engineers, construction planners, specialists, construction superintendents, field engineers, vendors, and subcontractors.

Some Interface Problems and their Solutions

Some Problems

In spite of formally accomplished coordination procedures and legally perfect contracts, the owner–contractor interface can often be less than satisfactory if not adversarial. There may be some problems. For example:

- Basic personality differences
- Interpersonal difficulties caused by language and culture*
- Managerial errors.

Interpersonal problems can often be simply overcome by role statements and role delineation. A good project coordination procedure ought to help here. Frequently, this procedure is not "customized," but a standard format that does not take the specifics of owner and contractor organizations into account. If they have not been able to influence the selection of effective owner–contractor team members before project begin, knowledgeable project executives will see the signs of trouble and call on "transactional analysis" specialists for help. In team building exercises – they have to be timely – owner representatives and contractor personnel are interacting in off-work situations such as in a weekend camp. The outcome is that future interfacing will be on the basis of enhancing and maintaining self-esteem of others, of listening to each other with empathy, and of asking for help in solving the problem at hand.

Managerial errors are frequently more difficult to overcome as the type of contractual arrangement will determine how a contractor is motivated. For example, on a lump-sum project the contractor has a strong motivation to sacrifice the project service factor goals to reduce cost and to improve his schedule. On a reimbursable cost project the contractor has nothing to gain financially. Here the contractor might have the tendency to overreact to the specifications and overdesign resulting in unnecessary high costs without any gain in ultimate plant availability.

Another obstacle to good owner–contractor relationships is the "cascading" of contracts. Here, one contractor has a subcontractor working for him, who in turn has another one working for him, and so forth. The results of these arrangements are invariably communication problems. Whatever the conditions are, it is the responsibility of the owner's representatives such as field engineers, plant engineers, and specialists to interface with the contractor in order to protect his or her company's investment, to get the best quality possible, and to assure project service factor goals are met.

---

* One of the authors had the opportunity to work on a project in France where the engineering contractor was from the UK, the construction company English and French, and many staff members hailed from other countries.

Some Solutions

The key to overcoming interface problems is the early assignment of experienced technical and maintenance specialists to the project. They are best suited to interpret project operating standards and specifications by dealing with contractor personnel. Further, large projects have been successful by assigning a coordinator between the future operating department, the owner project administration group, and the contractor organization. This person could typically be the future start-up leader, the senior equipment specialist or a former maintenance department head.

While this seems obvious, some of us have no doubt participated in projects where these assignments were not made at all; where they had been made they were either not timely or involved inexperienced personnel. This is especially true for the assignment of maintenance representatives. Frequently, project professionals fail to understand the contribution maintenance specialists can make because of a widely shared philosophy of first cost, not last cost plant facilities. Additionally, there is an urgency to get facilities designed, constructed, and on stream without finding the time to invite maintenance. As very few contractors are concerned with maintenance, having a maintenance specialist on board makes double sense.

Successful Interfacing Assures Uptime Goals are Met

In the plant equipment area, owner–contractor interfacing activities typically will occur during the following functional project steps:

- flow sheet and design specification review;
- vendor selection;
- pre-order review with vendors;
- vendor drawing review;
- inspection and test at vendor's site;
- review of spare parts and documentation;
- equipment field handling and storage;
- field installation and equipment turnover;
- pre-commissioning.

Two interface functions common to the steps just listed are of particular interest in our context. They are *review* and *inspection*. Both have a high potential for assuring project service factor goals are met. In the following, we are going to discuss these functions as they relate to contractor–owner interfacing regarding specifications, vendor selection, documentation, quality assurance, equipment field installation, and finally turnover.

Review functions are sometimes equated with audit activities. Reliability audits are defined as any rigorous analysis of a contractor's or vendor's overall design after purchase order issue and before equipment fabrication begins. Reliability reviews are defined as a less formal, ongoing assessment of component or sub-system selection, design, execution, or testing. They are aimed at insuring compliance with all applicable specifications. These reviews will also judge the acceptability of certain deviations from applicable specifications. Moreover, an experienced reliability review specialist will provide guidance on a host of items which either could not or simply had not been specified in writing. He will draw on his field of expertise and start-up experience when making recommendations aimed at insuring a successful plant start-up and safe, reliable operation of the equipment for years to come. Such a specialist must, by definition, work the interfaces between owner and contractor.

Priorities of the review and inspection effort must always be kept in mind as it is obvious that one cannot address all equipment items equally well. The degree of intensity of owner–contractor technical interfacing is determined by following the decision routine shown in Figure 14-6.

Flow Sheets and Specification

Flow sheets and specifications are the technical basis for our interfaces with contractors' representatives. What are technical specifications? They are quite simply documents which define in writing, together with drawings and flow sheets, plant and equipment which a purchaser wants a vendor to supply or conversely that which a bidder is prepared to offer to a buyer. It defines the technical conditions of a contract; it does not concern itself with the commercial contract conditions which are usually the subject of a separate covering letter or other documents between the two parties. Writing specifications is a skill of technical communication in print. The owner's representative must write a clear, precise statement of what he wants. The contractor or supplier must read, understand, and make an offer in line with the owner's requirements. It is also necessary that the supplier takes exceptions to specific requirements. For instance, an offer for a centrifugal pump might still be acceptable to a purchaser, if a statement is included, that, while most of his items comply with API 610, his bearing arrangement does not. His is a "face-to-face" design as shown in Figure 14-7 with internal clearance control based on a large number of successful installations. This purchaser, however, has the right to be annoyed, because he had to find out during precautionary inspection of the pump prior to start-up that he did not get what he asked for. How well was the interface between owner, contractor, and vendor handled in this case?

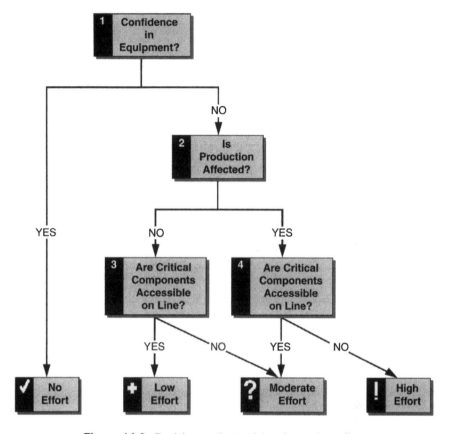

Figure 14-6. Decision routine to determine review effort.

The use of a specification requires careful consideration of needs. For example, simply specifying the maximum flow rate of a cooling water pump may be inadequate if the pump is expected to operate over a range of flow rates; obviously, cooling water system resistance changes and where a pump may work adequately in laminar flow, it may vibrate in turbulent flow and vice versa.

Specifications for single plant items such as a control valve, a motor, or a pump are prepared using a data sheet for each piece of equipment. The data sheet contains all necessary design information to specify the plant item. It is usually split up in sections presenting operating conditions, technical details, material requirements, and general information such as design standards, inspection, and shipping details. A typical pump seal data sheet is shown in Figure 14-8. A data sheet might also be accompanied by a short two or three page owner's standard equipment or

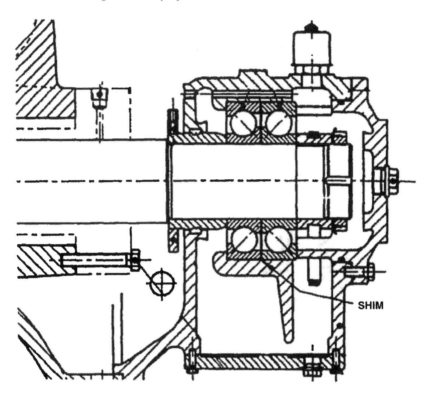

Figure 14-7. Face-to-face thrust bearing arrangement – incorrectly installed preload shim.

component specification describing special requirements, characteristics, preferences, and other needs based on standardization objectives.

By far the most effective method of specifying equipment is the single narrative document. Instead of using a series of disjointed individual specifications and stapling them into a stack of reference leaflets, the narrative document serves to blend all applicable references into a unified whole. In developing this single narrative, the responsible project engineer will use thought processes which tend to uncover oversights, weaknesses, and deficiencies in the procurement and design efforts for equipment installed in process plants. The single narrative pulls together only the truly relevant information whereas the cross-referenced individual plant specification approach tends to be extremely bulky. It puts the onus on the vendor or contractor to find relevant specification clauses and deprives the purchaser's project engineer from detecting oversights or other deficiencies. Experience confirms that additional cost incurred in developing single narrative specifications is often recovered before the plant starts to manufacture on-specification product at full capacity [5].

PAGE ___1___ OF ___2___

JOB NO. _____ ITEM NO. _____

PURCH. ORDER NO. _____ DATE _____

INQUIRY NO. _____ BY _____

REVISION _____ DATE _____

PUMP SEAL DATA SHEET
CUSTOMARY UNITS

1 APPLICABLE TO: ○ PROPOSAL ○ PURCHASE ○ AS BUILT

2 FOR _____ UNIT _____

3 SITE _____ NO. REQUIRED _____

4 SERVICE _____ PUMP SIZE, TYPE & NO. STAGES _____

5 MANUFACTURER _____ MODEL _____ SERIAL NO. _____

6 NOTE: ○ INDICATES INFORMATION COMPLETED BY PURCHASER □ BY MANUFACTURER ☒ BY MANUFACTURER OR PURCHASER

7 | ○ PUMP DATA | ○ LIQUID |

8 ○ SPEED _____ RPM
   ○ TYPE OR NAME OF LIQUID _____

9 ○ SHAFT/SLEEVE DIAMETER AT SEAL _____/_____ IN
   ○ CONCENTRATION _____

10 ○ STEPPED SLEEVE MINIMUM DIAMETER _____ IN
    ○ PUMPING TEMPERATURE

11 ○ STUFFING BOX I.D. _____ IN
    NORMAL _____ °F MAX _____ °F MIN _____ °F

12 ○ STUFFING BOX DEPTH _____ IN
    ○ SPECIFIC GRAVITY _____

13 CASING MOUNT:
    @ NORMAL TEMP _____ @ MAX TEMP _____

14 ○ CENTERLINE ○ NEAR CENTERLINE ○ FOOT
    ○ SPECIFIC HEAT _____ Cp (BTU/LB °F)

15 ○ VERTICAL ○ SEPARATE MOUNTING PLATE
    VAPOR PRESSURE

16 ○ IN–LINE ○ SUMP
    ○ @ NORMAL TEMP _____ PSIA

17 CASING SPLIT:
    ○ @ MAX TEMP _____ PSIA

18 ○ AXIAL ○ RADIAL
    VISCOSITY

19 IMPELLER
    ○ @ NORMAL TEMP _____ cP

20 ○ BETWEEN BEARINGS ○ OVERHUNG
    ○ @ MIN TEMP _____ cP

21 PUMP MATERIALS:
    ○ MELTING POINT (OR POUR POINT) _____ °F

22 ○ TABLE H-1 CLASS _____
    ○ CORROSIVE/EROSIVE AGENT _____

23 ○ SHAFT _____
    ○ CONCENTRATION (% OR PPM) _____

24 ○ SLEEVE _____
    ○ CHLORIDE CONCENTRATION (PPM) _____

25 ○ GLAND _____
    ○ H₂ S CONCENTRATION (PPM) _____

26
    SOLIDS CONTENT

27 | ○ SITE AND UTILITY DATA |
    ○ NAME OF SOLIDS PRESENT _____

28 UNUSUAL CONDITIONS: ○ DUST ○ FUMES
    ○ CONCENTRATION (PPM OR % WT) _____

29 ○ OTHER _____
    ○ PARTICLE SIZE _____

30
31 ○ UTILITY CONDITIONS:
    ○ TOXIC ○ FLAMMABLE ○ OTHER _____

32 STEAM: DRIVERS HEATING
    | ○ OPERATING CONDITIONS |

33 MIN _____ PSIG _____ °F _____ PSIG _____ °F
    ○ PUMP CAPACITY _____ GPM

34 MAX _____ PSIG _____ °F _____ PSIG _____ °F
    ○ SUCTION PRESSURE

35 ELECTRICITY: DRIVERS HEATING CONTROL SHUTDOWN
    MINIMUM _____ PSIG

36 VOLTAGE _____ _____ _____ _____
    RATED _____ PSIG

37 HERTZ _____ _____ _____ _____
    MAXIMUM _____ PSIG

38 PHASE _____ _____ _____ _____
    ○ DISCHARGE PRESSURE _____ PSIG

39 COOLING WATER:
    DIFFERENTIAL PRESSURE _____ PSI

40 TEMP. INLET _____ °F MAX RETURN _____ °F
    NPSH _____ FT

41 PRESS NORM _____ PSIG DESIGN _____ PSIG
    □ SEALING PRESSURE _____ PSIG PRESSURE @ THROAT

42 MIN RETURN _____ PSIG MAX ALLOW Δ P _____ PSI
    SERVICE: ○ CONTINUOUS ○ INTERMITTENT (STARTS/DAY _____)

43 WATER SOURCE _____
    CYCLING: ○ TEMPERATURE ○ PRESSURE ○ FREQUENCY _____

44 INSTRUMENT AIR: MAX/MIN PRESS _____/_____ PSIG
    REMARKS: _____

45 REMARKS: _____

46 _____

47 _____

48 _____

49 _____

50

Figure 14-8. Pump seal data sheet (page 1/2) – API 610.

Reviewing flowsheets and technical specification with the contractor will involve reviewing together the formal side and the technical content of the specifications as shown in Tables 14-5 and 14-6.

Table 14-5
Checklist of formal completion of a specification

1. Is the scope properly stated -
2. Is the extent of supply defined -
3. Are terminating points clearly identified -
4. Design basis and operating requirements clear -
5. Equipment description clear -
6. Manufacture, inspection, testing covered -
7. Packing, delivery, storage covered -
8. Site erection, commissioning covered -
9. General, i.e. hazards, quality assurance, painting, identification
10. Documentation (timing) requirements clear -
11. List of appendices -
12. Are there omissions -
13. Overstatements -
14. Duplications -
15. Vagueness -
16. Ambiguity -
17. Assumptions -
18. Language and terminology understandable -

Table 14-6
Checklist: Flow sheet and specification review (special purpose steam turbine: Partial example)

General
1. Is there a warm-up vent (at least 1 1/2") on the inlet line?

    (Yes or No)

2. Does the inlet block have a 1" bypass for line warm-up?

    (Yes or No)

3. Does the exhaust valve have a 1" bypass for warm-up?
    (Back pressure turbines only)

    (Yes or No)

4. Is there a trap and bypass upstream of the trip and throttle valve?

    (Yes or No)

5. Is there a trap and bypass on the steam chest of single valve turbines?

    (Yes or No)

6. Is there a trap and bypass on the low point of the exhaust casing?

    (Yes or No)

7. Is there a low pressure seal vent line on both seals?

    (Yes or No)

8. a) What devices cause a trip of the turbine other than the built-in ones?
   b) Have these been specified?

    (Yes or No)

   c) All special purpose turbines have a separate trip and throttle valve which is a shutdown device and can be actuated by an electrical signal. Do the process safety shutdowns utilize this device?

    (Yes or No)

[. . . and so forth]

Maintainability input is accomplished during the mechanical layout and flowsheet review by:

- Identifying special safety procedures and equipment necessary to assure inspection and maintenance along the guidelines issued with the project operating standards.
- Determining type and quantities of tools and equipment to maintain the facility.
- Suggesting to the contractor additional valving, blinding, and bypasses to allow for safe isolation of equipment during on-line and off-line maintenance.
- Determining to what degree the new equipment is complying with the plant's standardization policies – especially determining preferred mechanical seal vendor, reducing number of electric motor and valve types and so forth.
- Providing guidance to the contractor as to the best orientation of vessels, storage bins, and tower nozzles as well as external and internal man ways for ease of maintenance and inspection.

Vendor Selection

Vendor selection is based on several prerequisite steps of a bid evaluation process. The major equipment – high effort – bid evaluation has 11 distinct phases:

1. Pre-select vendors
2. Prepare inquiry document
3. Receive bids
4. Make preliminary evaluation
5. Make technical evaluation
6. Make commercial evaluation
7. Conduct pre-award meeting
8. Condition bids
9. Select vendor(s)
10. Conduct pre-commitment meeting
11. Award the order.

Depending on contractual arrangements, an owner–contractor team will be involved together in all these steps.

Let us look, for example, at the bid evaluation step. It is at this point where the owner–contractor interface must work harmoniously. Figure 14-9 is a typical technical bid tabulation form. It provides a checklist of general and common points to be considered during the

Service: _____

Flowrate-GPM Duty _____ Max Design _____ TDH-FT. _____

VENDOR			
Pump			
Size and Type			
Horizontal/Vertical			
Stage/Volute			
NPSHR			
Efficiency - %			
BHP at Duty Pt.			
Impeller Dia. -Bid/Max.			
Suction Specific Speed			
Slope of Curve - FT./GPM			
Duty Pt. Flow/BEP Flow - %			
Minimum Continuous Flow-GPM			
Suction Eye Velocity - FT./SEC.			
Shaft Size at Stuffing Box -IN.			
Shaft Size at Impeller-IN.			
Shaft Size between Brgs.-IN.			
Shaft Overhang - IN.			
Shaft Deflection - IN.			
Bearing Size - Radial/Thrust			
Brg.Lubrication Method			
Thrust Load			
Suction-Size/Rating/Flg.Face			
Disch. -Size/Rating/Flg.Face			
Location - Suction/Disch.			
Allowable Loading - Suct./Disch.			
Mechanical Seal			
Type			
Seal Flush Plan			
Baseplate			
Type			
Center Line/Foot Mount			
Coupling			
Type			
Spacer Length -IN.			
Driver Type			
Mfr.			
Power - HP			
Frame			
Class			
Efficiency - %			
Pricing			
Tests			
Pump Price, Each			
Driver Price, Each			
Freight, Each			
Duty, Each			
Total			
HP Evaluation - ($/HP)			
Maintenance Cost			
Total Cost			

Figure 14-9. Technical bid tabulation form.

technical evaluation. Once the owner–contractor team has agreed that the equipment is technically acceptable and meets reliability and maintainability goals as laid out in the specification, it has to look for additional benefits.

Examples are utility consumption, efficiency, superior quality of materials, spare capacity, uprateablilty, and other design benefits to the future owner. The team has to also look for hidden additional costs or savings. For instance, a well laid out modular piece of equipment may require fewer piping connections and better access for maintenance.

If little is known about a particular vendor, his experience level has to be evaluated. Table 14-7 shows a questionnaire that may be used. If equipment is procured in the international market, i.e. purchased in one country for installation in another, attention must be paid to whether or not it will be serviced once it is operating. It can be a vexing maintainability problem to see how the vendor's representative in the country of installation will not have anything to do with the equipment as it was not purchased through him. How well was the owner–contractor–vendor interface handled here?

Finally, in the pre-order review the owner–contractor team will interface intimately with the vendor and typically they will try to answer together the questions shown in Table 14-8. The team would certainly need to agree that the vendor should be selected based on the following criteria:

- Knowledge of the particular vendor's capability
- Reliable performance on previous installations
- Experience with various sizes of his product line
- Good relationship of local or regional service personnel
- The ability to obtain spare parts with reasonable delivery.

Table 14-7
Equipment vendor experience checklist

1. Start-up date (not ship date) _____
2. Run lengths between overhauls/inspection _____
3. Run lengths between failure _____
4. Total operating hours to date _____
5. Problems encountered during S/U _____
6. Major outstanding problems _____
7. Result of T/A inspections _____
8. Maintenance contact and phone # _____
9. Warehouse spare? _____
10. Installed spare? _____
11. What spare parts are stocked? _____
12. Would you purchase this equipment again? _____
13. With what changes? _____

Table 14-8
Checklist: Preorder review with vendor (special purpose gears:
Partial example)

---

1. Compare the gear service factor quoted with those in the latest version
   of AGMA 421. Is the quoted SF at least equal to the specified value?

   _____
   (Yes or No)

2. If the unit is a double reduction gear does it have a service factor of 2 or
   more?

   _____
   (Yes or No)

3. Check the rated HP
   a) If motor driven, the gear rating must be equal to the motor rating
      including any service factor. Is it? _____
      (Yes or No)
   b) If steam turbine driven, the gear rating must be the capability of the
      driver with maximum inlet steam pressure and normal exhaust. Is it? _____
      (Yes or No)
   c) If gas turbine or engine driven, the gear rating must be the capability
      of the driver with the lowest specified ambient temperature. Is it? (Yes or No)
4. Has the vendor adequate experience?
   a) Location of similar machine _____

   b) How closely does it conform to _____

   c) Have any similar machines experienced difficulty either on test or in
      the shop? If so, what were the characteristics of the problem and what
      steps were taken to correct it and to avoid a repetition? _____

[. . . and so forth]

Documentation Review

As mentioned earlier, reliability reviews are aimed at ensuring compliance with all of the specifications. To assist in fulfilling this task, the reviewing owner's engineer generally instructs the contractor or equipment manufacturer to submit drawings and other data for his review. For example, many of the vendor data and drawing requirements are tabulated in the appendices of applicable API standards and can be adapted to serve as checklists for the task at hand. Other checklists may have to be derived from the reviewer's experience.

Keeping track of the status of documentation reviews is best accomplished by first listing the vendor drawing and data requirements for a given equipment category. Figure 14-10 shows a typical listing for a special-purpose turbine purchase. Whenever possible, the listing should contain the data requirements of available industry standards. For instance, Figure 14-10 is derived from API 612.

Each "tracking sheet" is supplemented by three columns in which the reviewer can enter the review status (e.g., "Preliminary Review Completed"). Also, the "tracking sheet" shows seven columns which indicate a particular project phase during which the vendor is expected to submit

Project Title _____ Date Ordered _____

Project No. _____ Item No. _____

Purchase Order No. _____

Location _____ Unit _____ Inquiry No. _____

Service _____ Date _____

Submit to Owner Before:

Preliminary Review

Comments

Final Copy

Description	P.O. Issue	Coordination Meeting	Shop Test	Shipment	Field Erection	Field Test	Plant Startup		
1 Certified dimensional outline drawing and list of connections									
2 Cross-sectional drawing and bill of materials.									
3 Rotor assembly drawing and bill of materials.									
4 Thrust-bearing assembly drawing and bill of materials.									
5 Journal-bearing assembly drawing and bill of materials.									
6. Packing and labyrinth drawings and bill of materials.									
7 Coupling assembly drawing and bill of materials.									
8 Gland sealing and leakoff schematic and bill of materials									
9 Gland sealing and leakoff arrangement drawing and list of connections.									
10 Gland sealing and leakoff component drawings and data.									
11 Lube-oil schematic and bill of materials.									
12 Lube-oil arrangement drawing and list of connections.									
13 Lube-oil component drawings and data.									
14 Electrical and instrumentation schematics and bill of materials.									
15 Electrical and instrumentation arrangement drawing and list of connections.									
16 Control and trip system									
17 Governor details									
18 Steam flow versus horsepower.									
19 Steam flow versus first-stage pressure.									
20 Steam flow versus speed and efficiency.									
21 Steam flow versus thrust-bearing load.									
22 Extraction performance curves.									
23 Steam correction charts.									
24 Vibration analysis data.									
25 Lateral critical analysis.									
26. Allowable flange loading									
27 Alignment diagram.									
28 Weld procedures.									
29 Hydrostatic test logs.									
30 Mechanical run test logs.									
31 Rotor balance logs.									
32 Rotor mechanical and electrical runout.									
33 "As-built" data sheets.									
34 "As-built" dimensions.									
35 Operating and maintenance manuals.									
36 Spare parts recommendation and price list									

Figure 14-10. Example of a documentation control sheet (API 612).

certain drawings or analytical data for the owner's review. The decision as to when – i.e. which project phase – data are to be submitted is best made by mutual agreement and interfacing of all parties involved.

Problems in this area must always be anticipated. Since final documentation as built drawings and manual presentation by the contractor are frequently late in the project, it may happen that not enough resources are available to prepare the required documentation. Mediocre equipment files are the result and especially machinery maintainability and availability will suffer.

Supplier Quality Assurance

For those capital projects that cannot rely on a world-wide supplier quality improvement process [2], old fashioned quality assurance by checking and appraisal must prevail.

Contracting of quality assurance and inspection services makes for flexibility in the owner's or contractor's organization but can be fraught with problems. For example, the authors, in their role as owner's engineers, have seen contract inspectors come into a vendor's facility, take name-plate data of the equipment they were supposed to inspect and disappear. Others have no doubt made the same more or less distressing experience.

Successful interfaces with inspection service organizations should be a three-step approach:

1. Successful suppliers to the capital project should be seen to have a satisfactory in-house Quality Assurance organization. They should have sufficient resources to man and support the project.
2. In those cases where the owner wants to use an inspection service, he should himself interview the services personnel and not rely on resumés. So often a resumé does not reflect the real experience and education of a person.
3. An audit should be made that uses checklists agreed to by contractors' and owners' representatives as shown in Table 14-9.

Field Installation and Equipment Turnover

Field installation of the equipment must be followed on a day-to-day basis by the owner's representatives in close interfacing with the contractor's field supervisors. Some personnel considerations seem in order.

Table 14-9
Inspection services audit

- Does the quality audit checklist produced by the QA engineer include all relevant project requirements –
- Does the inspector carry a copy of the contract specification or the purchase order for the material or the equipment being inspected –
- Does the inspector carry copies of the latest revised drawings –
- Does the inspector carry or have access to up-to-date issues of the client's specifications and relevant national specifications –
- Is the inspector checking the client's test procedures –
- Is the inspector ensuring that the supplier's test equipment is properly calibrated –
- Is the inspector understanding and interpreting test results to ensure they comply with the client's requirements –
- Is the inspector ensuring that certification complies with the client's requirements –
- Do inspection reports issued by the inspector show that he is making sure that all client's requirements are being complied with –

We envisioned previously the involvement of experienced maintenance people in capital projects. At the stage of equipment field installation this involvement of maintenance is to become a must. Not only is early ownership of the equipment by the people who will operate the plant guaranteed, but reliability and maintainability of the plant will get a final boost. The authors found that the best person to interact with the contractor's personnel as mechanical field inspector is an experienced first-line maintenance supervisor or an experienced lead hand mechanic. While the owner's specialist field engineers are capable of "pinch hitting," they are far too busy to spend all their time in the field. Mediocre results may be achieved if field engineers are not supported by experienced maintenance personnel. Also remember, contractors often know very little about maintenance!

Tasks in the field installation phase are mainly to assure that:

- Equipment is protected during construction.
- Installations are executed in a reliable manner according to agreed-to procedures and standards.

Equipment protection is accomplished by the personal effort of a dedicated individual, the owner's mechanical field inspector. The basis for equipment protection during field storage is defined as part of the project coordination procedure.

Justified attention to detail – see Table 14-10 – by the owner's field inspectors will frequently lead to interface problems with the contractor's personnel. This should be anticipated and interface meetings should be

---

Table 14-10
Owner's field inspector interfacing with contractor

TO: Field Superintendant – Good Contracting Co.

SUBJECT: Pump installations

As you are getting into full stride on pump installations it seems necessary to draw to your attention several eversights or shortcomings in your installation methods. Some of these points have been discussed with the trades previously but the following poor practices still persist.

Oil and dirt on concrete bases:
Before a machine is moved from the fab shop millwrights should be asked to remove any preservative oil that may be in the machine. They may also remove any auxiliary parts that are required before the machine is placed outdoors. Oil has spilled on several bases where this has not been done. In these cases the concrete will have to be cleaned, because grout will not adhere properly to dirty concrete . . .

Pump flange blinds:
Our standards require that pumps be blinded off with gaskets on *each side of the blind.* They will then be filled with a preservative oil. In most installations so far this is *not* being done. It cannot be emphasized too strongly that pump, turbine and compresser flanges must be closed or covered at all times . . .

---

held during which contentious points are aired. Senior specialists will usually chair these meetings.

Installation of plant equipment is being followed as part of the project's quality assurance effort. Ideally, the equipment is readied for turnover and subsequent commissioning as a joint effort by both owner's and contractor's personnel. A typical quality tracking form is shown in Figure 14-11. A smooth transition into the turnover and pre-commissioning phase is achieved by following the steps described in Figure 14-12. This figure might, at first sight, appear to be a duplication of the checklist shown in Figure 14-11. However, since there is frequently a time lag between installation completion and commissioning, it cannot hurt to have certain steps of the previous phase repeated in the latter one. If owner–contractor interfaces have been harmonious, commissioning will be successful and project service factors goals will be most likely met.

Evaluation

As in all human endeavor, we would like to know how we did. "Feedback is the breakfast of champions," someone once said. It starts with a contracting firm worth its name being clearly interested in our topic. The company would want to know how it fared. Table 14-11 will give us an

RELIABLE CONTRACTORS		THE OWNERS			
Quality Control Plan	MECHANICAL EQUIPMENT BLOWERS	QCP **S**			

CONTRACT : PRCA690   Client :   Subcontractor:

EQUIPMENT Nr.:   DWG Nr. :   UNIT :
                                                                        Page 1/2

	Subcontr.		CONTR.		OWNER	
DESIGNATION	Self Control	QC	C	A	C	A
1 – BASEPLATE LEVEL AND PROPERLY GROUTED. JACKSCREWS REMOVED AND FOUNDATION BOLTS TIGHTENED.	X	X	X			
2 – REMOVE SHIPPING PRESERVATIVES FOR BEARINGS AND RELUBRICATE. (WHEN REQUIRED). LUBRICANT :	X	X				
3 – EXCESS PIPING STRAIN ALLEVIATED.	X	X				
4 – FINAL COLD ALIGNMENT TO MANUFACTURER'S TOLERANCE. NOMINAL COUPLING GAP :	X	X				
5 – CHECK BLOWER AND DRIVER FOR FREEDOM OF MOVEMENT.	X	X	X			
6 – DRIVER CHECKED PER GEAR UNIT CHECK LIST. (WHEN. APPLICABLE).	X	X				
7 – GEAR UNIT CHECKED PER GEAR UNIT CHECK LIST. (WHEN APPLICABLE).	X	X				
8 – BEARING TUBE SYSTEM CHEMICALLY CLEANED, FLUSHED AND OPERABLE. LUBRICANT :	X	X				
9 – CHECK MOTOR BEARING LUBRICATION. LUBRICANT :	X	X				
10 – BELTS INSTALLED OR COUPLING LUBRICATED AND CLOSED. LUBRICANT :	X	X		X		
11 – BELT OR COUPLING GUARD IN PLACE AND SECURED.	X	X				
12 – BLOWER AND DRIVER DOWELLED. (WHEN SPECIFIED).	X	X				
13 – CASING DRAINED.	X	X				
14 – VENTS AND DRAINS CORRECT.	X	X				
15 – INLET GUIDE VANES SET CORRECTLY AND OPERABLE.	X	X				
16 – AIR FILTRATION SYSTEM CORRECT.	X	X	X			
17 – ALL CONTROLS, PROTECTIVE DEVICES AND AUXILIARY SYSTEMS CORRECT.	X	X				
18 – PIPE SUPPORTS FITTED AND ADJUSTED.	X	X				
19 – LOUVRES SET AND OPERATING.						

APPROBATION DATE : SUBCONTRACTOR REPRESENTATIVE	APPROBATION DATE : REPRESENTATIVE:	APPROBATION DATE : REPRESENTATIVE:
NAME :	NAME:	NAME :
SIGNATURE :	SIGNATURE :	SIGNATURE :

SELF CONTROL   : BY WORKER HIMSELF
QC   : CHECKING CARRIED OUT BY S/C QC MAN
C   THE SC HAS TO CALL.   AND/OR   SUPERVISION – HE CAN CONTINUE TO WORK
A   HOLD PT – THE S/C CANNOT CONTINUE TO WORK WITHOUT FORMAL APPROVAL FROM
                                        A:\MECANIQU\ANGLAIS\QCPS2ENG.WK1

Figure 14-11. Example of an equipment installation quality tracking form.

BLOWER, PRE-COMMISSIONING

MANUFACTURER :　　　　　　　　　TAG N° :　　................................

MODEL N° :　　................................　　SERVICE :　　................................

SERIAL N° :　　................................　　DATE :　　　................................

		Pre-Turnover		Post Turnover	
		CONTR.	OWNER	CONTR.	OWNER
1.	Flush bearing housings with a solvent (SRB 7), then with ECA 7259.	X	W		
2.	Fill oil reservoir to proper level with ECA 9105.	X	W		
3.	Rotate shafts to check freedom.	X	W		
4.	Motor checked for rotation and run-in satisfactorily.	X	W/C,O		
5.	Check alignment of drive pulleys with straight-edge or string.	X	I		
6.	Install V - belts. Check that they are all matched and have equal tension. This should be 16 mm deflection per meter of center distance. See　　Op. / Mtc. Manual.	X	I		
7.	Install belt guards and insure clearance.	X			
8.	Instruments checked and simulated. Safety valves operational tested and certified.	X	I		
9.	Obtain hearing protection for run-in.	X			
10.	Run the unit for 20 minutes and check vibration, bearing temperature, hot spots, noise, and oil leaks.	X	W/O,C		
11.	If a hot spot develops, shut down immediately and investigate cause.	X	I		
12.	After 20 minutes run in, check belts and retighten if necessary.	X	I		
13.	Restart machine and continue to run for 24 hrs.				X
14.	Check belts for tightness				X
15.	Restart machine and continue to run for 200 hrs.*				X
16.	Inspect inlet filters. Clean or replace as necessary.				X

this is a "shake-down" run during day shifts.

X - Perform the work
R - Review the work or review plans to perform the work
W - Witness the work or sign-off that the work is completed
S - Supervise and direct the work of others
C - Construction
O - Operations
I - Inspect the work on an auditing basis only.

Figure 14-12. Equipment turnover checklist.

Table 14-11
Contractor rating questionnaire

to no extent		to some extent		to a great extent		
1	2	3	4	5	*(please circle number)*	

1. The engineering contractor field team (ECFT) made every effort to clarify its roles and interfaces with our organization.   1 2 3 4 5

2. It was clear to me how ECFT work fitted with my individual work.   1 2 3 4 5

3. ECFT members were consistently sensitive to the needs of others they worked with on a day-to-day basis.   1 2 3 4 5

4. It often seemed as though ECFT was an integrated part of the owner's project administration team (PAT).   1 2 3 4 5

5. I believe that integration between ECFT and owner's PAT was a desirable aim.   1 2 3 4 5

6. ECFT were rarely over-defensive on behalf of the contractor's design work.   1 2 3 4 5

7. The major decisions in the field were made by the owner's PAT.   1 2 3 4 5

8. The ECFT worked well together as a team.   1 2 3 4 5

9. The leadership of ECFT encouraged initiative and personal empowerment.   1 2 3 4 5

10. I felt that the response time to queries and questions was good.   1 2 3 4 5

11. I felt that any member of the ECFT would be able to handle any query I had.   1 2 3 4 5

12. I felt high personal trust in ECFT members' ability to deliver to my requirements.   1 2 3 4 5

13. ECFT encouraged an informal personal approach, which I appreciated.   1 2 3 4 5

14. I would be happy to work with such an ECFT again on a future project.   1 2 3 4 5

15. Overall ECFT completed their assignment to the satisfaction of my organization.   1 2 3 4 5

Any other additional comments or feedback:

idea of how such feedback could be accomplished. This effort and the question "How can we do it less wrong next time" is a good beginning for continuous improvement.

As for the owner's organization, a different review seems appropriate. Table 14-12 is an attempt to test where it stands in its development toward optimizing plant availability by successful project planning and execution.

Conclusion

In the foregoing, it was shown how new plant facility and particularly machinery service factors can be influenced by successful owner–contractor interfacing during the various phases of a capital project. It was demonstrated how an effort has to be made in the early phases of a project in terms of review and inspection as opportunities for reliability and maintainability input diminish with a project's progress. In several examples, it was explained how mutually agreed-to checklists can help maximizing new plant uptime. A questionnaire and an organizational matrix allowed the reader to review and evaluate his owner–contractor interfacing efforts and obtain an opportunity for improvement "next time around."

Quality Machinery Installation does not have to Cost More*

It is generally conceded that quality improvement programs require the participation of all plant members. What is less obvious is that obtaining a quality installation does not have to cost more, and we would like to explain why we try to make this point.

When visiting a refinery or petrochemical plant, it is not unusual to find a substantial portion of maintenance or operating personnel involved in day-long team building sessions. This approach, though well intentioned, often fails to result in substantial long-term improvements. Problems in areas such as labor relations, training, or management commitment are well documented. The following comments show how design of plant facilities can affect motivation and performance of operators and maintenance personnel. Judicious engineering and attention to the details of mundane tasks can prevent costly mistakes.

A high-quality machinery installation offers numerous rewards: greater safety, better operability, and improved maintainability. These factors

---

* By permission from U. Sela, Sequoia Engineering and Design Assoc., Walnut Creek, California.

Table 14-12

Maturity matrix: Optimizing plant uptime during capital project phases

Criteria	Level 1: Unaware	Level 2: Listening/understanding	Level 3: Policy introduction	Level 4: Continuous improvement
Definition				
• Philosophy • Goals and objectives	• No life-cycle cost concept • Availability ill-defined • Contractor–owner interfaces difficult • Don't know who is doing what • Cost and schedule are the rule	• Life-cycle cost concept explained • Operating philosophy published • Working relationships of owner–contractor defined • Basic rule is still safety and operability	• Philosophy/vision statement followed by R&M guidelines • R&M requirements defined qualitatively • Maintainability specifications issued • Beginning to focus on owner–contractor interfaces	• Life-cycle cost concepts well understood • Availability goals defined quantitatively • Owner–contractor partnership teams established • Basic rule: Meet project availability goal
System				
• Organization • Administration • Training	• No particular provisions for reliability/maintainability concerns • Reliability reviews hit and miss • Individual "go-getters" recognize problems but can't get things changed • Field inspection done by engineers training on the job • No training of project professionals in review/inspection management	• Reliability reviews by occasional specialist involvement • Introduce "R&M" change requests • Field inspectors are operators without role statement • Begin to train project professionals in reliability/maintainability concepts	• Team building owner–contractor encouraged • Early assignment of R&M specialists involved in reviews, audits, and inspection • Feedback from other projects is collected, disseminated, and used • Regular job-specific interface meetings with contractors	• Assign experienced mechanics as field inspectors who will operate and maintain the equipment • Have feedback mechanism for future projects • Have position (change agent) straddling operating department, owner's project administration group and contractor's organization • Review mechanism for maintenance specifications exist • Interchange on R&M issues within the company exists • Owner–contractor interface problems almost non-existent

Table 14-12

Maturity matrix: Optimizing plant uptime during capital project phases–cont'd

Criteria	Level 1: Unaware	Level 2: Listening/understanding	Level 3: Policy introduction	Level 4: Continuous improvement
Standards				
• Procedures • How problems are handled	• Designers know all • No procedures/field checks/tools • No repair/recourse of recognized maintainability problems – leave for after project completion	• Introduce installation checklists – not just "paperwork" • Reaction: "If we can't do anything about R&M problems now, at least let's try to feed back" • Identify R&M fixes needed and list • Some maintenance lifts, etc. are tried early	• Procedures, check lists for all functional project phases are collected, customized, and used • Designers ask for help about maintainability • Problems are handled in open and honest way	• Pre-job sessions with contractor are routine • Problems handled by owner–contractor teams
Measurement				
• Reviews/audits • Evaluation	• No evaluation because concept not understood • Specifications are assumed to cover everything	• No formal reviews/audits of owner–contractor interfaces • Evaluation of project in private conversations, i.e. "not good – not bad" • How do we compare with others? Not yet asked. • How could we improve next time? Not yet asked.	• Review/audit teams are formed to look at degree of owner–contractor co-operation at tools such as quality control and installation checklists • Fewer start-up problems • New plant service factor improving	• Reviews, audits infrequent and directed toward system and standard improvement • High new plant facility service factors • Life-cycle cost minimized

Note: R&M = Reliability and Maintainability

reduce operating costs and increase profits. Preventing design shortcomings that affect the work of maintenance and operations personnel does not always require costly investments. Rather, it is a question of avoiding common design errors that result in difficult working conditions. A facility layout that impedes access to rotating machinery is a common design problem affecting equipment maintenance.

As mentioned before, machinery train layout should take maintenance into account. Proper equipment access must be provided to facilitate inspection or repair tasks, i.e. maximize accessibility or surveillability. Typical design and construction shortcomings include the following design mistakes:

- Small piping (lube oil, seal oil, instrument air, process gas, etc.) often interferes with access to one or more sides of the machinery train. The problem becomes severe when electrical conduit, instrument transmitters, and explosion-proof boxes block access to the other sides of the train base plate as well.
- Often, junction boxes, transmitters, controls, and gauges are located to provide easy access for maintenance. However, little is gained when this "preferred location" blocks the mechanics' access to the machinery. Time required to remove and reinstall these obstructions is costly because the extra work can significantly extend machinery downtime. The hourly cost of instrument and electrical specialists pales in comparison to downtime production losses.

  Instrumentation removal and reinstallation work is not only time-consuming, but often results in additional damage to instrumentation and control systems. Instrumentation and controls not only must be located out of the way of the maintenance personnel, but they must be "mechanic proof." In other words, the design must take into account that maintenance personnel will step and lean on electrical conduit and small piping during the course of their work.
- Thoughtless equipment layout is not confined to large special-purpose machinery trains. It is fairly common to see large multistage pumps where access to mechanical seals or couplings (the likely maintenance items) is blocked by a seal pot stand, small piping (lube oil, process fluid, or seal flush), or electrical conduit.

Another result of a sloppy facilities design is that it creates resentment and barriers between operating personnel and engineering. The message it sends to mechanics and operators is that the design engineer has no concern for their work conditions. This kind of message discourages team work in the plant.

Improper facilities design also impacts reliability. Instrument stands, which seem to be preferred by some plant instrument departments, can sometimes adversely impact machinery train reliability:

- Besides occupying prized floor space and restricting access to the machinery, these stands also require long tubing and electrical conduit runs. These long tubing runs are likely to be damaged by maintenance personnel and become a source of leaks.
- Such stands, though fairly rigid, are often bolted to grating or checkerboard-type floors. This type of installation often results in a weak support system that tends to amplify machinery-induced vibration, thus leading to early instrumentation failure.

Safety hazards that can be eliminated. Rising incidences requiring first aid deteriorates the plant's overall plant safety record. There is no excuse for creating tripping hazards by locating small piping or electrical conduit below knee level (thus not easily seen). Gas or oil leaks due to vibration-induced failure of improperly installed tubing or small piping are also avoidable incidents.

Operability is adversely affected by poor equipment layout. Ultimately, the most significant impact of poor plant layout is on equipment operability. Consequences of the following errors affect the operators' effectiveness as well as their motivation. The latter is perhaps the most significant damage.

- It is not unusual to find that conduit and junction boxes are located so as to hide lube and seal oil drain sight glasses from view. Such a situation guarantees that operators will neglect periodic checking of lube and seal oil flow. This is a serious problem, especially during machinery train startup. Another unfortunate result is that it leads the operator to disregard an important step in operating instructions and, ultimately, to the practice of operating "by the seats of the pants."
- In some plants, electrical stub-outs are located 2–3 ft away from the electrical motor. The electrical conduit must then be routed knee-high above ground between the motor and the stub-out, thus creating an unnecessary obstacle. Similarly, cooling water stub-outs are sometimes located to block access to other facilities. Such obstacles are more than just minor safety hazards; they prevent the access required to permit the operator to inspect machinery condition.
- Problems caused by a sloppy facilities design go beyond causing the operators to neglect some of their duties. For example, a steam turbine is normally provided with a manual trip device that can shutdown the machine by dumping oil pressure that keeps the trip and throttle valve open. Thus, in an emergency, the operator can shutdown a machinery train even in the case of total electric power

failure – an event that might disable local control panel operation. To do so, the operator has to be able to reach the trip/reset lever, which is normally located near the turbine outboard bearing. Unfortunately, access to this spot is often restricted by instrument air tubing and electrical conduit! Such a design shortcoming can prevent a timely shutdown of the train.

Machinery control panel location is crucial. Local panels are meant to provide information to operating personnel. This information is critical, particularly during a turbine startup. Thus, efforts must be made to locate and orient each panel so the operator can see all relevant information at a glance during startup. In some poorly designed installations, it takes an extra operator to read the tachometer and call out speed readings to the operator located at the turbine trip and throttle valve.

Common local panel design errors. Design of the local panel should help the operator acquire needed information fast. During an upset or a difficult startup, the operator should not have to read labels on a dozen randomly arranged identical instruments to locate the one with the needed data! Lack of concern for basic ergonomic considerations is illustrated by the following examples of common errors:

- Using identical-looking local panel gauges regardless of function.
- Lack of consistent identification label location, e.g. the display label is sometimes located above the gauge and at other times below.
- Odd gauge ranges (e.g., 70–220 psig) make it more difficult to interpolate between gradations and slow interpreting gauge readings.
- No apparent order in information display; temperatures, pressures, speed, and flow readouts for both turbine and compressor are randomly located on the panel.
- Lack of machinery vibration data near the machine. It is not acceptable to have to call the control house at each step of a turbine startup to determine if vibration levels are normal; especially if each time the board person must get up and go to another section of the control room to take the readings.

Local panel design is crucial for quick operator response. A turbine-driven train startup requires that information on the machine's status be readily available to the operator. During steam or gas turbine warm-up and slow roll, the operator has to continuously monitor variables such as:

- Pressures
  lube and seal oil supply;
  filter pressure drop (Δp);
  compressor gas inlet;
  compressor gas discharge;
  steam turbine exhaust.

- Flowrates
    process gas including side streams;
    steam to turbine including extraction rate;
    fuel gas to gas turbine;
    steam injection to gas turbine;
    buffer and purge gases;
    emission control system.
- Temperatures
    lube and seal oil supply;
    lube and seal oil drains;
    process gas inlet and discharge;
    steam inlet;
    gas turbine exhaust;
    gas turbine injection steam.

During emergencies or when the machinery malfunctions, the need for quick and clear information is crucial. To achieve this objective, the local panel must be engineered in accordance with good ergonomic design practice. A well-thought-out ergonomic design uses spacing, color, shape, grouping, and plain common sense to convey information clearly. During the 1960s, automobile dashboard designers violated these guidelines by providing rows of beautiful and identically shaped control knobs; however, one often turned off the headlights when attempting to switch on the windshield wipers.

In an ergonomically correct design, proper use should be made of display instrument type (analog versus digital), shape, color, and location to differentiate between functions. As an example, pressure gauges should be analog to quickly convey approximate values from a distance. These gauges (which can be of vertical design to save space) should be arranged in a logical order:

- Turbine steam pressures from inlet to exhaust
- Compressor inlet and discharge pressures
- Auxiliary subsystem (lube oil, seal oil, buffer gas) pressure gauges should be separated from turbine and compressor data.

    - Temperatures, which vary slowly, can be displayed digitally, perhaps using a switchable shared readout to save space.
    - Flow data should be displayed in engineering units and should not require one to use a calculator! Digital output is a must.
    - Turbine speed should preferably be indicated in two ways: by an analog instrument (highly visible circular gauge) for quick information and by a digital readout for precision.

Attention to local lighting conditions should be considered when selecting digital readout devices. Liquid crystal displays (LCDs) are sensitive to temperature extremes. Electro-fluorescent displays can be difficult to read in bright lighting, while LCD displays often need supplementary lighting.

In the same vein, the local panel annunciator should indicate which parameter tripped the unit or prevents startup. There should be a clear indication of alarm urgency. This helps control house personnel (board person) to distinguish between common trouble priorities and determine if the operator should respond immediately.

Local panel design and location have an importance that transcends the previous considerations. Provisions that keep the operators inadequately informed might create a dangerous attitude (the mushroom syndrome): "As usual, I am kept in the dark. Therefore, this machine is not my concern. Let the board person worry about it." This attitude is dangerous because it defeats the efforts to improve quality by empowering the operator.

Floor grating elevation can make a difference. Small bore piping layout on a compressor train can be affected by platform floor elevation. Raising the grating (or checkerboard) above the machine base plate can markedly improve access to the machinery. Sometimes significant improvements can be achieved by routing some of the piping through the middle of the base plate. Such routing improvements can be achieved by timely design review with the machinery vendor.

Get your money's worth out of your instrumentation. A lot of valuable machinery data can be accessed by the process control computer. This information is important for optimizing operation, reducing energy consumption, monitoring machinery health, and failure troubleshooting. However, many potential benefits will be lost unless steps are taken to extract "all the juice" out of the lemon. The following will help the plant realize more profit from the investment.

- Check data error such as determining when the data are out of range. As an example, when a transmitter range is 400–1000 psig and the computer displays a 397-psig reading, it is not necessarily obvious to the control board person that the instrument is out of range and that the actual pressure is only 160 psig.
- Flow meter data must be corrected for temperature or pressure changes. Flow data could even be corrected for gas mole weight changes by smart algorithms such as using centrifugal compressor performance as a simplified gas analysis tool.
- Instrument stands are often in the way of machinery maintenance. These devices are fairly flimsy (less rigid than many machinery structures such as pump seal pot stands) and require longer tubing and conduit runs than otherwise necessary. It makes little sense to install a seal pot pressure switch on a stand located at grade when

the same pressure switch can be mounted on the hefty (and already paid for) seal pot stand.

- There is no need for differential pressure gauges. These devices are fragile (Bourdon tube failure is a real hazard), expensive, and difficult to stock because each application differs in range and pressure rating. To be seen by the operator, this type of gauge often requires long individual tubing runs that are fragile and can be improperly connected. A much easier approach is to use differential pressure transmitters (these are required anyway in most cases), to display data on standard, low-cost 4–20-mA gauges placed in the local panel. Pressure transmitters should be mounted on the compressor or turbine feet and could be readily accessible for instrument maintenance without impeding access to rotating equipment mechanics. The same case can be made for all pressure gauges in general whenever a pressure transmitter is in place. This approach can vastly reduce clutter on machinery trains, save money, and reduce the chance of leaks from high-pressure oil or process gas tubing.

- Data from all machine parameters (including bearing, steam and compressor gas temperatures, pressures, and data from auxiliaries such as lube and seal oil consoles) should be routed directly to the process control computer. This approach is not only cheaper than using supplementary "smart data acquisition systems," but it ensures that all parameters are recorded at the same rate and are time stamped by the same computer clock. This feature will facilitate future failure troubleshooting.

The preceding comments attempted to show that a quality plant design does not require more costly equipment, just more attention to details. Unfortunately, many of the design shortcomings described are the result of "back-end" engineering (or the lack of it). The plant machinery engineer can normally be trusted to review the compressor rotor-dynamic analysis with great care. Similarly, the instrument engineer will carefully select the surge control system, and have very specific requirements about the type of transducers acceptable to the plant. On the other hand, these engineers cannot be bothered to rout small piping, which is often left to the discretion of a draftsman or, more often, to be field routed by the construction contractor.

Cleaning machinery and process piping is an engineering task. Lube and seal oil systems, and steam and process gas piping require a thorough cleaning before a new machinery train can be commissioned. Cleaning seal and lube oil systems is also often required during a machinery turnaround. In practice, however, cleaning these systems rarely gets the attention deserved. The reasons are multiple; following are some typical stories:

- The process department claims that the process gas piping is already clean because it was flushed. Translation: "Someone dumped the water after the hydrostatic test!"
- Blowing steam lines is left to the process personnel without any practical engineering guidance. As a result, the lines are blown for long periods or until the noise cannot be tolerated any longer. Targets are not installed, flow rates are inadequate, and flow meter orifices left in the line are damaged; yet no one is certain that the lines are really free of damaging debris.
- Cleaning lube and seal oil lines proceeds for days, but the gauze pads still remain full of metallic particles. Shortcuts are then taken when the project schedule is jeopardized.
- Small but crucial lines are forgotten. Buffer gas supply lines and balance lines are often ignored. Yet these lines feed gas that must flow through restricted passages or through labyrinth seals. Small metal shavings or weld slag left over from piping fabrication can cause considerable damage when caught between rotating and stationary parts.

Most of these problems could have been avoided if an engineer would have taken time to follow-up on the job rather than to leave these seemingly mundane tasks to untrained personnel. Cleaning lube oil lines sometimes involves a variety of methods depending on the situation. Just circulating oil is usually inadequate. Metal particles, for example, tend to get trapped at the bottom of vessels, in check valves, and behind baffles. Calling in a contractor who uses a chemical cleaning method will not help if the problem is caused by a fluid velocity that is too low to entrain the metal particles. In this instance, bubbling air or nitrogen at the bottom of the vessel or behind the baffle can sometimes be sufficient to dislodge metal particles in a short time.

Is a Quality Design Really so Difficult to Implement?

The plant owner must convey the right message to the equipment supplier and engineering contractor. Initial cost, energy consumption, reliability, installation costs, and performance are not the only factors to be considered at the time of equipment selection. Selecting and designing large, critical process pumps should be given almost as much attention as that of special-purpose compressors. Careful specifications (not just API standards) and detailed follow-up by machinery and instrumentation specialists during the "quality control" phase (after award of order) can yield an improved product at little or no additional cost.

If the operating facility lacks the in-house standards needed to guide projects, there is added incentive that the plant's most experienced engineers be long-term residents in the engineering contractor's offices. The role of resident engineers is to provide guidance otherwise provided by engineering standards and to closely supervise contractor personnel to ensure that the proposed design meets plant requirements.

Baseplate Grouting*

An average refinery has approximately 1000 process pumps of various sizes. As the backbone of sound pump operation, every baseplate must be properly installed. Recent experience demonstrates that pre-grouting baseplates reduces pump installation and commissioning costs yields higher integrity and performance than conventional on-site grouting.*

Preparing a new horizontal pump for operation is a complex process. The goal of a properly aligned pump, operating efficiently and predictably in short order, is not usually achieved. Instead, refineries and process plants experience a high level of unpredictability and increased costs due to on-site problems related to the mounting, leveling, and grouting of baseplates, especially for API 610 pump-driver sets. These problems can include:

• Warpage caused by stress relieving that occurs in transit from factory to refinery or during on-site handling.
• Poor bonding of grout to the underside of the baseplate when poured on-site.
• Reworking mounting pads in the field because they do not meet coplanar flatness and parallel specifications.
• Interference of site environmental conditions, particularly extreme temperatures, to meeting installation specifications.

These are some of the counteractive remedies engineers have tried:

• Strengthening baseplates with additional structural members and thicker top plates to improve stiffness.
• Specifying extremely tight tolerances for machining pump and motor mounting pads.
• Specifying non-grouted baseplates.

---

*Courtesy of R. Hasselfeld, Jacobs Engineering, Cypress, California. F. Korkowski, Flowserve Pump Division, Vernon, California.
* Ibid.

- Heating baseplates in ovens to relieve stress.
- Sandblasting and priming with inorganic zinc silicate to improve grout bonding.

In response, engineers have devised another solution – pre-grouting the baseplate – which moves the work from the uncontrolled environment of the site to the controlled environment of the factory. In the case of a California crude desalting project, the result was lower total cost and faster installation and commissioning.

When engineers for a major oil company's crude desalting project in Wilmington, California, specified 37 new pumps, including 195–1000 hp (3.5–750 kW) API units, one major goal was to reduce installation cost and time. That is why they chose to have the vendor pre-grout the baseplates at the factory. The largest measured 5 ft (1.5 m) by 12 ft (3.5 m) and weighed 21,000 lb (9525 kg), with over 7000 lb (3175 kg) of grout.

Here are the actual benefits this project realized:

- Only one small site pour of 1–2 in. (25–50 mm) in depth per pump was needed to bind the pre-grouted baseplate to the foundation. This took about two-thirds less time than the conventional method.

    A conventionally grouted baseplate requires at least two pours plus time to find voids after the grout has cured. This can occur no sooner than 12 hr after the last pour. With traditional baseplate grouting, the voids must be methodically "sounded out" with a small ball peen hammer, then marked. This can be tricky with cross braces and reinforced areas often producing confusing sounds when struck with a hammer. Then vent and injection holes are drilled above each void and grout is injected carefully, taking care not to over-pressurize the area.*
- Pre-grouted baseplates travel better and arrive at the site flat and aligned, just as they left the factory. Their structural integrity is better because numerous large grout pour holes are not needed. The process includes five stages, all done under controlled conditions before shipment to the site.

    1. Fabricate the baseplate.
    2. Stress-relieve it.
    3. Pre-grout it – see Figures 14-13–14-15.
    4. Allow the forces from bonding to become static.
    5. Machine the baseplate so that it is flat.

---

* Finite Element Analysis (FEA) by grout manufacturer Escoweld confirms that forces created by on-site grout application can cause serious distortion of API 610 baseplates.

Figure 14-13. Pre-grouting a pump base.

Figure 14-14. Pre-grouted pump base (bottom view).

Figure 14-15. Finished pre-grouted pump base (bottom view).

- Pre-grouting the baseplate in the upside down position reduces epoxy grout volumetric shrinkage that causes baseplate distortion. Few, if any, pre-grouted baseplates require even, minor field machining to re-establish flatness.
- Grouting at the factory can be done quickly under controlled conditions. In one instance, it took less than an hour to mix grout, fill three baseplates, and clean up. On site, workers encounter dirt, weather, and other impediments to efficient grouting.
- Pre-grouted baseplates can be installed with pumps and drivers remaining mounted. Pumps with 1000 hp (750 kW) drivers were installed, leveled, and grouted to the foundation with everything attached to the baseplate. The pre-grouted baseplates were so stable that most of the factory alignment readings were held and no adjustments were needed.

In summary, these are the advantages of pre-grouted baseplates:

- They increase lateral and longitudinal rigidity during factory machining of mounting pads.
- Protect against damage in transit.
- Prevent twisting during installation.
- Reduce separation from grout.
- Reduce installation and shaft alignment time.

- Maintain shaft alignment design specifications.
- Improve vibration dampening through greater unit stiffness.
- Increase pump, driver, and mechanical seal reliability.
- Reduce total life-cycle costs of the pump set.

Of course, no method is all positive, and pre-grouting is no different, though the negatives are few. The costs to ship each pre-grouted baseplate, pump, and motor combination ranged from 50–75% more per unit because of the extra weight. Handling the units on-site could be more difficult, again because of the extra weight. As a preventive measure, the weight was clearly marked on each pump skid and on all paperwork; no problems were experienced in either transit or on-site. The smaller pumps were moved with normal capacity forklifts and larger ones with slings and cherry pickers. Though some vendors might not be able to machine pre-grouted baseplates, again because of their extra weight, this was not a problem encountered on the referenced project.

Analysis of Savings and Benefits*

From their experience, the engineers involved in the project concluded that pre-grouting saves money overall when installing 5–1000 hp (3.5–750 kW) units, ranging from $2000 to 8000. They analyzed the installation of a 50-hp unit in depth, comparing its actual pre-grouted cost of $7700 to the average $9800 cost of on-site grouting. This 21% cost reduction supports the view that savings of 10–30% can be achieved with pre-grouting.

The contractor was initially concerned that the final pour between the pre-grouted baseplate and the foundation would be difficult but found that the epoxy bonded satisfactorily, with only minor final cleaning of the factory poured grout surface. The contractor's representative stated: "My preference would be to work with pre-grouted baseplates, mainly because of the added stability of the unit, resulting in less field adjustments or rework to meet alignment and pad coplanar flatness." A longer-term benefit of pre-grouted baseplates expressed by the engineers is that they can help raise equipment reliability to the 7-year average interval between planned overhauls.

Based on the experience with this project, the authors believe the back-end savings in installation and commissioning more than offset the added cost at the pump vendor's facility and added transportation cost.

---

* Similar findings, illustrations and checklists together with rigorous analyses have been published by Stay-Tru® and can also be found in a comprehensive reliability improvement text, ISBN 0-88173-472-7, Bloch/Budris "Pump User's Handbook: Life Extension."

Given that the backbone of sound pump operation is the baseplate, it must be properly installed. This experience demonstrates that pre-grouting baseplates, in fact, reduces pump installation and commissioning costs and yields higher integrity and performance to specifications than on-site grouting. That, in essence, translates into extended machinery uptime and a more beneficial operating environment – our next chapter.

References

1. Kletz, T., Plants should be friendly, *The Chemical Engineer*, March 1988, p. 35.
2. Dubner, P. M., Supplier quality improvement process, *Proceedings of the 8th International Pump Users Symposium*, Texas A&M University, March 5–7, 1991.

Bibliography

British Standards Institution, BSI Handbook 22, *Quality Assurance*, 1983.
Bloch, H. P. and Geitner, F. K., *Machinery Component Maintenance and Repair*, Gulf Publishing, Houston, Texas, 1990.
Bloch, H. P., *Improving Machinery Reliability, Practical Machinery Management for Process Plants*, Volume 1, 3rd edn, Gulf Publishing, Houston, Texas, 1998.
Special Purpose Steam Turbines, API Standard 612, 5th edn, April 2003.
Monroe, T. and Gregg P., Stay-Tru® Pre-Grouted Baseplate Technology, Houston, Texas.

Chapter 15
The operational environment

Process Safety Management (PSM): An Important Component of Machinery Uptime

The PSM in the hydrocarbon processing industries is a systematic approach to minimizing the occurrence and adverse effects of operating incidents to people, environment, assets, and production. Process safety management is based on the understanding that incidents are rarely caused by a single equipment failure, human error, or environmental condition. Rather, it is held that most incidents occur due to one or more failures in the safety management system to adequately anticipate, prevent, and mitigate incidents. Process safety management is defined as a program or activity involving the application of management principles and analytical techniques to ensure the safety of process facilities. It is not occupational health and safety, industrial hygiene, environmental management or preventive maintenance (PM), although all of these should integrate with PSM efforts. An effective PSM system will cover the 12 elements shown in Figure 15-1.

What does Figure 15-1 tell us? In a database compiled by the Canadian Chemical Producers' Association from 1998 to 2000, over one-third of total process-related incidents were attributed to the Process and Equipment Integrity element. Within this element the contributors cited most often were PM and maintenance procedures. "This involves understanding what equipment is truly critical, establishing risk-based PM frequency, having procedures to carry out PM that can pick up faults, and ensuring PM findings are recorded and analyzed for trends on a particular piece of equipment or 'family' of equipment. Maintenance procedures include Permit to Work procedures, ensuring that equipment is properly

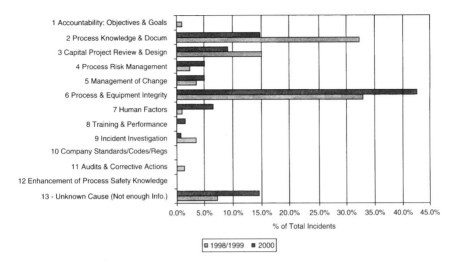

Figure 15-1. PSM cause elements [1].

prepared and ensuring that personnel are well trained in carrying out tasks" [1].

Without trying to take exception to this somewhat global explanation we believe that adherence to details in the operations–maintenance interface will avoid machinery equipment caused incidents. Here are some points from our experience we would like to stress. First, the rush factor would have to be mentioned. There is usually not enough time to:

- Allow equipment to heat up at a reasonably slow rate. When steam turbines are warmed up too quickly, some internal parts cannot undergo uniform expansion, and binding or rubbing may occur. Also, wet steam may seriously damage equipment internals.
- Make sure compressor train coupling alignment is being checked just prior to start-up after extended downtimes, such as overhauls and repairs.
- Align piping to the machinery in a stress-free manner. Pipe stresses exerted on fluid machinery suction and discharge nozzles will frequently cause equipment-internal misalignment and decrease the life of couplings, mechanical seals, and bearings.
- Ascertain that all emergency shutdown devices are functional. An equipment restart is often the only chance to verify the integrity of critically important instruments, yet trip testing of remote devices is often overlooked.

- Provide proper grout support of pump baseplates. This has often resulted in abnormally high vibration or out-of-alignment operation of driver and driven equipment.
- Apply proper torque values to such critical bolts as used in reciprocating compressor valve caps, piston rods, high-speed couplings, high-pressure reciprocating pumps, etc. There are many literature references pointing out costly and even fatal incidents attributable to careless bolting procedures.

Then, there are the oversights or well-meant practices that could be characterized as the "more is better" syndrome:

- Reciprocating compressor cylinder cooling water temperatures are maintained too low. Saturated gases are allowed to enter the compressor cylinder at a higher temperature and moisture condensation will cause equipment distress.
- Overlubrication of bearings may cause heat rise and rapid oxidation of the lubricant. Statistics show more grease-lubricated bearings fail due to excessive lubrication than due to insufficient lubrication!
- Steam turbine trip and throttle valves must be exercised at regular intervals if their functional integrity during emergencies is to be assured.
- Control oil accumulators can and sometimes must be checked and refilled with the equipment running if steam turbine reliability is important.
- Auxiliary lube and/or seal oil pumps on turbomachinery must be tested on a regular schedule.

Our experience shows that sites that have convinced their operators to do these tasks are running with greater safety and overall dependability than sites just paying lip service to these routines. Further, there is always the occasional site which allows shortcuts and second-guessing, such as substitute lubricants and materials. Finally, lube oil analysis and dewatering should be taken more seriously since the economics of performing these tasks will inevitably win out over the ultimate cost of not performing them.

Other issues relate to accepting compressor surges as normal occurrences, tolerating pump cavitation, excessive vibration, lack of thermal insulation, and many other situations where a well-informed operations department can have a surprisingly beneficial effect on the safety and hence overall profitability of modern process, pipeline or gas companies [2].

A Reliability Policy – Do You Have One? – Why Not?*

Reliability for any business starts with management. Management uses policy statements to address major issues. What does the reliability policy for your company say to the organization? Does your management know enough about reliability to endorse a reliability policy, which you, as the reliability professional, would propose for their signature?

Management needs clues about reliability details to achieve lasting reliability improvements for products and processes to increase profits. Reliability professionals must help educate managers about reliability in short, tight, clear, attention-getting sound bites to sell the organization on the benefits of reliability. Effective communications from reliability professionals must be short, to the point, and business oriented to emphasize money and time.

As the reliability professional in your organization, can you give a 60-second sound bite to state your view of a reliability policy? Or, do you require a non-productive lengthy, tedious presentation? Remember you must sell reliability – not preach – so ask how your verbal presentation would sound on the 6 p.m. evening newscast? Would your sales pitch be viewed as a turn-on or turn-off?

Deliver your management sales presentations in listener train of thought:

1. cost
2. time
3. alternatives – including the datum cost of doing nothing for improvements
4. benefits. (Tutorials for management cannot drone on about mind bogging minutia concerning probabilities and confidence limits for which management has zero tolerance.)

Effective reliability professionals must help management communicate reliability to the organization with a clear reliability policy statement. Policies can mobilize actions for considering cost of alternatives to prevent or mitigate failures, which require knowledge about times to failure, and failure modes, found by reliability technology.

Modern organizations have:

- A safety policy that generally says "We will have an accident free working environment."
- A quality policy that generally says "We will ship a defect free product that meets the needs of the customer.

---

* Contributed by P. Barringer, P.E., www.barringer1.com.

- A reliability policy that generally says "- ——"! You have the opportunity to fill the void by saying "Our products and production processes will be failure free in use for a pre-established and specified period of time when performing the intended function with correct operation in the proper environment." If you do not have a reliability policy, then get one – in the end, it is all about money and time!

Of course, the simplest statement for managers to adopt would be consistent with quality and safety policy such as "We will have a long defect free life for our products and processes." (Yes, I know you want to tell me a paragraph or maybe a chapter about why this is not precisely correct but the issue is too big and too important for too many small details!)

Managers rarely write their own policy statements. Reliability professionals must mobilize and persuade (sell) management to do the right thing with a clear policy proposal. The proposal must address the need for improvements using a top-down thought process (avoid bogging down in a bottom-up argument as management will not be interested). One way to motivate a policy is to address the cost of unreliability [3].

Reliability professionals must do the staff work for their managers and sell the need for improvements in a persuasive manner. Thus, reliability professionals must think about their sales strategy – if you cannot sell your boss, how are you going to sell the organization? Think about what is in it for people to change their thinking process for reliability?

Communications for reliability improvements require knowing:

1. when things fail
2. how things fail
3. conversions of failures into statements about time and money.

Reliability engineering principles help define when and how things fail. The principles provide evidence for making life-cycle costs comparisons. Reliability details provide evidence for the lowest long-term cost of ownership driven by a single estimator called net present value, which converts hardware issues and alternatives into money. If you are not speaking in money/time terms involving the lowest long-term cost of ownership then you are missing the boat for management conversations.

Once you have a reliability policy, you can then set reliability procedures. Procedures provide guidance for how the policy should be implemented. Pre-think the procedures so the organization can make monetary decisions in an un-emotional manner as a business fact rather than becoming immobilized with indecisions. The procedures can address the "facts of life" for getting problems resolved rather than promoting endless discussions about what is a failure and how much does our failure cost.

Rules for reliability follow the procedures. Want an example? Think about your safety programs. Safety programs have a policy, which sets the accident-free environment. The policy drives procedures for how the policy will be implemented. The procedures drive the safety rules to take or not take action. Good effective safety programs have been demonstrated to be both altruistic and cost-effective. Should you expect anything different for a reliability program? Go sell your programs on this basis.

If you have a top-down reliability program driven by a policy statement, procedures, and rules, then it is easy to perform a reliability audit. What would an audit show for your facility in substantiating your reliability program? Avoid the procedural and bureaucratic ISO-9000 audits, and go for a value audit.

Would a value audit show you are doing what you promised for the corporation in moving toward a long failure-free life? Most organization would, by knee jerk reaction, focus on product hardware. However, the lost money in most organizations is in the failure of the process to produce the product. This requires an assessment of process reliability and quantification of losses. Get the process under control and then product reliability control will follow. The last place most reliability professionals investigate is control of the process [4]. Tight process control for reliability is an important place for profit improvement, which is unseen to your competitors.

Have you quantified the cost of reliability failures for your organization by products and processes? If you want management's attention, present issues in terms of time and money! Avoid glibness. Set down details in simple terms so you can sell your position from a professional reliability perspective. Sound simple? It is not! Get the facts. Make your sales pitch. Remember the sales challenge: Salesmen do not really start selling until the customer says NO!

Reliability professionals – it is your job to sell a reliability policy to your organization. If it is not your job, tell us whose job is it? And what is it we can learn from our peers, the folks that have been successful in selling these policies to their respective managements and have subsequently implemented the requisite procedures and work processes?

What We Can Learn from Others

Performance Benchmarking

In June 1998, Lee H. Solomon, CEO of Dallas/Texas-based Solomon Associates Inc., gave an overview presentation on "Pacesetter Performance and the Role of Reliability" to an audience in Houston. In the years since 1982, his company has performed numerous benchmarking and maintenance performance studies on all six continents. Likewise, he and his staff continue to be involved in performance improvement consulting and performance measurement in all functional areas.

Expectations and Reality

Studies performed by Solomon Associates focus on a clearly definable business. They address all issues that impact profits. They examine historical facts not plans. And since the company is staffed by experienced personnel, these individuals had certain expectations:

- Little variation in performance – after all, we live in the space age and everyone has access to modern technology.
- Similar results for the affiliates of world – renowned companies.
- Differences in performance due to such physical issues as size, age, location, and unionization.

But, what they found surprised even those seasoned individuals:

- Wide variations in performance, in spite of access to modern technology. Profitability, expressed as return on investment (ROI) of pacesetter, best of class (BOC) companies was typically in the vicinity of 16% and exceeded that of the low performers by 12%.
- No affiliation synergism. Top quartile plants and bottom quartile plants had the same owners.
- Weak correlation to physical factors. Big plants at the top, big plants at the bottom; small plants at the top, small plants at the bottom. Old plants at the top, old plants at the bottom; and the same with plants in this hemisphere or on that continent, unionized or non-unionized.

How Companies Differ

Both management practices and employee culture at leading companies differed from those that did not score well. Companies with high unreliability also reported high maintenance costs (Fig. 15-2). Among the top companies, total annual refinery maintenance costs per unit of refinery capacity related to capacity and complexity, in 1992-dollars, amounted to $17/EDC (equivalent daily capacity). On the opposite side of the spectrum, low performers spent more than twice as much, $36/EDC in 1992.

As indicated in Figure 15-3, rising reliability trends were experienced in the 1986–1992 time frame. Costs went up as mechanical availability declined from 1986 to 1992 (Fig. 15-4). Improved mechanical reliability is thus *not* related to the amount of maintenance effort.

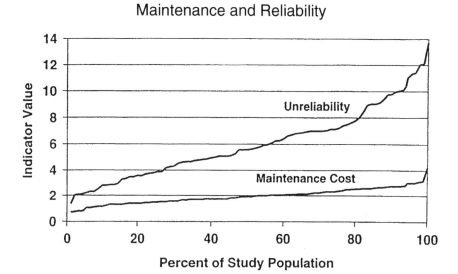

Figure 15-2. Plants with higher maintenance costs are less reliable.

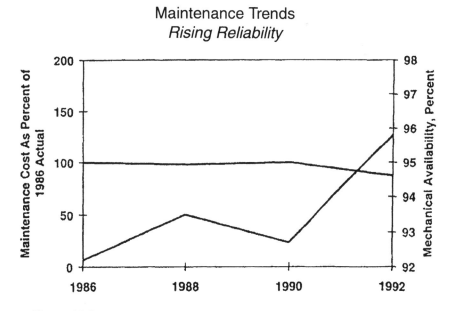

Figure 15-3. Rising reliability trends experienced in the 1986–1992 time frame.

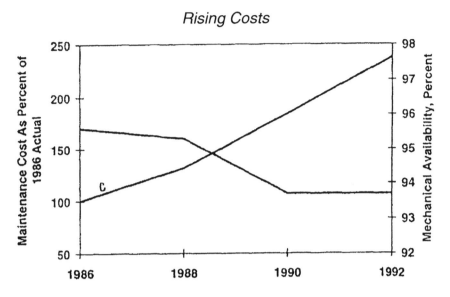

Figure 15-4. Rising cost trend as mechanical reliability declined, 1986–1992.

In the refining industry, the pacesetters BOC companies enjoy interesting and immensely valuable advantages, as tabulated in Figure 15-5. Our own observations largely parallel those of Solomon Associates. We have observed that pacesetters share a number of values. These companies are reliability-focused, not repair-focused. In the mid-1990s, the highest (maintenance cost) quartile's craftspeople inevitably worked for a repair-focused organization, whereas the lowest cost quartile craftsmen worked for a reliability-focused organization.

Workload and Planning

Statistics show that the lowest (maintenance cost) quartile's craftsmen had four times more pieces of rotating equipment per person than the highest cost quartile. Those in the highest cost quartile are kept busy repairing failures and have no opportunity to examine the causes of these failures. They thus cannot participate in the formulation and implementation of action plans to make permanent repairs or to devise preventive or predictive remedies.

It was further shown that the consistently high performers base management decisions on real data. They adhere to the plan and deal with all deviations. They always focus on economics, optimize revenue and expense, take responsible risks. We know they record events and

> 5% Higher Throughput
>
> 15% Less Energy Consumed
>
> 20% Lower Maintenance Expense
>
> 5% Higher Reliability
>
> 20% Lower Operating Cost
>
> 40% Higher Financial Return

Figure 15-5. Advantages enjoyed by pacesetters (BOC Companies).

thoroughly investigate all causes. These profitable plants follow through by revising their planning to avoid repeat events. Moreover, solid performers seek sustainable excellence and most decidedly engage their employees.

Where Operations, Maintenance, Technical and Organizational Relationships Mesh

In pacesetter companies, there is unconditional acceptance of the fact that facilities, maintenance, and organization are an interdependent continuum. This implies that there exists a commendable level of communication, cooperation, and consideration among virtually all job functions in the plant. A good example would be a petrochemical company with not only a management committee, but also a steering committee that gives guidance and actively elicits feedback. This latter activity is structured to give visibility to the efforts of every competent worker.

Facilities with first quartile capability are almost certain to engage in LCC. They will view every maintenance event as an opportunity to upgrade and will base the decision on the findings of a rigorous root cause failure analysis (RCFA). Combined with LCC of the various remedial options, these BOC companies have positioned themselves to capture financial credits from the chosen course of action.

Best of class, implying first quartile companies, thus perform reliability-centered maintenance (RCM) in a thoughtful, results-oriented manner, quite unlike their fourth quartile peers for whom RCM is often a laborious, costly, and largely procedural effort. Many of the low performers have at one time tackled RCM simply because it had been viewed as the cure-all or magic bullet. Seeing their efforts frustrated, they have since abandoned it and have gone back to their old, and ineffective, ways of doing things.

Who Does What, and How

Extensive use of predictive tools and monitoring instruments is found in companies ranging from top to bottom. However, where BOC performers use the operator to determine if a deviation exists and to then report this deviation to the highly qualified condition-monitoring technician for detailed analysis and follow-up, the low performers waste the trained professional's time by compelling him to collect reams of data on equipment exhibiting no deviation. In essence, the high performers execute maintenance as a mutual effort involving operations and maintenance as equal partners. Just as in modern aircraft, this maintenance approach will inevitably include both preventive (time based) and predictive (condition-based) methods. Subscribing to either one to the exclusion of the other has been shown to be flawed.

In any event, all maintenance decisions taken by high performers are based on real data and not on tradition or hearsay. Craftsmen constitute the primary point of control and there is individual accountability. To the extent that self-directed work teams are employed, they are not only empowered, but enabled. Enabling through proper and truly relevant training is viewed as a prerequisite to empowering. At first quartile facilities, supervisors and technical personnel are less involved in decision-making than elsewhere. Instead, they are used as a resource by craftsmen and operators. Their time and talent are optimized by allowing them to focus on longer range plans. At these pacesetter companies, there are fewer levels of management. The primary function of the managers is to enforce standards and, as was stated earlier, to actively sponsor all cost-justified reliability improvement targets. The steering committee is asking technician-workers to discuss methods and accomplishments, thus giving visibility to the grass-roots efforts.

Twelve Common Attributes Examined

Best of class companies share a large number of work practices, reliability engineering, implementation concepts, and organizational alignments. Some of these have been highlighted earlier, whereas others are implicit and follow a logical pattern of progression. While their numbers could be easily expanded by simply going into greater detail, we are limiting our listing and examination to 12 of these shared attributes. Please note the similarity to so many of the points and observations made earlier in this text as regards BOC companies:

1. They have on their bidders' lists only top-notch vendors and manufacturers. They recognize that reliability comes at a price and that competent suppliers are entitled to a reasonable profit.
2. They specify and procure equipment and components on the basis of life-cycle cost studies, having built reliability and low maintenance into the equipment specification.
3. They engage in pre-award (pre-procurement) reviews, design audits, and selective, systematic, pre-delivery engineering quality control. First quartile companies know that this effort, while requiring an up-front investment, will always result in pay-back.
4. They pay attention to detail, work toward perfection, and understand that 'business as usual' and 'hurry up, we want it running again this afternoon' attitudes are intolerable impediments to the achievement of reliability improvement in any plant.
5. They are obsessed with doing every job right the first time. To that end, they develop, acquire, use, and consistently invoke written checklists and procedures across all job functions.
6. They treat every maintenance event as an opportunity to upgrade, with a view toward run-length extension. The decision on how to proceed is again based on life-cycle cost considerations.
7. They use major elements of PdM to determine when and how to perform PM. They are aware that optimized, bottom-line cost-justified maintenance is a composite of both PM and PdM.
8. They will not tolerate the employment of highly trained machinery condition analysts (vibration monitoring technicians) to periodically acquire data from 1000 equipment bearings, only to find out that 970 of these bearings show no signs of distress. Instead, they train their operators to perform equipment surveillance. Operators report deviations from normal equipment behavior to the condition analyst for follow-up and definition of remedial action, component upgrading, etc.
9. Cross-functional teams perform true RCFA and monitor their overall progress by maintaining accurate failure statistics. Top performers have a reasonably good idea as to what improvements are feasible and, in fact, achieved elsewhere. Their *repeat* failure events show significant downward trends.
10. They recognize the virtual impossibility of acquiring expertise in *all* fields of major equipment component design, application engineering, and component optimization. This realization prompts BOC companies to teach and actively pursue *resourcefulness* by maximizing all aspects of vendor/manufacturer experience. Vendor and

manufacturer statistics are extensively consulted and design reviews performed whenever applicable and cost-justified. Electronic or conventional reference libraries are maintained and consulted by reliability professionals.

11. They have implemented an extremely close working relationship between the production (operations), maintenance, and technical (reliability/project engineering) functions. The "services" concept with its wait-until-called-upon connotation has been abandoned in favor of a support and partnership concept that demands self-activation, contribution, and participation at all levels and across all functions. More than mere lip service is paid to this critically important issue!

12. They take results-oriented training seriously. Reliability professionals are given guidance and direction through company-devised training plans, progress reviews, and easy access to mentors.

As was mentioned before, many subsets exist beneath these principal attributes shared by BOC companies. However, a process plant striving to excel in a highly competitive world economy would do well to give priority to our 12-point listing. And, while it is reasonable to proceed one step at a time, all of these items will ultimately have to be implemented if a company wants to measure up to the challenge.

However, we see even more when we look at BOC or BPs companies. First, though, a few words that define the terms.

What are Best Machinery Practices?

Best practices pertain to procedures, methods, and hardware, which have proven to be efficient and reliability enhancing. They are a process, technique, or innovative use of resources that have a proven record of success in providing significant improvement in cost, schedule, quality, performance, reliability, safety, environment, or other measurable factors which impact the health and profitability* of an organization. Establishing and monitoring the consistent use of BPs documentation forces us to consistently do the right things well. Best of class companies have over 100 BPs related to machinery. In BOC plants, there is a rigorous procedure for the development and endorsement of BPs.

The "mechanical technical services" (MTS) within the machinery department in an organization usually maintains a register† of such

---

* Profitability must not always be an objective.

† Meaning document control.

practices. It reviews their application at every opportunity such as new and small improvement projects and turnarounds. Typically BPs are tied to the overall site reliability improvement planning process and have their credits established a *priori*. Figure 15-6 represents our attempt to structure machinery BPs. Probably the most important BPs are the operating directives. They specify the condition for which the plant or unit should be operated to achieve high product quality and efficiency.

Operating directives are developed by the technical organization and executed by the operating groups. The use of operating directives allows the best knowledge of the organization to be used in specifying key variables. They are also a way of improving the standards of control by specifying the ranges within which operations must be maintained. Standards can be raised by gradually narrowing those ranges, but perhaps the most important contribution of operating directives is the clarification of

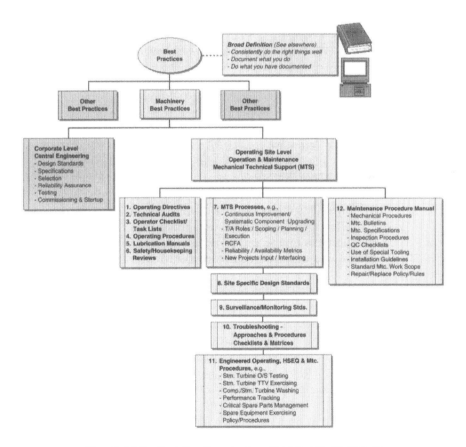

Figure 15-6. Best Practice structure for process machinery.

responsibility between the operating and technical groups. If an operation condition is not within operating directive range, it is a deviation which requires corrective action. This in itself sets priorities on what is to be done; it helps to focus the entire organization on key problems. Use of operating directives is fundamental for making sure-footed technological progress.

Regular and routine auditing of an operation by the technical group helps to close the information loop in the ideal operating system. The technical audit is useful both to determine whether or not directives are being followed and to determine if the directives are working properly. The technical audit is an essential part of the ideal control system in that it assures the follow-up and management attention to the operating directive system.

Operator checklists and the task system are two complementary subsystems. The operator checklist describes certain tasks to be done each shift on an operator post. The task system describes tasks to be completed on less than a daily frequency. The purpose of both the operator checklist and task system is to assure that tasks needing periodic completion are done on a timely basis. Activities which go into an operator checklist or task system are those pertaining to safety and housekeeping, PM, surveillance and monitoring, process control, and administrative functions. In regard to assuring operating safety, operator checklist and task system are the most important of the subsystems. There are just too many items needing periodic attention left to chance.

Lubrication manuals are really operating procedures. However, lubrication is such an important subject that it deserves to be set aside in greater detail and emphasis by having a separate lubrication manual for the each unit. The lubrication manual is the reference for the process technician to use for specifying lubrication procedures, type lubricant for each piece of equipment on the unit. Mechanical, technical personnel supporting a unit provide the main source of knowledge for the lubrication manual.

Operating procedures are somewhat similar to operating directives in that they tell how things should be done. Whereas operating directives are more concerned with specifying conditions and targets to be achieved in operating process, operating procedures specify how to get those targets and conditions. Operating procedures should be consistent with the operating directives, checklists, and task system. They should be detailed enough to be a useful resource. They should contain enough information about the equipment to help troubleshoot that equipment when the need arises. Operating procedures should give enough information to allow the average person to safely operate the equipment in question. Operating procedures are most important in facilitating good operator training.

Routine safety and housekeeping inspections should be conducted by the operating management and/or with participation from outside organizations on a periodic basis. Routine safety and housekeeping inspections are very similar to a technical audit in that they provide some feedback on what is actually happening on the unit. It is generally important for the housekeeping inspections to be documented in a punchlist form which makes correction of deficiencies easy to follow-up on safety and check up on all the systems in use. For example, adherence to operating directives, maintenance of updated operating procedures, the following of lockout, and other safety procedures or for handling of fire fighting equipment can all be spot checked during the safety and housekeeping review in addition to the general unit housekeeping. The safety and housekeeping reviews also afford the department head and operating supervisors a fine opportunity to emphasize areas they think important or needing closer attention. They are also a show of interest by the management team in safety and housekeeping. It communicates by action to the process technician management's emphasis on operating safety.

The Mechanical Procedure Manual should contain:

- *Mechanical Procedures* document repetitive jobs of a general nature involving more than one department. Written procedures will provide ready reference for new supervisors and staff personnel and will provide recall of information on repeated jobs.

  These procedures are not mandatory except in those cases where they reiterate procedures outlined in the Safety Rules, Administrative Letters, Administrative Instructions, and similar management publications. They should, however, be followed unless an equally safe or safer and (not or) equally efficient or more efficient procedure is devised. In the event that better procedures are devised, the official procedures should be revised to take advantage of this experience.
- *Maintenance Bulletins* are issued by MTS personnel to provide technical information on maintenance of process equipment.
- *Maintenance Specifications and Maintenance Inspection Procedures* together with checklists refer to troubleshooting and inspection procedures usually involving one craft.

Operator Training will Contribute to Uptime

In Chapter 1 we talked about the role of the operator. We believe that there is indeed no reliability without operator involvement! Just as the most well-designed and best-maintained automobile will fail in the hands of a thoughtless or inexperienced driver, the best and most reliable machine will not perform optimally if the operator lacks training, care, or motivation. We accept the responsibility of viewing the dashboard instruments of a modern automobile; similarly, the operator in a modern process plant must accept equipment surveillance as his or her prime responsibility.

According to the Abnormal Situations Management Consortium (ASMC), US process plants lose over $20 billion a year from abnormal situations; $8 billion (40%) is directly attributable to human error. The losses are caused by insufficient employee knowledge, and operator and maintenance worker errors. Further analyses indicate that most of the personnel-related causes are due to a lack of:

- properly designed jobs
- properly structured training
- performance support.

Recently we were asked to assess the need for what our client called "Advanced Operator Training" (AOT) around large compressors and their drivers. We decided to interview the operators, both field and control (DCS) operators in their facility. We used a 30-min audit interview format as shown in Figure 15-7.

We also asked the operators simple questions such as: How do you start-up a centrifugal pump or how do you recognize surge on a centrifugal compressor?

After we had convinced ourselves that we had identified most of the training needs of this organization, we sat with the managers in a debriefing session where we listed three of the most urgent needs:

1. There are gaps in knowledge and skills concerning machinery among process operators.
2. Skill erosion is no doubt a factor affecting training and needs to be considered
3. Continuity and uniformity of training is not sufficiently guaranteed. Sound familiar?

When the manager asked us how his operation ranked among those others we had contact with, we said that our standard was determined by the licensed operating engineers [5] environment in Canadian (Ontario) process plants. This standard calls for competencies and competency testing of personnel operating machinery. Compared with this standard, the manager's organization was at the leading edge with their competency-based training program soon to be implemented.

Now, we used to suffer the odd ridicule from our colleagues outside Canada: "There they go again, talking about their licensed operating engineers – we also have good operators!" Admittedly, we all have capable

A. Tell us about your job related training courses, prior experience and current responsibilities.

B. Describe your job function, who provides your day-to-day operations training? What is your area of operations responsibility?

C. How much of your job is organizational, supervisory, technical? Which part do you like best? For which part do you think you need more time to do your job better?

D. Is there a clear statement of your job function? Is your performance measured against it? Who provided you with operations technical support?

E. Where do you see the greatest training needs? At your level? At other levels?

F. What would you like to know more about (train in) to make the performance of your job easier, safer and/or enhance your confidence?

G. Current training received. Do you view current training as adequate or sufficient to do your job? Is current training relevant to your job? What areas can current training be improved in? What additional training is required?

H. When a new project makes changes to the base system, where and how are training and training needs documented?

I. What would you describe as your major problem areas? Technical? Operational? Equipment related?

J. Safety and reliability. Where do you see areas that have potential impact on the reliability of your plant operations? What does not work and needs fixing?

K. Where do you go for help to deal with operational problems?

L. What areas do you see as future problem areas?

Figure 15-7. Operator interview questionnaire.

and efficient operators around process machinery without licensing them, but what we must do, something common to both concepts, is train them and check them out.

Process machinery operators must have a set of competencies which would most likely be applicable to all process operators. There are two types of competencies:

- Core competencies which include adaptability, critical thinking, the desire for continuous improvement, communication skills, teamwork ability, and problem-solving abilities.
- Technical competencies which comprise a sense for loss prevention and control, equipment and process understanding, troubleshooting, monitoring, equipment feeding, care and maintenance – especially

in an environment where more and more integration of maintenance and operations is being promoted.

What does all this mean in the context of machinery uptime? Competent process machinery operators pay attention to details such as:

- Subtle signs of impending trouble, for example, changes in noise, vibration, and other indications noticeable by using their five senses to avoid or reduce consequences of equipment failures.
- Lubrication requirements of each lubrication point.
- Check oil reservoirs for discoloration, contamination, and presence of water.
- Feel or measure bearing surfaces for temperature.
- Report oil leaks and determine cause.
- Clean oil bubbles, level glasses, and gauges as required.
- Check seal flush lines for proper operation.
- Check cooling lines for effective operation.
- Check for proper operation of heat tracing.
- Check if steam traps operate normally.
- Check operation of automatic level-controlled blow devices by manual bypass in compressor suction drums and interstage heat exchangers.
- Steps of initial start-up routines.
- Points of normal shutdown routines.
- Phases of emergency shutdown response.
- Startup routines following an emergency shutdown.
- Safety procedures.

There are certainly more items that can be added to this list – they all should be part of a competency inventory to be verified by field observation and actual checkout.

Effective Machinery Monitoring: Getting the Most for Your Investment*

Design of many machinery monitoring systems relies on four implied assumptions. These assumptions are not necessarily spelled out, but nevertheless such concepts have influenced the design of most process machinery monitoring systems. These concepts lead to designs that often fail to achieve their objective because plant operators have been left out of the picture. Thus – instead of discussing Fourier transforms or the beauty of a Nyquist plot – why, when, where, and how a plant should

---

* By permission from U. Sela, Sequoia Engineering and Design Associates, Walnut Creek, California.

spend the effort and investment to monitor process machinery will be clarified. The four implied assumptions are:

1. Careful machinery vibration monitoring is sufficient to yield substantial returns. Monitoring (data collection) is done by operating technicians.
2. Buying vibration monitoring equipment from a reliable vendor guarantees effective machinery monitoring.
3. Competent and dedicated vibration analysts are the key ingredients to getting maximum returns from the investment in monitoring equipment. Analysts are not mere "data collecting personnel."
4. The last, which is really a corollary of the previous one, is that the vibration monitoring equipment should be located in the plant control house.

It is difficult to justify adding a machinery monitoring system to an existing plant. Usually, it takes a severe accident or catastrophic losses to re-emphasize the need to provide a comprehensive machinery protection and surveillance system. However, a convincing case to justify machinery-monitoring investment cannot be made unless there is a clear understanding of why it is needed and what its potential benefits are. In addition, system designers have to ensure that the monitoring system will be put to good use after installation. This means full acceptance of the system by operating personnel and management.

Why Spend Money to Monitor Machinery?

The most obvious reason is for machinery and personnel protection, to minimize consequences of machinery failure by reducing damage, thus decreasing repair cost and shutdown duration. A timely shutdown will also reduce danger to personnel and minimize environmental impact (e.g., by shutting down a vibrating compressor before a compressor seal failure causes a gas release).

Another reason is to detect gradual deterioration and allow for a planned response, for example changing the unit operating mode to decrease the problem's severity (e.g., a speed or load reduction). It will give machinery specialists the time needed to capture analysis data and perform a diagnosis. It will also enable the plant to schedule downtime and plan maintenance, e.g. order parts and schedule equipment and personnel. In some cases, it might provide the opportunity to line up alternate product sources or adjust the plant process.

A machinery monitoring system should also provide information. This will permit process improvements for higher yields or better product quality. The information can also be used to optimize plant operations to save energy. Most importantly, it gets O&M personnel involved. These people are key to good operation; they must have ownership and pride in what they are doing.

What Type of Equipment Should be Monitored?

There are many reasons why cost of monitoring a machine is justified. Candidates for monitoring should include machines where a failure can affect plant safety. Machines that are essential for plant operation and where a shutdown will curtail the process should also be considered prime candidates. Essential machines include unspared machinery trains and large horsepower trains. Also consider machines that are very expensive to repair (e.g., high-speed turbomachinery) or machines that take a long time to repair, such as large gearboxes. Perennial "bad actors" or machines that wreck at the slightest provocation – such as certain packaged high-speed air compressors – also warrant such an investment. Finally, consider machinery trains, where better operation could save energy or improve yields.

Is Not Vibration Monitoring Sufficient?

Setting up a machinery monitoring system involves more than just ordering vibration sensors and monitors from a reputable vibration monitoring equipment supplier. A number of parameters can give effective information about machine condition. In addition to shaft and seismic vibration, the obvious variables include shaft position and clearances, component temperatures such as bearings and auxiliary systems including buffer gas or sealing systems, process temperatures, process, and auxiliary system pressures, lubricant conditions (e.g., water and solids), flows (process and auxiliary systems), speed(s), power consumption (electricity, steam, or fuel gas), and corrosion rate.

Do We Really Need Operators to Watch the Machinery?

Some less obvious parameters are more difficult to instrument. Hence, these are the ones that the operator should monitor. Such indicators include clues given by machinery noise, appearance or color of machine elements (such as burned paint on a bearing or smoke coming from a vent), and an unusual smell (e.g., an H_2S smell, which indicates a process gas leak). Even the stain left by an oil drip can give a valuable clue to an observant operator. Yet, there are no readily available detection sensors for most of these parameters. The operator should be considered an integral part of the monitoring system and the system should be designed to reflect this understanding.

In brief, buying good vibration monitors and sensors will give the plant a system that can protect against damaging vibration or excessive bearing temperatures. The plant engineers must still design the rest of the system to ensure that all pertinent variables are adequately monitored.

What are the System Design Objectives?

To recap, the main purposes of a machinery protection and surveillance system are to permit safe equipment operation and minimize unscheduled downtime. The latter requires that, in the event of a failure, consequential damage be reduced by a timely shutdown of the machine while at the same time avoiding nuisance trips. The rapid rate at which many types of machinery failures can progress may well preclude timely reaction by an operator and requires some sort of automatic shutdown device. During the past 25 years, there has been marked progress in monitoring hardware reliability. The industry has also become more sophisticated about designs that minimize nuisance trips. Using "two-out-of-two" and "two-out-of-three" voting logic is becoming popular. Triple modular redundant (TMR) systems are now readily available.

Other purposes of the system must be to provide information necessary to monitor machine operation, trend changes in machinery condition, provide a basis for PdM, and allow for optimizing operating conditions. In addition, history recall features must permit effective analysis of events surrounding a failure or unplanned shutdown. The following performance attributes should be considered to meet the stated objectives.

Protective System Requirements

Effective machinery protection. Ensure that trips occur when needed. Easily replaceable parts such as bearings are considered sacrificial, but non-spared components such as a gear case or turbine diaphragms should not be damaged because of the extended downtime required for repairs. Therefore, we find that the objectives of plant operations management are sometimes quite different from those envisioned by the vibration monitoring equipment supplier.

Minimum nuisance trips. Consequences of an unscheduled shutdown vary and must be weighed against the possibility of extensive machinery damage. Sophistication of shutdown logic must reflect individual situations (two-out-of-two voting and TMR systems).

Operator friendliness. The importance of this design requirement is seldom understood. Vital data about equipment condition must be clearly presented and the operator's attention must be quickly drawn to the important facts. As an example, visibility of the vibration monitor display under all lighting conditions must be considered. Local temperature display devices must supersede obsolete and ill-maintained strip chart recorders. Eliminate the grouping of important alarms into common trouble displays in local panels.

Hardware reliability. Design of the recent generation of monitoring equipment tries to improve long-term reliability. However, digital devices and displays such as LCDs might be more sensitive to ambient temperatures than older, analog-type hardware. These devices require more attention to proper packaging than old-fashioned analog instruments.

Information system requirements. Machinery status messages should be integrated with other process information to provide rapid response by control supervisors ("single window to process" concept). Machinery data must be easily related to process data to facilitate troubleshooting. The same data must also be stored for long-term trending and history recall. In addition to vibration, machinery information must include parameters such as temperatures, speed, flowrates, and pressures. Furthermore, there should also be a high-speed (i.e., once/second) critical data stack to record machinery trips. Machinery data should be easily accessible to machinery specialists in their offices as well as from their homes. The latter feature should be of immense help in responding to 3 a.m. phone calls by the shift supervisor. ("The hydrotreater recycle compressor bearing temperatures went up again. What should we do now?")

The system must provide for data flow to a "number crunching" computer to allow for time-consuming optimization calculations that cannot be performed by the data acquisition computer. This could be another dedicated computer such as a mini-computer, engineering work station, or even a personal computer. Reliability of the data cruncher should not impact operation of the data acquisition computer. In other words, if the number cruncher crashes, it should have less impact than a process control computer crash. Hence, data should flow through the process control computer and then to the number cruncher.

Information that the system design must consider. The information system design should take into account hardware maintenance needs, impact of hardware obsolescence, and need for future software improvements. Respective areas of responsibility of the Computer Support Group and the Maintenance Department Instrument Group might need to be redefined to account for the growing amount of digital data acquisition systems located in the field. For example, Should a digital input/output device located near a compressor train be maintained by instrument or computer technicians? The answer might require additional personnel training or organizational changes.

System design must minimize initial installation costs (hardware and software). Trade-offs between custom design and off-the-shelf hardware, as well as choices between commercially available software versus in-house developed proprietary programs, will affect initial costs and future maintenance costs.

Data acquisition computer constraints must be considered. The process control computer is often limited in data acquisition rates and calculation capabilities. Even though all data do not have to be sampled at the same rate, the impact of adding 1000 or 2000 machinery data points on process control computer operation must be considered. Addition of data acquisition modules, disk drives, and computation modules might be required. Use of a separate computer for data acquisition presents a host of other problems:

- Cost of an extra computer, system installation, and software development.
- An additional computer to maintain (hardware and software). Traditionally, the machinery data acquisition computer gets a much lower maintenance priority than the main process control computer.
- Interfacing between operating systems.
- The additional difficulty of events that are tracked by separate computers.

Routing of data communications is no trivial issue. Several options are available to bring data from the field to the computer. These include underground versus overhead wiring. The latter is cheaper, but more likely to be damaged during a fire. This approach is acceptable only for "informational" data, not data needed by control loops. One option is to install hardwired underground data cables from the control house to the machinery platforms. This is usually an expensive option that can be justified only in conjunction with other projects. For reduced costs, consider adding only the cables required by a bus system. The bus can be controlled by dedicated programmable logic controllers (PLCs). Fiber optics, instead of conventional copper cables, should be evaluated for technical feasibility and cost. Fiber optics are not affected by radio frequency interference and can be routed in common trenches with power cables.

Local panels: Monuments to the designer's incompetence! On most machinery platforms, the local panel (if available) has been perversely designed to be out of the operator's sight when the machine is started up! Turbine-driven trains, in particular, require that the local panel be located near the turbine startup controls to provide the information needed for a safe turbine startup. Experience has shown that some wrecks could have been avoided if the operator had been aware of the high vibration data displayed at the other end of the platform. Repeater-type displays should be added near the operator station for variables such as turbine speed if the main panel digital speed displays cannot be easily seen from a distance.

Sometimes the local panel is located near the platform stairs but away from the compressor train. This design might sound wonderful to save steps for the operator when he or she goes up to look at the panel. However, this design also prevents the operator from walking alongside the machinery train and noticing problem signs such as oil leaks or burned paint on a bearing cap, feeling the floor vibrate, hearing a gear screech, or smelling the H_2S, which might indicate a leaking compressor seal!

Plants where all the information is only in the control room have in effect taken responsibility and ownership away from the operator! It implies the belief that all the important information is in the control room. The result can demotivate the operator. A local panel must be located where the operator can see all relevant information at a glance during critical times such as machine startup or an emergency.

Local panel design should take ambient lighting conditions into account. Electro-fluorescent displays are difficult to read in bright light, even if the panel is in the shade, because the human pupils sense ambient lighting and close down. On the other hand, LCDs require good lighting to be seen. Viewing angle is also critical. Thus, display elevation off the floor is critical to allow viewing by people from 5 to 6.5 feet tall! Local panel instruments should be designed to permit fast identification of each variable. This means that each variable should be displayed by an instrument that gives immediate clues to its function by shape, type (analog or digital), location, and color. In other words, all local panels should have a similar layout and identical instruments (which also simplifies maintenance). The following guidelines are suggested:

- Speed read-out should be digital (near the operator station). Consider also an analog speed display on the local panel (which is easier to track during turbine acceleration).
- Temperatures should use digital displays.
- Flows should be in engineering units.
- Pressures should use analog gauges arranged in process order.
- Vibration monitors should be laid out in the same order as the equipment and gaps should be used to provide grouping.

- Labels should be in plain language (avoid abbreviations if possible).
- Annunciator panel alarm lights should be grouped by function.
- First-out and reflash should be available.
- Use standard 4–20-mA line powered displays.

In addition, design of new panels might have to provide cooling required by the displays and heat loads of the new generation of digital systems (e.g., compressed air-powered vortex coolers). Selection of display devices and instruments must, of course, be done after review and approval by the instrument maintenance group. Not only will these specialists offer valuable feedback about potential application problems, but their "buy-in" is crucial. A device bought without that stamp of approval might not work out as well as the one favored by the service personnel.

Selecting Vibration Monitors

The advent of a new generation of lower cost modular monitoring systems requires a re-evaluation of equipment selection practices. The new designs offer more flexibility in setting options, are claimed to be cheaper to maintain and more reliable, and have new features such as computer interface ports and digital displays. The following points should be considered during selection:

- Compliance with API Standard 670 requirements. Some offerings might reflect a design philosophy that does not meet plant requirements such as shutdown voting logic implementation.
- Ease of operation of the monitors is crucial. Often, operators are unable to access needed information (such as switching to another vibration channel) because of an unfriendly design.
- Equipment reliability, maintainability, ease of installation, and area classification are also factors to be reviewed.
- Ensure that there is full compatibility between monitor design and third-party sensors that you might want to integrate into the system (such as high-temperature accelerometers).

Equipment installation: The neglected child. API Standard 670, third edition covers many of the details to be considered during vibration and temperature monitoring system installation. Sections 3 and 4 and Appendix C offer many important installation pointers with one important omission: How to locate electrical conduit and junction boxes without blocking access to the machinery for maintenance and operation! A common error is to rely on draftsmen to determine electrical conduit routing. Often, conduit and instrumentation tubing are left to be field routed, leaving the plant at the mercy of a contractor who only knows that the cheapest and shortest path from one device to the other is somewhere between ankle and waist height! A few hours of careless installation can badly compromise access to a machine.

Vibration system shutdown philosophy. One of the most consistently asked and debated question is how to set vibration shutdown limits. A variety of well-known approaches exists.

For turbomachinery shaft vibration:

- A commonly accepted method is to set the trip at a value somewhat less than bearing clearances (say 80%).
- Base the trip setting value on machine speed and the degree of concern. For example, consider a high-speed packaged plant air compressor running above 40,000 rpm that normally operates at 0.24–0.30 mil p-p (6–8 microns p-p). This machine can wreck at 1.5–1.6 mil p-p (38–41 microns p-p) and, therefore, a trip setting of 1.0 mil p-p (25 microns p-p) might be reasonable.
- Determine a normal operating range and use a multiple of it as a trip setpoint. For example, if a machine runs at 1.0–1.4 mil p-p (25–36 microns), use a × 4 multiplier and trip at 4.0–6.0 mil (100–150 microns p-p).
- From operating records and problem information, find out what the machine can survive and use this as a guideline. In the above case, one might set the trip at 5.0–7.0 mil p-p (125–175 microns p-p).

As a matter of fact, there is no single shutdown setpoint that protects against all eventualities! One must consider the failure mode. As an example, consider a compressor that can survive a 6.0-mil p-p (150 microns p-p) vibration level caused by an oil whirl. The same compressor will rub and wreck itself in two or three minutes with less than a 4.0-mil p-p vibration level if it is running at its first critical speed. This, of course, is because at resonance the shaft bow results in mid-span vibration levels well in excess of labyrinth seal clearances! The logical conclusion is that what we really need is a "smart shutdown" system capable of performing an overall data review, including vibration analysis, before tripping a machinery train.

For bearing cap and casing vibration (absolute vibration). Raw acceleration signals are not recommended for gear box shutdown. Acceleration signals should be integrated to velocity if a high vibration trip is desired. Location of measurement points for gear boxes should be carefully selected. The previous comment also applies here. Vibration level can vary widely with failure mode. Very slow grinding of the gear teeth can result in substantial noise and vibration levels (well in excess of

50 G pk). Yet, it might take weeks to achieve significant damage. Thus, the decision to shutdown can be made at leisure. A broken gear tooth might result in vibration levels only one-third as high yet require immediate shutdown before the gear case gets distorted when a tooth fragment is caught between the rotors.

Vibration shutdown acceptance. A real problem might still exist even after installing a modern machinery monitoring system. Plant management or individual process unit owners might be reluctant to commission the vibration monitor danger/shutdown feature. This reluctance is based on the fear of nuisance trips and a justified concern about the impact of a sudden machinery shutdown upon the process itself. An automatic machinery trip also takes the shutdown decision away from management, perhaps the most difficult hurdle. Ultimately, total cost of a machinery train failure (including process losses caused by an extended shutdown) should be the major factor that brings about acceptance of the danger/shutdown feature.

The secret ingredient: Operator cooperation. The system should be designed with front-end input from the operators. They are the ultimate customers and, thus, they must "buy-in" and it must serve their purposes.

- Strive to achieve "operator friendly" facilities to ensure that people in the field use the tools effectively.
- Train all operators during system commissioning. Listen to their comments and use this feedback for further improvements.
- Convey the message that the monitoring system is for their benefit first and that it is not a toy for the engineers only.

Consideration of these points may lead to cost and effectiveness optimizations that might place your plant in the "pacesetter" category.

A Spare Equipment Policy is Needed

How should you treat your spare machinery? We should treat our spare machinery conscientiously. Our company has invested a considerable amount of capital in spare equipment and it would stand to reason to take good care of this asset. So what? We often find that operators do not pay attention to their spared equipment with the result that standby equipment ultimately is not available when needed in an emergency. Practices around spare machinery installations can range from total neglect to following standby or spare operational policies as described in Table 15-1.

Table 15-1
Spare equipment operating policy choices

Stand-by policies	Changing on failure	Assigned spare	Periodic change (swinging)
	Machine B starts running on failure of A and stays running until it is its turn to fail	Machine B does not run until A has been taken out for repair	For example every 2 weeks or 15 days
Advantages	• Not many "maneuvers" • Good MTBF* if the spare stays is in good condition and switching is without risk • No start-up unless the machines have been repaired like new	• One strives to always have a spare in good condition	• If well managed, the chance of having an unscheduled outage is small
Disadvantages	• Start-ups often take place in a "catastrophe" scenario • The spare machine could degrade	• The spare must be kept in good condition	• Multiplies various risks of swinging machines • Risks having two failures in a short-time frame
Improvements	• Test the spare periodically	• Try out the spare periodically	• Before each change, "perform diagnostics," i.e. condition monitoring • Run unequal time intervals

* MTBF = Mean time between failure

All the above policies have advantages and disadvantages. No matter what, a common trait is the need for judicious periodic servicing or exercising of the spare; there is just no other way to ascertain the condition of a "fail-to-danger" device such as a critical standby pipeline pump, an electrical emergency generator, or a spare steam turbine.

We prefer a machinery standby policy as described in the fourth column of Table 15-1. Our basic approach, however, is that simple equipment such as general purpose centrifugal pumps in a standby condition do not deteriorate provided the bearings are kept coated with lube oil. Such a coating can be applied either by rotating the shaft at approximately weekly intervals or by oil mist. The shaft may be rotated by hand, by steam if it has a turbine driver, or by a single bump if it has a motor drive. Of the three we favor hand rotation where this is practical. Frequently, it is neither necessary nor desirable to bring the unit up to full speed due to the inherent risk of damage at start-up. On the other hand, we feel that on machinery units equipped for automatic start, the advantages of periodic testing the full instrumentation loop outweigh the potential damage.

What are the consequences of not looking after our spares in regular intervals? Besides an obvious inherent unreliability, we have frequently seen that some sites lack confidence in their running "spares" after the "main" machines have failed. As a consequence, maintenance forces are being expedited for a quick and often compromised repair. A vicious circle! It consists of hasty fixes followed by short running times causing a high percentage of priority work orders and overtime in the repair shops. Earlier we defined the relationship between machinery MTBF, unspared running time, and availability.* For example, in order to maintain an availability of 99%, a spared centrifugal pump that has failed in a population with an MTBF of 16 months could stay out of service for 5 days – more than enough for a thorough repair. With other words, there is only a 1% chance of loss of service during the indicated outage time.

Why are existing standby policies sometimes not followed? We believe that the answer is not so much operator reluctance than a lack of built-in accessibility and operability. Many spared equipment arrangements, where spare exercising is indicated, require complicated swinging maneuvers which may result in a unit upset or even a loss of production. Technical management must recognize these situations and change the design or provide whatever hardware is needed to make the testing of spare equipment possible if not just easier. For example, in a critical spared centrifugal pump installation, where swinging pumps would cause a "bump," one should think about installing controlled bypass lines or even smart pump technology to assure spare pumps can be started up without impacting unit operation.

Here are some more improvement ideas. Personnel responsible for machinery reliability should strive to establish operational confidence by enforcing or, where a policy does not exist, by introducing a spare machinery care and exercise program as part of the plant's PdM and

---

* Refer to Chapter 5, p. 83

monitoring concept. Why is PdM involved? The bottom right hand box of Table 15-1 points to a very important activity connected with critical spares exercising. There is a need for condition monitoring of both units, of the running machine about to be shutdown, and of the standby machine as it is being brought on line. Condition monitoring would mean the deployment of suitable weapons in our PdM arsenal, for example vibration analysis and performance parameter checks, such as motor current, process flow, suction and discharge pressures on a critical pump. This condition assessment will tell us whether or not we are shutting down a good piece of equipment and bringing on a similarly healthy standby machine. We are trying to answer the question: Does either of the units require maintenance attention at this time? The foregoing is an example showing how O&M personnel can work together by exploring the connection between equipment standby practices, reliability, and maintenance costs.

References

1. del Luz, D. and LaLonge, L., Making it safe: process safety management in Canada, *Engineering Dimensions*, May/June 2001.
2. Bloch, H.P. and Geitner, F.K., Series practical machinery management for process plants, Volume III, *Machinery Component Maintenance and Repair*, 2nd Ed., Gulf Publishing Co., 1990.
3. http://www.barringer1.com/cour.htm.
4. http://www.barringer1.com/may01prb.htm.
5. Operating Engineers Act, Province of Ontario (Canada), www.tssa.com.

Chapter 16
The maintenance environment

Introduction

How to Achieve Quality Machinery Maintenance

Mentioning maintenance was unavoidable in several of the preceding chapters. Maintenance was mentioned in conjunction with avoiding routine work and optimizing routine work. The term was also used in conjunction with quality. Indeed, as in many human endeavors, the quality of our activities should take precedence over quantity. In this regard, machinery maintenance is no different from maintenance of any other equipment; the quality focus must be accomplished through preventive maintenance.

The needs of quality operations demand uptime and trouble-free functioning of production equipment. To achieve these requirements, it has long been recognized that maintenance activities designed to anticipate and avoid failure have been, and continue to be, a sound investment in the overall maintenance strategy. Many companies are investing in software programs to improve or optimize maintenance strategies. Yet, no computer system can help a maintenance department unless the basic elements of a preventive maintenance program are in place.

In both small and large facilities there are identifiable components of the preventive maintenance program that can be generally described by 7 elements:

1. facilities management;
2. inspection routines;
3. predictive or diagnostic activities;
4. integration of maintenance within the production activity;
5. insurance activity;

6. corrective activity;
7. continuous improvement.

Facilities Management. Simply stated, each significant piece of equipment and its related components are identified uniquely and logically, in order that all maintenance activity can be related and selected history maintained. Once each piece of equipment has been identified, equipment manuals should be acquired. Consultation of these documents together with the maintainer's own equipment knowledge will assist in developing quality preventive maintenance. Machinery maintenance classifications are shown in Figure 16-1.

Inspection Routines. Establishing quality routines and meaningful frequencies requires thorough study of the manufacturer's documentation, consultation with manufacturer's representatives (where necessary), and careful study of the equipment environment. Once this has been completed, the established routines and frequencies must be subjected to ongoing refinement and adjustment. It is important that the routines ensure specificity and, where possible, quantifiable measurements should be utilized. For example, changing product quality, pressure temperature, vibration, and noise are all relevant indicators for the analysis of equipment health. This approach coupled with visual checks by experienced personnel will contribute toward a quality program.

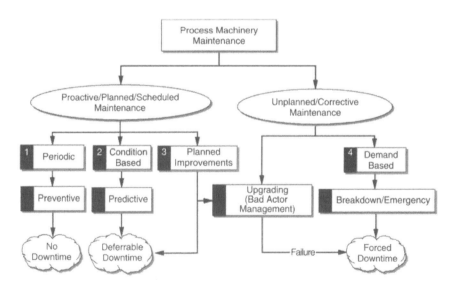

Figure 16-1. Process machinery maintenance classification.

This philosophy of specificity will help to control excessive frequencies which contribute to significant, unnecessary costs. A systematic approach toward the preventive maintenance activity will help to decrease preventive maintenance costs which can be as much as 30% of the overall maintenance labor cost.

Predictive and Diagnostic Activity. Significant results are achieved through quantifiable preventive maintenance. Recognition of these results has created a demand for technologies to meet preventive maintenance needs. A quality program should include one or more of the following diagnostic techniques:

- vibration analyses
- thermographic techniques
- spectrographic oil analyses.

If these techniques are not in use, a re-evaluation of progress in the preventive maintenance activity may be in order.

Integration with the Production Strategy. Most process operations have recognized the need for a complete integration of the maintenance activity within the overall production strategy. Scheduled periodic shutdowns, based on maintenance needs that have been identified through preventive maintenance, are mandatory. It is essential that scheduled equipment downtime be available in order to achieve the benefits from the corrective activity of preventive and predictive maintenance.

Insurance. Spare parts and repair facilities play a significant role in the preventive maintenance strategy. Facilities knowledge of what needs to be done and the scheduled shutdown to get it done are only half the activity. A balanced capability must be in place to ensure the corrective repair. Repair facilities and spare parts must be evaluated to ensure the right balance. Statistical analysis of failures and preventive maintenance findings can, and do, play a significant role in determining the right mix of inventory levels and repair facilities.

The Corrective Activity. Are your reliability professionals working on fully quantifiable cost-benefit projects? An excellent base for prioritizing the maintenance engineering activity, for example, are history records containing information on the facilities management activity, quantifiable cost data from preventive maintenance, and other maintenance activities. Sound, corrective designs (while requiring significant investment) can achieve enormous cost-benefits in parts and labor.

Everyone should evaluate their present programs within the framework discussed here. You may very likely have a quality program. If not, formulate a plan for change. The results will be gratifying and cost-effective.

Asset Management: The Essential Foundation for Plant Optimization*

Introduction

In the competitive world of production enterprises, all participants recognize the necessity of improving performance, effectiveness, and shareholder value. Many have tried to increase profit by cutting costs. Those who have followed this course eventually find that degrading performance overruns the temporary effects of cost reductions. The real solution is a process of improvements, specifically targeted to increase financial performance. Directed toward optimizing the performance and effectiveness of capital assets, the process is popularly called Asset Management.

Today, for example, most power generating utilities are actively considering the necessity and benefits of Asset Management to derive greater profit and return from capital assets. Some recognize that Asset Management is a competitive imperative and are implementing programs to improve asset utilization and effectiveness. An enlightened few understand that an Asset Management process is the foundation and key ingredient for success in many other optimization initiatives including production scheduling, supply chain (logistics), and maintenance management.

Automation Research Corporation (ARC) continually points to their concept of Asset Management (Plant Asset Management [PAM] and Enterprise Asset Management [EAM]) as a necessity and mutual opportunity for industrial production enterprises and suppliers.

The challenge is that Asset Management, as a business process, is not yet well defined. In addition, assembly and organization of the practices and technology to gain success is even more obscure. Some view the current situation as just another of many buzz word programs that eventually will be abandoned to obscurity. Others see the necessity, real value, and an opportunity to define the term and process, and lead industry with a "must have" solution.

Many potential providers of Asset Management solutions, notably process control, maintenance, and management information system suppliers, seem to be following very narrow strategies to exploit the opportunity. None appear to be addressing full requirements for gaining optimal performance, effectiveness, and value from the broad range of equipment

---

* Contributed by John S. Mitchell.

and structures that comprise capital assets in a typical production facility. Many of the suppliers seem constrained by a business culture that views opportunities only as an extension of their core business and fails to recognize significant changes that may be required to meet real market expectations. None appear to be pursuing the comprehensive Asset Management process that will cause a prospective customer to exclaim, "that's exactly what I've been thinking about and must purchase!"

A number of service providers have introduced offerings that are promoted under a heading of Asset Management. As in the case with the process control, information and maintenance management suppliers, most of these offerings reflect the suppliers' culture and main business. All lack the range and scale needed for a comprehensive Asset Management solution. Likewise, there are a number of smaller system and service suppliers offering products and services under a banner of Asset Management. For the most part, all are partial solutions that address one aspect of Asset Management but not the comprehensive whole.

Opportunity

With a clear, unfilled market demand there is an opportunity to develop, introduce, and implement a comprehensive Asset Management process that will gain maximum value from capital assets. The Asset Management process and all implementing practices and technology must be tightly linked to financial performance with every improvement initiative directed to gain a return that will be highly attractive to management and financial executives.

In the following, we are going to attempt to demonstrate how Asset Management contributes major value to a customer's bottom line and outline a solution that will be highly attractive to prospective customers. While the process and principles of Asset Management contribute substantial value during the full lifetime of capital assets, we want to address only the operation and maintenance phase of asset lifetime. We assume that the justification, design, procurement, and installation phases are completed for capital assets and the end-of-life, retirement phase is still well in the future in a typical operating production facility.

Definition

Asset Management and the resulting asset optimization are essential elements of business improvement. It is fair to state that a production enterprise cannot gain full business value and return without an effective Asset Management program.

In most modern production enterprises, Asset Management, and the resulting asset optimization, offer the largest, perhaps the only, remaining opportunity to increase business value, return, and profitability.

Asset Management establishes the optimum balance of production, cost, and capital effectiveness for a given set of market conditions and production capability. A balance that will assure fulfillment of delivery commitments, maximize the ability to exploit market opportunities (e.g., spot market sales), minimize risk of accidents and lost production – at an optimum cost.

Elements of Asset Management

As illustrated in Figure 16-2, Asset Management makes a strong contribution to business optimization. Production operations, cost and capital effectiveness and risk management all gain significantly from Asset Management and optimization.

Production Effectiveness. Production effectiveness, a primary value producer within Asset Management, is measured in terms of the familiar OEE,

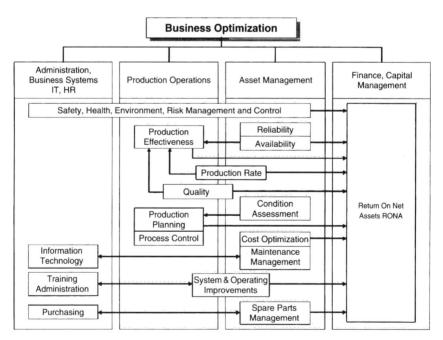

Figure 16-2. Asset Management as an essential component of business optimization illustrating cooperative elements and contributions.

Overall Equipment Effectiveness. Of the three terms that make up OEE, availability and production rate are key contributors to production effectiveness. (The third term, quality, is more a function of production operations than Asset Management.) Reliability, an essential product of Asset Management, produces availability. All four elements make a strong contribution to financial results (produce more from the same assets).

The condition assessment and lifetime estimation elements of Asset Management (described later) are invaluable for production planning. Is capacity available at a defined future date to deliver a commitment on time and cost?

In addition to a strong influence on availability and production rate, Asset Management improves operating effectiveness through system, component, and procedural improvements. Improvements are implemented by training and improvement projects that make systems more efficient, easier to operate, and less prone to human error. In these three roles, Asset Management makes a large and powerful contribution to the value generated by production operations.

Cost Effectiveness. The cost-optimization element of Asset Management is primarily directed to improving the maintenance component of O&M expenditures. Within the Asset Management process a concept of business-centered reliability is directed to optimizing expenses per unit production considering market, delivery, cost, and quality considerations. The approach to optimizing maintenance costs demands an optimum level of reliability for each production asset and system based on market and delivery requirements. Optimum reliability is achieved by eliminating/minimizing problems that cause downtime and drive maintenance costs combined with providing both information and the technical means to safely extend overhaul intervals.

Increasing production with costs held essentially static is a primary objective of Asset Management. Meeting this objective in a market capable of absorbing additional production with no erosion in price improves production effectiveness by reducing costs per unit output. There are stories in the literature where industrial producers following a process equivalent to Asset Management have achieved as much as a 60% increase in production with a 10% increase in O&M costs.

Cost effectiveness is improved by an array of comprehensive improvement initiatives including:

- Prioritized RCM to develop and implement an optimized maintenance process emphasizing condition assessment and condition based maintenance (CBM).
- A structured reliability improvement program including RCFA.

- Improved maintenance management including an optimum computerized maintenance management system (CMMS), optimized PM, and improved planning and scheduling.
- Optimum spares management system including outsourced spare parts and optimum stocking levels.
- A system of performance metrics to assess improvements and identify financial return.

Capital Effectiveness. Asset Management improves capital effectiveness defined as greater output and profit from the same assets. Improving availability and production rate often delays investment requirements for additional capacity in an expanding market. Capital effectiveness metrics such as return on net assets (RONA) and return on capital employed (ROCE) demonstrate the improvement in performance.

Optimizing stores and spare parts (logistics) management (mentioned in the last section) is another area where an Asset Management program improves capital effectiveness. Reliability improvements gained through RCFA and life extension initiatives allow a safe reduction in spare parts stocking (reduced capital). With adequate warning of requirements, reducing capital by outsourcing spare parts is far more practical with greatly reduced risk. Extending the warning period before work commencement or an outage greatly improves the effectiveness of the maintenance planning and scheduling process. It also allows least cost delivery of repair parts through normal channels with minimal expediting (cost effectiveness). (It has been stated that the best way to improve the effectiveness of the planning and scheduling process is by reducing the number of work requests, particularly emergency requests that utilize time and resources so inefficiently.)

Reduced Risk. Finally, effective Asset Management greatly reduces the financial, safety, and environmental risks associated with a production facility. Everything associated with production operations becomes more stable and predictable.

Condition Assessment. Within the Asset Management process, condition assessment is an essential core practice. The cost benefits of condition assessment and CBM are well known – greater than a 50% improvement in cost effectiveness compared to reactive, run-to-failure maintenance. Far more important, condition assessment identifies problems that could lead to an outage in adequate time to minimize safety and environmental risks as well as the financial consequences of outright failure and lost production. The ability to accurately predict capacity to meet future delivery requirements and safe operating lifetime is a key, if less well recognized attribute of condition assessment. This aspect of condition assessment can potentially contribute far greater business value than the ability to reduce costs.

Process Overview

Asset Management is a top-down process directed to improving business performance, effectiveness, and profitability. All subsequent initiatives and actions for improved process, technology, and practice connect and contribute to this overall objective.

The Asset Management process begins with an audit of current conditions to identify potential improvements in the three primary areas of Asset Management: production, cost, and capital effectiveness (Fig. 16-3). In each area, current performance is compared to industry benchmarks. The difference or Gap represents opportunities for improvement. Improvement opportunities are valued based on contribution to business results for current and anticipated market conditions and prioritized for implementation.

Audit Process

The initial Asset Management audit is conducted to identify specific opportunities for improvement in the areas of production, cost, and capital effectiveness. The initial audit may be conducted on an entire enterprise, a single facility, or even an individual production unit.

Assessing audit results requires a financial model capable of attaching a valid business value to each opportunity for improved effectiveness. An example of a financial model may be found in [1]. Overall Asset Management metrics to be compared to industry benchmarks for an indication of current performance, identification, and financial assessment of opportunities for improvement include:

- Production effectiveness
  - Heat rate, including average heat rate as a percentage of best attained.
  - Availability, percent, variations in month-to-month availability over the past 2 years.

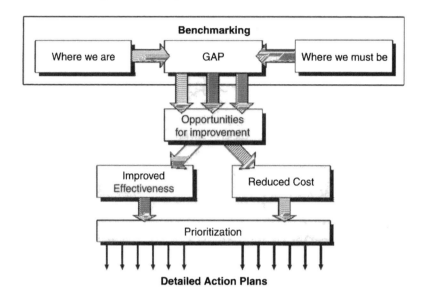

Figure 16-3. The Benchmarking Process.

- Production rate, percent of maximum sustained, variations in month-to-month production rates over the past 2 years.
- Operating and maintenance cost per unit output; MW, ton, barrel, etc., variations in month-to-month costs over the past 2 years.

- Cost effectiveness

   - Maintenance costs per unit output.
   - Preventive and CBM as a percentage of total maintenance.
   - Planned (1 week ahead) as a percentage of total work.
   - Overtime as a percentage of total time worked.
   - Percentage rework (repeat work within 2-month period).

- Capital effectiveness

   - Spare parts value as a percentage of, for example, replacement asset value (RAV).
   - Work delays due to lack of parts.

Identifying Improvement Opportunities

Values of the preceding measures of performance are subjected to a business analysis. The business analysis is directed to determine the greatest opportunities for increased value that can be gained by an ambitious, but attainable, percentage improvement. It must include market conditions. For example, increased availability has full value only when the extra production can be sold at full price. If the reason for reduced availability is lack of demand, full value cannot be credited for an increase. The same holds true for production rate. Only de-rating caused by a mechanical or operating problem, where the lost production could be sold, counts toward additional value.

The general opportunities for improvement identified in the audit must be further refined to identify individual projects. Using availability and production rate as examples, what are the specific causes for outages and de-rating? A Pareto analysis is very useful for this task and provides an excellent tool for visualization. In order to conduct the analysis, the exact causes of outages and de-rating as well as the time consumed must be either known or developed from plant records. In most cases two Paretos will be required. The first to separate cause of lost availability by system, i.e. pulverizer, boiler, fans, condensate, feed, generation, balance of plant, etc. From there, the cause should be further refined to specific equipment and finally root cause.

In evaluating potential improvement initiatives, it is necessary to look at both absolute values and statistical variations. For example, annual maintenance costs that are essentially equal month-to-month lead to quite different conclusions and opportunities for improvement compared to essentially equal annual maintenance expenditures that vary greatly from month to month.

From the perspective of selling Asset Management within an operating organization, each improvement must connect to an improvement in asset effectiveness measured by RONA, ROCE, or some other capital effectiveness metric. In other words, proposed improvements must be shown to deliver business results!

Strategy for Selling Asset Management Technology and Products

In our opinion, selling asset optimization products as a fully integrated, comprehensive Asset Management solution offers the greatest opportunity for success and maximum sales revenue. With customers increasingly demanding a comprehensive solution, a single source supplier capable of meeting the broadest range of requirements (the SAP model) has substantial advantages over suppliers with a more limited offering who must "knit" questionably "integratable" components into a foreign system.

Individual optimization products may have established markets, an annual sales volume, and predictable growth. In the author's opinion, the only way to significantly improve established trends is to bundle the optimization products into a comprehensive Asset Management solution that is clearly superior to competition who appear focused on segments rather than the whole. Properly executed, a mutually reinforcing strategy of this type should gain results that are substantially greater than the sum of the parts.

Overall Strategy

The proposed strategy begins with a detailed, comprehensive audit process to identify and assign full business value to primary improvement opportunities. For maximum credibility with a prospect, technology, product, and service, proposals should all connect to gaining an audit-based business value. For example, condition monitoring, CBM and improved planning and scheduling are justifiable to improve specific gaps in a facility with maintenance costs above benchmark values. Likewise, process control optimization might be justified in a facility with greater than benchmark heat rate and/or sub-optimal control indicated by excessive variations from an average.

This leads to a second element – the requirement to provide linked products and services for the strongest offering. Again, using condition monitoring as the example, products alone are not sufficient. Many who have purchased state-of-the-art condition monitoring equipment end up with failed programs due to inadequate post-purchase priority and support. To assure a level of success that will provide enthusiastic user recommendations, some entity, presumably the supplier of condition monitoring instruments and technology, must be able to establish and commission a CBM program, train the purchaser to take over operation, and monitor results to assure greatest value recovered. (There are major benefits for a supplier. In addition to invaluable user recommendations, the supplier gains valuable knowledge from the process that is highly useful in product marketing, design, and improvement.)

The ability and processes necessary to establish and implement programs of RCM and RCFA to organize and train in CMMS, planning and scheduling, and other maintenance optimization methods are imperative for the strongest and most attractive solution offering.

Audit Process

A solid audit process, capable of determining a prospect's primary requirements and attaching credible business value to reasonable improvements, is crucial to the success of an Asset Management offering. This is the first stage in the engagement with a prospect and must simultaneously demonstrate an attractive business return as well as a highly credible capability for implementation (prospect is convinced the supplier can deliver expected results). In the current economic climate, a prospective Asset Management supplier must be willing to develop, propose, and accept objective-oriented risk and profit-sharing contracts.

Thus far, this paper has implied a single audit – in fact there should be two audit procedures. In addition to an in-depth audit procedure that might take several experienced people 2 or more weeks to gain a detailed understanding of a prospect's operations, there is a requirement for an abbreviated "quick look" procedure that will provide a good indication of opportunities with minimal investment for both the prospect and potential supplier. The "quick look" audit could rely on a review of public information plus a simplified subset of comprehensive audit questions that the prospect could be asked to answer. This information would qualify the prospect and determine whether business opportunities justified a full, comprehensive audit. Also, a pre-audit to assess opportunities might make a paid full audit more attractive to the prospect.

Proposed Implementing Strategy

In addition to the audit process described earlier, the winning implementing strategy must include the ability to bring, install, commission, and possibly operate advanced methods, practice, and technology to a prospect – all prioritized by financial return. Asset Management methods, practice, and technology include:

- Financial model of the prospects business.
- Process, including heat rate, optimization technology, and system.
- Reliability centered maintenance.
- Total productive maintenance for OEE methodology and calculation. Note that some elements of Six Sigma provide a more rigorous and detailed process for evaluation.
- Root cause failure analysis.
- Computerized maintenance management system.
- Condition assessment and CBM technology and methods.
- Optimum planning and scheduling methodology.
- Optimum stores management process.

It is very attractive to offer one or more automated expert systems to supplement a prospect's in-house expertise. Potential automated experts include rotating machinery condition assessment, problem diagnosis, and lifetime estimation and process optimization systems. Expert systems must be promoted and sold on the basis of assisting human experts and making them more effective – otherwise the very people within a prospect who should be enthusiastic supporters of an Asset Management initiative will feel threatened and may impede its adoption.

The implementing strategy must address a cost-effective method for integrating measurements and information from existing plant systems as well as adding additional measurements and information required by the Asset Management system. The MIMOSA–OPC interface should facilitate the integration task. All systems in an Asset Management structure must be fully integrated with a common operating "look and feel."

Suggested Action Plan

1. Select a broad-based solution-oriented team with in-depth knowledge of Asset Management and machinery condition assessment as well as requirements for control optimization and condition monitoring methods:

   - Develop a matrix list of all capabilities and products required to offer a comprehensive utility industry asset optimization solution.
   - Identify products and applicable expertise that are available immediately and how they could best be integrated into a comprehensive Asset Management solution.
   - Prioritize requirements for immediate product integration and medium-term process and product development to fill gaps and satisfy the majority of utility requirements for asset optimization.
   - Formulate a strong product, sales and marketing plan to capitalize on current strengths, market position and broad range of asset optimization products.
   - Gain approval for integration and development work necessary for immediate sales.

2. Formulate a strong, value- and benefit-oriented marketing message and plan that features a broad range, comprehensive solution:

   - Emphasize broad capabilities, especially those that point up weaknesses of potential competitors, e.g. value of full integration, range of capabilities compared to other likely offerings.
   - Showcase and promote the flagship methodology and products, demonstrate how they produce value at the core of a utility Asset Management solution.

3. Form a team to develop, test, and "procedurize" a utility audit process:

   - Develop an audit process directed to uncover opportunities for product and technology insertion.

- Test the initial procedures with established utility customers who are willing to critique the process and results.
- Prepare detailed audit check offs, comprehensive instructions for use.

4. Introduce (roll out) the Asset Management solution:

- Identify high priority prospects; known interest in Asset Management open to improvements in reliability, maintenance management, and condition assessment.
- Provide marketing and sales support.
- Train added personnel to assure quality, consistent results.

5. Develop support agreements with specialized third-party suppliers as necessary.
6. Promote the Asset Management solution through papers and articles:

- Joint Power Generation Conference.
- EPRI utility conferences.

About Maintenance Strategies

Back in the early 1970s, we listened to the maintenance consultant responsible for the ideas in Figure 16-4. This was a good while before we heard about RCM and all the other "M's." After the seminar we considered this concept of "TPM" not exactly revolutionary but as something we could take back to our plants and use as a strategy which has served us well ever since.

In reviewing the decision flow sheet shown in Figure 16-4, we start with *step number one* which reminds us to register all defects and failures around our equipment in order to come up with the best counter measures.

The *second step* is to test each failure as to its normalcy. Here we would consider normal the attainment of a MTBF benchmark set either by ourselves or by others. A good example would be a mechanical seal for a pipeline pump application. For instance, we hear about MTBF values of some 6–8 years for mechanical seals in hydrocarbon (HC) service.

If the seal failure is considered normal with respect to a benchmark MTBF, we check in *step three* if we could possibly extend seal life by PM. Now, there is a large body of evidence that suggests that well selected and designed seals tend to exhibit random rather than time- or cycle-dependent failure behavior. Therefore, they do not respond well to PM measures. The old maintenance adage "leave well enough alone" applies. This is contrary to what some other experts seem to convey. They

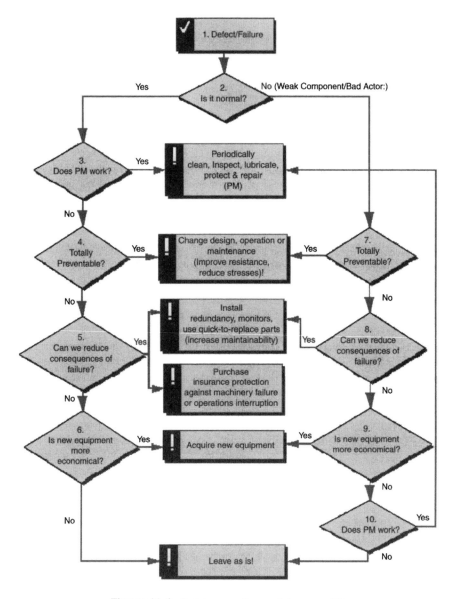

Figure 16-4. Total preventive maintenance [2].

prefer to use an index such as mean-time-between-planned-maintenance which suggests that mechanical seals should receive periodic inspections, overhauls, or other invasive maintenance attention. We would like to state that this is not being borne out by our experience in the HC industries.

If we have been visiting our mechanical seals periodically, it has been in 5-year or so turnaround intervals applied to major turbomachinery.

We would now come to *step four* in which we ask ourselves if the failure could be totally eliminated by altering the design, operation, or maintenance mode.

If we are convinced that we have already optimized these parameters, we proceed to *step five* that forces us to investigate the possibility of reducing the consequences of this failure that will occur regardless of what we have done. Here we consider protecting ourselves by such measures as introducing redundancy features, i.e. dual seals or spare pumps, monitoring devices such as incipient failure indicators, quick-change-out assemblies, i.e. cartridge designs; we could even think about taking out machinery failure and business interruption insurance.

After being unable to come up with any ideas in the previous step, we are faced in *step six* with the question as to whether different components or new equipment are more economical. If we do not see any way of justifying new and more reliable equipment we have come to the "bitter end" and must leave things as they are.

In case our example mechanical seal does not meet the normalcy test in step two, we go to *step seven*. This means that we are faced with a weak link or a "bad actor." We must test, as in step four, if we can eliminate it by the measures already described. To stay with our example, a hands-on failure analysis should lead to a thorough investigation of the suitability of the seal for the service it is in. Startup and operating procedures as well as maintenance and repair standards should be reviewed.

If step seven yields a no, we go to *step eight* to find out if the actions indicated in step five would work.

If we have not been able to find any possibilities to apply the measures mentioned in step eight, we investigate if a new investment is economical in *step nine*.

Arriving at *step ten* means that our only possibility for tackling these types of failures would be preventive maintenance measures as described in step three. Most likely preventive maintenance and increased vigilance would be of some help. However, they should be considered a last resort in the case of a frequently failing seal. If PM is yielding no results, we must resign and live with the problem.

How Often should we Inspect Our Machinery? We were discussing CBM programs with our maintenance first line supervisor. We wanted him to start using predictive tools such as portable vibration data terminals to be deployed on a regular basis – as an enhancement to inspection visits his mechanics were making based on their five senses around the plant's pumps and compressors. The response was "no, we do not need any sophisticated tools – we know when a pump is about to fail. Just let us do what we have been doing all along."

Rather than continuing the discussion, we decided to invite the representative of a company engaged in vibration monitoring on a contractual basis. When we started to discuss the frequency of potential rounds and data taking we were surprised that the contractor seemed to have no scientific or non-intuitive approach to determining this frequency. He indicated that his survey frequency was going to be around 4 weeks – this was how everybody else was performing periodic monitoring; this would suffice to catch equipment problems in time. We found this puzzling as it was quite obvious that some problem equipment, sometimes referred to as bad actors, would naturally require more frequent visits. The question arose, how do we set inspection frequencies?

It is not uncommon to see plants determine their equipment inspection frequency based on a criticality analysis. At first view, it may seem reasonable to use criticality of the equipment, but the foregoing discussion already indicated that there might be some other criteria that should govern inspection frequencies.

Let us suppose we are setting the inspection frequency for a simple component such as a V-belt. We assume the belt is part of a drive train for a fan associated with a large pipe-ventilated induction motor. The inspection method would probably involve the on-line use of a stroboscope which would reveal any wear progress. The motor and the fan are very critical according to the criticality study. Inspection frequency for the highest criticality score is often recommended to be one inspection every shift.

Intuitively, we recognize that it does not make any sense to inspect a belt every shift. Why? Because the inspection frequency must be based on how long, on average, it takes to develop a failure in a component. The belt in our example will not fail from one shift to another unless there is a completely random event. The most likely failure is that the belt will wear over a period of 6–8 months. We should therefore, determine the inspection frequency according to the component's time to defect limit or failure developing period (FDP) [3]. After estimating the FDP we will set the inspection frequency to *FDP*/2. In our example, we estimate an FDP of 8 months and set the inspection interval to 4 months.

Hopefully, the reader will have understood that our example pertained to a fairly time dependent, non-random failure pattern. There are, however, many pieces of equipment in our plants that fail in a random mode – just like the pumps and compressors we mentioned above. Here we must optimize our predictive maintenance program, and we return to the question of how to determine inspection or checking frequencies in face of predominant randomly occurring defects.

The answer could lie in using the MTBF value pertaining to the equipment population in question. In many operating units, this number has now become a rough yardstick for equipment reliability or sometimes even a key performance indicator (KPI). Let us recall from previous chapters that MTBF is a variable in the reliability function:

$$R = e^{-\frac{t}{MTBF}} \times 100 \qquad (16.1)$$

where R = reliability (%), e = base of the natural logarithm, t = mission time.

By calling *R* the capture rate (CR), i.e. the probability of coming across faults around our equipment covered by periodic inspection programs, and knowing our MTBF we are now able to determine the inspection interval, *t*, when solving Equation 16.1 for *t*. Thus:

$$t = MTBF \times \ln CR \qquad (16.2)$$

Figure 16-5 shows inspection intervals as a function of equipment MTBF and various CRs. It points us to the fact that, as our equipment's MTBF increases, we could also prolong inspection intervals. Here is an example. If our machinery population covered by predictive monitoring techniques had an MTBF of 36 months and we wished to have a capture rate of 95%, our inspection frequency should be 7 weeks. As we might improve the reliability of our machinery to, say, an MTBF of 45 months we are prompted to increase our inspection interval to 9 weeks. This also means, however, that all along there is a probability that we will not be able to capture 5% of the occurring problems; we are going

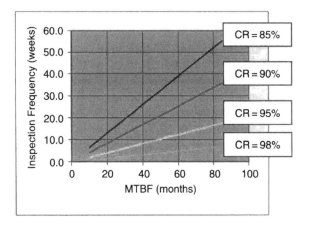

Figure 16-5. Inspection frequency versus MTBF.

to miss them because we are not there to inspect. This is borne out by the fact that we experience unexpected machinery break-downs in spite of periodic monitoring programs – much to our management's disappointment. Only continuous monitoring will reduce if not eliminate that risk.

Use Selective PM and PdM for Your Compressors. Perhaps you, too, are working for a company that is asking challenging questions or demanding certain implementation strategies for which they have not laid the necessary groundwork. Such seems to be the case at a well-known refinery. One of their staff wrote:

> Regarding process gas compressors, is it possible to use predictive monitoring and no longer perform preventive maintenance on set intervals? Management has stated that the previous method of performing periodic overhauls during planned turnarounds is not acceptable. We are asked to use state-of-art predictive equipment to determine when a failure will occur and then plan an outage accordingly.
>
> Maybe I was mistaken but I thought one did the periodic preventive tasks to ensure that the process would not be affected during planned run times. The state-of-art predictive routines can still be used to minimize the impact of a premature failure, or to prevent off-design operation such as improper rod loading.

To provide an answer to the gentleman's questions, we must direct our attention at a number of facts, conventions, and scenarios.

Preventive and Predictive Maintenance Explained. Preventive maintenance encompasses periodic inspection and the implementation of remedial steps to avoid unanticipated breakdowns, production stoppages, or detrimental machine, component, and control functions. Predictive and, to some extent, also preventive maintenance is the rapid detection and treatment of equipment abnormalities before they cause defects or losses. This is evident from considering lube oil changes. This routine could be labeled preventive if time based, and predictive if done only when testing shows an abnormality in the properties of the lubricant. Without strong emphasis and an implemented preventive maintenance program, plant effectiveness and reliable operations are greatly diminished.

In many process plants or organizations, the maintenance function does not receive proper attention. Perhaps because it was performed as a mindless routine or has, on occasion, disturbed well-running equipment, the perception is that maintenance does not add value to a product. This may lead management to conclude that the best maintenance is the least-cost maintenance. Armed with this false perception, traditional process and industrial plants have underemphasized preventive, corrective, routine maintenance, not properly developed maintenance departments, not properly trained maintenance personnel, and not optimized predictive maintenance. Excessive unforeseen equipment failures have been the result.

Correctly executed, maintenance is not an insurance policy or a security blanket. It is a requirement for success. Without effective preventive maintenance, equipment is certain to fail during operation. In today's environment, effective maintenance must be selective. Selective PM (selective preventive maintenance) results in damage avoidance, whereas effective PdM allows existing or developing damage to be detected in time to plan an orderly shutdown.

Compressor Maintenance in Best Practices Plants. Four levels of effective compressor maintenance exist. Although there is some overlap, the levels of maintenance are:

1. *Reactive, or breakdown maintenance.* This type of maintenance includes the repair of equipment after it has failed, in other words, "run-to-failure." It is unplanned, unsafe, undesirable, expensive, and, if the other types of maintenance are performed, usually avoidable.
2. *Selective preventive maintenance.* Selective preventive maintenance includes lubrication and proactive repair. On-stream lubrication of, say, the admission valve control linkage on certain steam turbines should be done on a regular schedule. In this instance, anything else is unacceptably risky and inappropriate.
3. *Corrective maintenance.* This includes adjusting or calibrating of equipment. Corrective maintenance improves either the quality or the performance of the equipment. The need for corrective maintenance results from preventive or predictive maintenance observations.
4. *Predictive maintenance and proactive repair.* Predictive maintenance predicts potential problems by sensing operations of equipment. This type of maintenance monitors operations, diagnoses undesirable trends, and pinpoints potential problems. In its simplest form, an operator hearing a change in sound made by the equipment predicts a potential problem. This then leads to either corrective or routine maintenance. Proactive repair is an equipment repair based on a higher level of maintenance. This higher level determines that, if the repair does not take place, a breakdown will occur.

Predictive maintenance instrumentation is available for both positive displacement and dynamic compressors. It exists in many forms and can be used continuously or intermittently. It is available for every conceivable type of machine and instrumentation schemes range from basic, manual, and elementary to totally automatic and extremely sophisticated. Not knowing the size of the questioner's compressors and if the owner employs such sparing philosophies as installing three 50% machines, two 100% machines, or perhaps only one 100% machine, it is not possible to make firm recommendations as the most advantageous level of monitoring instrumentation, shutdown strategies, etc.

However, PdM instruments are available from key vendors in the USA and overseas. We have dealt with some of them over the years; we recall, among others as shown in Table 16-1:

An Internet search will uncover competent manufacturers of monitoring equipment; some of these are discussed in *Reciprocating Compressors: Operation and Maintenance* (ISBN 0-88415-525-0).

Certainly, a predictive maintenance expert system can monitor machine vibrations. By gathering vibration data and comparing these data with normal operating conditions, an expert system predicts and pinpoints the cause of a potential problem. The trouble is that detecting vibration is different from eliminating vibration. An intelligent, but highly selective preventive maintenance program may lead to actions that prevent bearing distress and thus prevent vibration from occurring in the first place. Needless to say, a selective preventive maintenance program may well be a more cost-effective program than the program that waits for defects to manifest themselves. This fact establishes that a sweeping management edict disallowing all manner of preventive maintenance does not harmonize with the principles of asset preservation and best practices.

Table 16-1
Listing of PdM literature

Web deflection monitoring (Hydrocarbon Processing, "HCP," 10, '97)
Spectrum, FFT, and cepstrum analyses ("HCP," 9, '97)
Reciprocating compressor wear detection ("HCP," 12, '97)
Improve selecting valve unloaders ("HCP," 4, '98)
Melding preventive and predictive maintenance ("HCP," 8, '98)
Apply engineering quality control ("HCP," 9, '98), also third-party inspection ("HCP," 10, '99)
Impact-echo techniques determine foundation soundness ("HCP," 11, '98)
Start with a "clean lube program" ("HCP," 12, '98)
Infrared condition monitoring ("HCP," 9, '99)
Consider machine condition inspection systems ("HCP," 5, '00)
Machine scanning is coming of age ("HCP," 8, '00)
Upgrade your compressor piston rods ("HCP," 11, '00)

Traditionally, industry has focused on breakdown maintenance and, unfortunately, many plants still do. However, in order to minimize breakdown, maintenance programs should focus on levels 2 through 4.

Emergency repairs should be minimized to maximize uptime. Plant systems must be maintained at their maximum level of performance. To assist in achieving this goal, maintenance should include regular inspection, cleaning, adjustment, and repair of equipment and systems. Repairs events must be viewed as opportunities to upgrade. In other words, the organization *must* know if upgrading of failed components and subsystems is feasible and cost-justified. On the other hand, performing unnecessary maintenance and repair should be avoided. Breakdowns occur because of improper equipment operation or failure to perform basic preventive functions. Overhauling equipment periodically when it is not required is a costly luxury; upgrading where the economics are favorable is absolutely necessary to stay in the forefront of profitability.

Regardless of whether or not PdM routines have determined a deficiency, repairs performed on an emergency basis are three times more costly in labor and parts than repairs conducted on a preplanned schedule. More difficult to calculate, but high nevertheless, are costs that include shutting down production or time and labor lost in such an event.

Bad as these consequences of poorly planned maintenance are, much worse is the negative impact from frequent breakdowns on overall performance, including the subtle effect on worker morale, product quality, and unit costs.

Effectiveness of Selective Preventive Maintenance. Selective preventive maintenance, when used correctly, has shown to produce considerable maintenance savings. Sweeping, broad-brush maintenance, including the routine dismantling and re-assembling of compressors is wasteful. It has been estimated that one out of every three dollars spent on broad-brush, time-based preventive maintenance is wasted. A major overhaul facility reported that "60% of the hydraulic pumps sent in for rebuild had nothing wrong with them." This is a prime example of the disadvantage of performing maintenance to a schedule as opposed to the individual machine's condition and needs.

However, when a selective preventive maintenance program is developed and managed correctly, it is the most effective type of maintenance plan available. The proof of success can be monitored and demonstrated in several ways:

- increased plant uptime;
- higher equipment reliability;
- better system performance;

- reduced operating and maintenance costs;
- improved safety.

A plant staff's immediate maintenance concern is to respond to equipment and system functional failures as quickly and safely as possible. Over the longer term, its primary concern should be to systematically plan future maintenance activities in a manner that will demonstrate improvement along the lines indicated. To achieve this economically, corrective maintenance for unplanned failures must be balanced with the planned selective preventive maintenance program. Every maintenance event must be viewed as an opportunity to upgrade so as to avoid repeat failure.

Know Your Existing Program. The starting point for a successful long-term selective maintenance program is to obtain feedback regarding effectiveness of the existing maintenance program from personnel directly involved in maintenance-related tasks. Such information can provide answers to several key questions, and the answers will differ from machine to machine and plant to plant. Your in-plant data and existing repair records will provide most of the answers to the seven questions given below. A competent and field-wise consulting engineer will provide the rest:

1. What is effective and what is not?
2. Which time-directed (periodic) tasks and conditional overhauls are conducted too frequently to be economical?
3. Which selective preventive maintenance tasks are justified?
4. What monitoring and diagnostic (predictive maintenance) techniques are successfully used in the plant?
5. What is the root cause of equipment failure?
6. Which equipment can run to failure without significantly affecting plant safety and reliability?
7. Does any component require so much care and attention that it merits modification or redesign to improve its intrinsic reliability?

It is just as important that changes not be considered in areas where existing procedures are working well, unless some compelling new information indicates a need for a change. In other words, it is best to focus on known problem areas.

To assure focus and continuity of information and activities relative to maintenance of plant systems, some facilities assign a knowledgeable staff person responsible for each plant system. All maintenance-related information, including design and operational activities, flow through this system or equipment "expert," who refines the maintenance procedures for those systems under his jurisdiction. He or she re-shapes preventive maintenance into selective maintenance.

Maintenance Improvement. Problems associated with machine uptime and quality output involve many functional areas. Many people, from plant manager to engineers and operators, make decisions and take actions that directly or indirectly affect machine performance. Production, engineering, purchasing, and maintenance personnel as well as outside vendors and stores use their own internal systems, processes, policies, procedures, and practices to manage their sections of the business enterprise. These organizational systems interact with one another, depend on one another, and constrain one another in a variety of ways. These constraints can have disastrous consequences on equipment reliability.

Program Objectives. An effective maintenance program should meet the following objectives:

- Unplanned maintenance downtime does not occur.
- Condition of the equipment is always known.
- Where justified, preventive maintenance is performed regularly and efficiently.
- Selective preventive maintenance needs are anticipated, delineated, and planned.
- Maintenance department performs specialized maintenance tasks of the highest quality.
- Craftsmen are skilled and participate actively in decision-making process.
- Proper tooling and information are readily available and being used.
- Replacement parts requirements are fully anticipated and components are in stock.
- Maintenance and production personnel work as partners to maintain equipment.

Why Training of Machinery Maintenance Forces is Important or No News is Good News

We had been talking with some mechanics in a major petrochemical process plant about how their major machinery inspection and overhauls were being executed. They insisted they would always have the OEM's representative present. We asked them why and they felt the vendor's rep would assure a quality job; besides, their management deemed the

OEM's presence necessary to satisfy warrante and insurance requirements. As we sat with them, several of the men were summoned out of the meeting. A major centrifugal compressor train had just shut down, and they were required to immediately attend to the problem. It turned out a compressor's kick-back valve* had acted up and the train had shut down on a thrust bearing high temperature and position indication. The men worked two 12-h shifts to open up the thrust bearing on a multi-stage barrel compressor and to replace the thrust collar together with the badly damaged entire thrust bearing assembly. When we visited the machine later in the week it was running nicely attesting to the skills of the people who had worked on it – without a vendor's representative who could not be scheduled in on a short notice from a distant location.

While we recommend to employ skilled field-service personnel of a reputable compressor manufacturer, we would like to point out that we should observe some subtle points regarding the interfacing of OEM personnel with our mechanics. Our first objective should be to make sure our people are trained for exactly the contingency described above.

Point one: The OEM's field service representative should seldom touch – just look.

Point two: Instruct the client's personnel by show and tell – demonstrating but letting the site mechanics do the actual work such as disassembly and final reassembly together with the necessary dimensional control and documentation tasks.

Where does this topic fit in? It fits into the frequently neglected area of machinery repair quality assurance. The old adage "no news is good news" applies when a machinery repair has been successfully performed, often nobody cares to mention it. However, when it has not been successful, "high tech" reliability jargon is used to identify causes ranging from lack of machinists' training to poor inherent equipment maintainability. This is the bad news.

Yet, we believe there is no need for recriminations had we thought about an appropriate basis for our machinery maintenance and repair routines beforehand. If we want to be successful in the repair business, we must define the "repair cycle" as follows:

- Assess the damage by taking failure mode inventory.
- Analyse and identify the cause of the failure leading to the repair.
- Execute the repair by dimensional checks and proper parts replacement.
- Follow a checklist format for a quality field installation if the equipment had to be removed for repair.

---

* Also known as recycle valve for surge control.

- Perform and supervise an equipment run-in or post-repair acceptance routine.
- Follow-up to eliminate failure root cause all the while documenting what you have done.

This repair cycle is applicable to all types of machinery, be it special purpose or general purpose equipment such as, for instance, a small centrifugal pump. Once all phases of the repair cycle are defined and understood, they need to be supported by providing appropriate training, tools, and procedures. In the training area, we would see that our mechanics are regularly updated in on-site seminars or toolbox discussions on subjects covering shop and field practices such as:

- alignment (e.g., the use of Laser alignment)
- anti-friction bearing fitting
- coupling assembly
- lubrication
- piping to machinery fit-up and alignment
- seal assembly
- function of machinery and components.

In the tool area, provide tool lists and kits for important jobs. Keep mechanics updated on the use of CMMS input tools, such as portable data terminals (PDTs), in order to free them from tedious and frequently meaningless documentation tasks. Management must develop incentives for shop and field personnel to supply the what, when, where, why, how, and who together with material and component changes into the repair history.

Finally, maintenance engineers must help mechanics to adhere to correct practices by leading the development of realistic and easily accessible procedures, for instance procedures for each specific machine on-site with sections covering:

- safety precautions;
- detailed inspection, maintenance, repair, and overhaul (IMR&O) procedures;
- rotor and spare check-out;
- clearance tabulation forms (e.g., "as found," "as repaired");
- bolt torquing for important joints (e.g., casings, reciprocating compressor valve caps, bearing caps, couplings, flanges, and others);
- sealant and adhesive use (e.g., an approved listing of S&As);
- alignment instructions (e.g., identification of operating shaft centre-line location);
- up-to-date spare parts list;
- tool list;
- digital photo file of last IMR&O.

Many companies have had success with sequestering-experienced shop and field mechanics to structure and write these procedures themselves thus achieving the necessary site acceptance and buy-in. We thought we saw some positive indications of what we just mentioned at the plant we visited. In the following, we are going to deal with planned major overhauls of process machinery.

Proven Turnaround Practices*

In process industries, plant and process unit turnarounds are major undertakings and can have significant impact on the plant's annual maintenance budget and future operating and maintenance performance metrics. A successful turnaround is one where safety, environmental compliance, cost, and duration are within expectations and benchmark performance is achieved. To be successful, turnarounds require careful planning and scheduling, which typically is done by skilled and experienced personnel utilizing proven practices that integrate and minimize the work scope. However, since the interval between turnarounds is generally long, the lessons learned from past turnarounds are sometimes lost because of personnel transfers and poor documentation.

We are going to describe some of the practices that have been successfully applied in planning and executing turnarounds that resulted in reduced turnaround cost and shorter downtimes. The key to an effective and efficient turnaround is proper and early planning. An effective work plan is achieved by early development of an overall milestone plan called "Planning the Plan." The areas covered by this plan will be discussed, including timing for each of the activities. The key to keeping turnaround cost under control, or within benchmark objectives, is to ensure the work scope only includes those items needed to achieve the current business objectives. This requires that a strict and consistent approach be adhered to for approving work list items.

Generally, capital projects are scheduled with turnarounds. Complete integration of the activities associated with the capital project into the overall plan and schedule is paramount if an efficient turnaround is to be executed. Planning, scheduling, contracting, safety, and other activities need to be integrated into a single plan to ensure the shutdown time will be minimized and the cost held to acceptable levels. Some practices in these areas will be discussed.

Another technique for improving the effectiveness of a turnaround is to conduct cold-eye reviews by an outside team of experts at critical

---

* By permission from Robert J. Motylenski, P.E., Maintenance Consultant, Woodland Park, CO USA.

times during turnaround preparation. This technique has cut duration time by up to 6 days and resulted in savings of several million dollars from reductions in work scope. The reviews ensure good practices and procedures are being used to plan the turnaround. The timing and scope of these reviews will be presented.

Introduction. In refineries and chemical plants, plant and process unit turnarounds are major undertakings and have significant impact on plant operating expense and future operation. The cost of a turnaround is the single biggest maintenance expense a process plant can expect to encounter. The cost of the single event can vary from less than one million dollars to over 35 million dollars, depending on work scope, and can represent, on an annualized basis, up to 50% of the annual maintenance budget. In addition to the maintenance expense, there is lost revenue because of process unit or plant shutdown. This can be anywhere from a few days to over a month, again depending on the units involved. In order to minimize the business losses, it is important that the planning, scheduling, and execution are done in an efficient manner.

A winning turnaround can be characterized by ten success factors:

1. absence of personnel injuries;
2. no facility incidents;
3. no environmental impact;
4. schedule objectives are met or exceeded;
5. target cost is not exceeded;
6. facilities are successfully commissioned;
7. equipment achieves planned run length and operating conditions;
8. workforce performed outstandingly;
9. no contractor claims;
10. improved contractor and vendor relationships.

One method of attack to ensure success is to use a disciplined approach similar to that used in developing and executing projects. The major difference between the two types of work is that turnarounds occur over a very short-time period and thus require more coordinated planning and scheduling. Some of the key similarities for managing a turnaround and a project are:

- use of a planning team;
- early identification of the work scope;
- a contracting strategy;
- agreement and adherence to timetables and milestones;
- site commitment to the objectives.

Practices Overview. To be successful, turnarounds require careful planning and scheduling which, typically, are done by skilled and experienced personnel, utilizing proven practices that integrate and minimize the work scope. Since the interval between turnarounds is generally long, the lessons learned from past turnarounds are sometimes lost because of personnel transfers and poor documentation. However, there are many other areas that need to be effectively developed when optimizing the work list and assembling an integrated plan and schedule. All the areas that need to be considered when preparing for a turnaround are:

Management	Milestone Plan
Pre-turnaround Reviews	Work Scope
Cost Estimating	Planning
Contracting	Scheduling
Inspection	Process Operations
Materials	Projects
Organization	Safety/Health/Environment
Communication	Logistics
Security	Execution
Closeout	

For each of these areas, there are demonstrated practices that support a world-class turnaround. The practices for some of the areas will have a greater impact on turnaround performance, such as safety, work scope, planning, contracting, scheduling, and execution. However, that does not mean the other areas should be downgraded in importance because, without implementation of the practices for those areas, the turnaround will not be a success.

The following will review the specific practices for some of these areas. The practices that will be presented have proven to be successful in lowering turnaround cost and reducing turnaround duration. The practices for any one area are a composite of practices that have been successfully used by several refineries and are not those of an individual refinery. No single refinery has implemented all the practices, although some refineries have been successful in implementing a large majority of the practices and are striving to implement all of them. The practices were collected and observed by the author when he led 16 refinery pre-turnaround reviews for an international oil company.

Practices for the following areas will be presented in the following sections:

- Management
- Milestone plan
- Work scope
- Projects

- Material
- Process operations
- Pre-turnaround reviews.

Management Management provides the underlying guidance and support needed by the organization to ensure a successful and effective turnaround. To achieve these ends, management needs to be an active participant in all phases of the turnaround, particularly during early development. Some of the key practices management must implement are:

- Establish guiding principles for the turnaround if they are not already part of the plant normal management practices. The principles should include:

  - clear turnaround objectives;
  - business objectives for the upcoming operating period;
  - a fully integrated team for planning, execution, and closeout;
  - an optimized process for identifying and executing work;
  - complete alignment of all personnel (contractor and own) involved in the turnaround.

- Form a Turnaround Management Committee (Fig. 16-6) to guide the turnaround team, and monitor and steward progress from early planning to execution and closeout. Representatives from each of the key plant areas (operations, mechanical, and technical) should be on the committee, plus the plant or site manager. The Committee should conduct regular meetings to steward progress, with meeting frequency increasing as execution approaches.

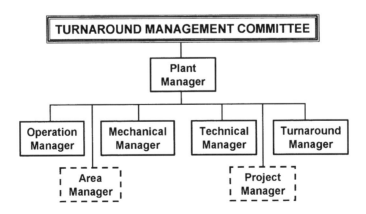

Figure 16-6. Turnaround organization chart.

- Draft specific turnaround objectives consistent with the guiding principles and overall plant objectives. The objectives should address budget, anticipated upcoming run length, downtime duration, starting date, estimated man-hours, and any other appropriate items. The Turnaround Management Committee generally is responsible for these.
- Prepare an overall milestone chart that outlines the planning process and includes the major key milestone dates.
- Form a single organization to do the planning, scheduling, contracting, and execution for the turnaround, including any project activities.
- Decide if pre-turnaround reviews will be conducted and, if so, indicate them on the milestone plan.
- Appoint a Turnaround Manager who will report to the Turnaround Management Committee and has the authority to organize an integrated team for planning and execution.
- Develop a contracting strategy for the turnaround that is consistent with the local contracting situation and experience.
- Develop screening criteria for primary contractor selection based on past performance, turnaround objectives, key mechanical work, safety, quality, and cost.
- Make arrangements to complement the turnaround team with experienced outside personnel to supplement existing manpower.
- Develop a plan for ensuring personnel needed for the next turnaround will be retained, or are being trained, so that valuable lessons and skills learned are not lost.

The impact good management practices have on a turnaround is difficult to quantify, but without it the turnaround team will have difficulty being effective. Maintaining a consistent turnaround planning organization supported by written procedures, computer-based equipment and cost files, and frequent communication to everyone involved is vital for successful turnaround management.

Milestone Plan The milestone plan is a time chart that identifies all the key turnaround planning activities that need to occur prior to execution. It is a high-level time line and is generally called "Planning the Plan." It serves as a key document that the Turnaround Management Committee and turnaround organization use to steward planning progress. The milestone plan is the first document prepared since it establishes the timeline for strategic turnaround activities and their interrelationship. The milestone plan can take different forms. The plan can be a T-minus chart, Gnatt chart, or a table. The key is that they outline the major tasks involved in planning against a timeline.

Figure 16-7 represents a partial milestone plan showing the activities that are discussed in this paper followed by some key milestone plan characteristics:

- The milestone plan provides an overview of strategic activities in the turnaround and is a key document in communicating planning progress. The plan must be developed early, gain full support and commitment from those responsible for performing the activities, and used by the Turnaround Management Committee for ongoing stewardship.
- Typically, there are 14 different areas represented on a milestone plan:

work scope	cost control
planning	scheduling
contracting	materials
inspection	health/safety/environmental
organization	communications
logistics/Security	project
process operations	execution.

- A unique milestone plan is prepared for each turnaround.
- The milestone plan indicates all the major activities that need to be developed by process operations, maintenance, inspection, project, and management, prior to turnaround execution. The plan should also show execution and post-turnaround activities.
- Two other areas that have significant impact on the planning process may be included on the milestone plan. These are pre-turnaround reviews and key meetings, especially of the Turnaround Management Committee.
- If the practice of conducting pre-turnaround reviews is adopted, the review timing needs to be included on the milestone chart and scheduled to conform to the availability of key information.
- The amount of detail included in the plan will depend on turnaround size. Generally, 100–150 k man-hours are the cut-off between a very detailed milestone plan and an abbreviated plan. Very small turnarounds can usually be managed using routine maintenance procedures.
- Separate and more detailed timeline plans need to be prepared for each of the key areas included in the overall milestone plan.
- The milestone plan is initially endorsed by the Turnaround Management Committee, reviewed at each meeting, and used as a tool to steward planning progress and overall schedule.
- The time range used for the milestone plan and the starting time used is generally dependent on turnaround size, i.e. man-hours. Table 16-2

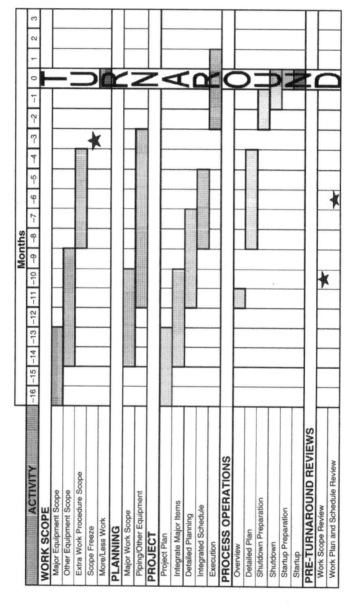

Figure 16-7. Partial milestone chart example.

Table 16-2
Turnaround man-hours and planning time

Turnaround man-hours	Milestone planning time – Months
Large – >300k	−24 to −18
Typical – 100k to 300k	−18 to −14
Small – <100k	−14 to −8

gives some reasonable time ranges, in months, based on estimated turnaround man-hours.

- To help assess progress and identify schedule delays, actual progress should be overlaid on the milestone plan. This is an effective tool for the Turnaround Management Committee.

Work Scope Work scope is the process of identifying mechanical, process, and project work that is required to be performed during the turnaround. The selection process for identifying work needs to ensure risk is managed and accepted for the upcoming run-length. The approved work scope excludes work items that can be performed more effectively and efficiently outside of the turnaround (normal maintenance).

Most plants have found that it is more costly to perform normal maintenance type work during a turnaround because of the additional supervision and overhead. Thus, their objective is to minimize scope and eliminate all routine type maintenance from the work list. A few plants have found it cost-effective to include routine maintenance type work in the turnaround, providing it does not lengthen the downtime. Each plant needs to analyze their data and make the appropriate decision.

Work scope is the single area that has the biggest impact on cost and downtime. The development of the scope needs to start early in planning and be prepared with care and consistency. Eliminating work that is routinely included on the turnaround work list because of tradition or just a look-see without any basis can result in savings of millions of dollars on large multi-unit turnarounds. This is a paradigm shift for most plants and will take considerable discipline to enact. The following are some practices that have proven to be successful in preparing the work scope:

- All requests for turnaround work should be screened, challenged, and approved by the process, mechanical, and technical teams responsible for the process unit(s) involved in the turnaround. The objective of the work screening procedure is to approve the minimum work required to support the business objectives.
- The primary sources of work are the individual equipment maintenance plans, previous turnaround closeout reports, regulatory

inspections, process shutdown and startup preparation, process work activities, and process unit improvement projects.

- A summary sheet should be used for each work request. The sheet should include work scope and justification, risk matrix, probability of work being required, estimated man-hours, and a good-quality cost estimate.

- Regulatory inspection requirements should be verified and challenged when they seem unreasonable or significantly increase the scope of the turnaround.

- A screening process that is risk-based and consistent across the plant should be used to screen all work requests. The timeframe for assessing risk should be either based on the upcoming run-length or two run-lengths.

- The work scope for fixed equipment should be identified early and closed to additional work about a year prior to the turnaround.

- Any potential work item that requires pre-turnaround inspection to confirm the need for the work must be identified early and the inspection completed prior to closing of the work list.

- The process unit operating team (process, mechanical, and technical) is responsible for endorsing the final work list and takes ownership of the list. The turnaround team, especially the planners, must NOT become involved in work justification. The turnaround planners are NOT the owners of the work scope.

- A timeframe, supported by management, should be established for collecting work list items. An extra-work procedure needs to be developed that will be used for reviewing and approving any work item submitted outside the timeframe.

- A "cold-eye review" of the finalized work list should be conducted to ensure the work list only includes the minimum work required to achieve the business objectives. The team, preferably, should be from the outside.

- Any work that is not clearly defined should be identified and a probability of doing the work assigned to it. Contingency plans for major unidentified work items need to be prepared. Table 16-3 provides guidelines for how probable work should be handled in the various phases of planning.

- The contracting strategy used for the turnaround will influence the cut-off date for defining work scope. The higher the percentage of lump sum/fixed price bidding, the earlier the work scope cut-off dates.

- An extra-work procedure needs to be developed for assessing additional work once the work list is closed. The procedure needs to include authorization criteria.

Table 16-3
Potential work guidelines

Probability	Turnaround planning phase				Execution
	Planning	Scheduling	Estimating	Contracting	Inspection
≥50%	• Take off materials and determine availability • Identify special tool requirements and availability • Assess executability • Determine how much preparation is needed	• Include in schedule with probability identification • Make available early in turnaround	• Estimate and include in overall estimate	• Include in contract as defined adder	• High priority inspection • Perform as soon as possible in turnaround
<50%	• Review materials and special equipment requirements	• Assess impact on schedule if work required	• Not included in estimate • Use contingency	• No provisions in contract	• Normal inspection priority

- The turnaround team should not become involved in the justification or approval-of-work requests after the work list is closed and the additional work procedure is in place. Only additional work approved by the procedure should be forwarded to the planners for incorporation into the overall work plan and schedule.
- Long-delivery material needs should be identified early in the planning stage.

Work scope is the area that can have the biggest impact on turnaround cost and schedule. Thus, systematic assessments of the work scope and the procedure for developing the scope can usually result in considerable reductions in the scope and savings of several days' duration.

Projects Almost every turnaround involves some capital project work, either minor capital improvements, or large revamps, or expansion projects. In either case, in order to have a successful and effective turnaround, a single integrated team with an integrated plan and schedule is required to manage both the turnaround and the project work. This also means using a contracting strategy that complements both activities. Projects are the one area, if not integrated into the overall plan and schedule, that can cause major extensions of downtime. Table 16-4 shows the timing relationship for some turnaround and project activities.

The following are some practices that should be followed in integrating a project into the turnaround:

- Integrate the project and turnaround teams into a single team so that there is one organization responsible for the planning and scheduling of turnaround and project work. This also means a single manager.
- Combine the turnaround milestone plan and project gate process into a single event milestone plan. Ensure that the timing for project activities is consistent with the turnaround schedule.
- Merge any planned project reviews with pre-turnaround reviews.
- Only project work that requires the unit to be shutdown should be executed during the turnaround. All other project-related work should, if at all possible, be done outside the turnaround, i.e. pre- or post-turnaround. Maximize the amount of project work that can be performed outside the turnaround.
- All project activities that are to be performed during the turnaround need to be integrated into the overall turnaround plan and schedule, and follow the contracting strategy.
- For larger projects, a project representative should be on the Turnaround Management Committee.

Materials Materials refer to the complete process of bidding, purchasing, delivery, and storage of supplies needed for the turnaround.

Table 16-4
Turnaround timing relationships

Approximate timeframe – Pre-turnaround				Post T/A
>20 months	−20/−14 months	−14/−9 months	−9/−6 months	+1/+3 months
Turnaround activities				
	• Appoint T/A manager • T/A objectives • Milestone plan • Contracting strategy	• Major work scope • Initial cost estimate • Critical path identified • Functional milestone plans • Contract plan	• Work scope frozen • Integrated schedule • Critical/sub-critical path plans • Mobilization plan	T U R N A R O U N D • Critique • Update files
Project activities				
• Project team appointed • Project objectives • Contracting strategy • Budget estimate	• Design basis • P&IDs available • Cost estimate	• Appropriation • Detailed engineering • Procurement • Construction	• S/D and S/U plans • Turnover plan • Testing and commissioning	• Commissioning

Specifically, it is a system of ordering against a pre-determined specification, developing competitive bids, selecting a supplier based on price, quality and delivery, expediting for assured quality and delivery, transportation to site, receipt and confirmation against purchase orders including PMI (positive material identification), appropriate storage and delivery to prearranged locations at specified times.

Materials can account for 25–35% of a total turnaround cost. Therefore, the effectiveness of this activity is vital to achieve a successful turnaround. The following are some practices to improve procurement of materials:

- An experienced materials coordinator needs to be assigned to the turnaround team or report to the Team. The coordinator is responsible for purchasing, contracting, expediting, and storing turnaround materials. Typically, the coordinator will be from the plant material or warehouse organization.
- The materials coordinator should be full-time for large turnarounds where there are many material requisitions. For smaller turnarounds, the coordinator may be part-time, depending on the number of material orders. However, in either case, their main responsibility will be to the turnaround organization.
- A regular communication link needs to be established by the materials coordinator between the turnaround team and the site materials/ purchasing function. This is because the turnaround can represent over 50% of the warehousing activity prior to, and during, the turnaround.
- The materials coordinator is proactive during the turnaround planning to identify long delivery items, specialty and unique items, special packing and delivery needs, availability of consumable items from local vendors, storage and delivery restrictions, and off-hour material receipt procedures.
- Typically, the existing purchasing and expediting function will be used for the turnaround. The key to its success is to identify early in the planning stage critical needs and any significant deviations imposed by the turnaround. If any deviations are noted, changes to existing procedures need to be developed and resources assessed so there are no bottlenecks developed during the turnaround.
- Plant versus contractor-supplied materials need to be defined, e.g. safety equipment, tools, etc., including the availability of free bins. In some cases, it may be advantageous to have the plant furnish individual safety items to prompt contractor usage.
- A separate milestone plan needs to be prepared for materials area, showing the major activities (Fig. 16-8).
- Procedures need to be in place for materials pickup and delivery for all turnaround activities to minimize ineffective use of contractor and own manpower. Specific areas of concerns are:

Item No.	Description	PR No.	PO No.	LPR Issued Date	Enquiry Issued Date		Quotation Receipt Date		Quotation Summary Date		PO Issued Date		Source Inspection		Ex. Work date		ETA at Port		ETA at Site	
					P	A	P	A	P	A	P	A	P	A	P	A	P	A	P	A
1	F-601 Elbow	1530-40376	1530-41031	Jun-98								Aug-98							Jan-99	Jan-99
2	F-101/F-102 transfer lines	1530-40417	1530-41381	Jul-98								Nov-98							May-99	May-99
			1530-41382	Jul-98								Nov-98							May-99	May-99
			1530-41383	Jul-98								Nov-98							May-99	May-99
3	T-101 Lining	1530-50150	1530-41031	May-98																
4	Chlorine Injection Line	1530-50132	1530-51113	May-98					Aug-98	Aug-98	Aug-98	Aug-98							Dec-98	

Figure 16-8. Material milestone plan (partial example).

– Entry of plant and contractor personnel into the warehouse during normal and off-hours.
– Receipt and lay down in the field of materials during off-hours.
– Pickup and return of special tools.
– Control of consumable items and personnel expendable equipment, e.g. gloves, flashlights, hoses, etc.

Process Operations Process refers to all activities undertaken by plant operations as part of the turnaround. These include preparing the process unit for the start of mechanical work, identifying any process work activities that need to be performed during the turnaround, developing an optimum shutdown and startup plan, and, lastly, supporting the mechanical work of the turnaround by providing suitable supervisory manpower. These activities have a tremendous impact on the overall turnaround plan and schedule.

• Early in planning, identify all process-related work that needs to be performed during the turnaround. All work requested by process operations needs to be justified, similar to mechanical and other types of work requests.
• Integrate all process work activities into the overall turnaround plan and schedule.
• Assign a senior process supervisor, who is a full-time member of the turnaround team, as process coordinator for the turnaround. The coordinators will be the prime contact between the process organization and the turnaround team.
• Prepare a process shutdown plan and startup plan that makes equipment available for timely start of mechanical work.
• Prepare a process shutdown plan that makes critical equipment, and equipment with a high probability of needing additional work, available early in the shutdown.
• Early in turnaround planning, review the blanking requirements and availability of blanks. Prepare a blanking procedure which includes a tracking system.
• Review the existing work permit system and identify any procedures that will cause delays in issuance of permits. Consider using area permits to minimize the number of permits issued.

Pre-turnaround Reviews Pre-turnaround or cold-eye reviews are a management tool for assessing the status and completeness of turnaround planning and scheduling at strategic times in the overall planning process. The type of reviews conducted depends on turnaround scope, size, and previous turnaround experience. The reviews are conducted to assess the effectiveness of the work selection process and planning, and ensure alignment and preparedness of the organization to plan and execute the turnaround. An internal or external team can be used, but an outside team will be able to bring to the organization lessons learned and proven practices used by others.

Pre-turnaround reviews have been successfully used by many organizations. The typical result from using an outside team of maintenance professionals to review the work scope and turnaround execution is a savings of three to 6 days downtime and a reduction of 10–20% of the budgeted turnaround cost. The actual savings are dependent on the starting point of the plant's turnaround planning efforts.

- If the practice of conducting pre-turnaround reviews is adopted, the review timing needs to be included on the milestone chart and scheduled to conform to the availability of key information.
- Two different types of reviews are most frequently used to improve turnaround effectiveness, work scope and work planning and schedule. The timing of the review depends on the availability of key information.
- The work scope review checks to ensure the work scope was generated using a systematic approach (RCM or equivalent), which included risk, and that the work scope includes all static equipment, major piping, special-purpose machinery work, and any major project work. The review generally also checks on the progress of the overall planning process and ensures it is proceeding according to developed strategies.
- The work planning and schedule review assesses the organization's readiness for execution, especially with regard to the detailed planning and schedule for the critical and sub-critical paths. In addition, the review will check the status for managing different aspects of the turnaround execution, such as cost control, contracting, inspection plan, communication, logistics/security, project, safety/health/environment, etc.
- Some organizations conduct specialty reviews focusing on specific turnaround issues that have been identified from previous turnarounds or because of unusual circumstances included in this turnaround.
- External teams of three to four people experienced in turnarounds typically conduct the reviews. This gives the receiving site the opportunity to capture lessons learned from outside organizations. The team make-up for each review may differ depending on the specific review objectives.
- Timing of the review is dependent on turnaround size and availability of key information. Table 16-5 outlines the approximate timing, duration, information required, and typical participants for each review.

Table 16-5
Review duration and timing

	Work scope	Work planning and schedule
Timing	−15 to −10 months	−9 to −6 months
Duration	2 to 4 days	2 to 4 days
Key documents	Major work scope and initial cost estimate	Critical and sub-critical plan and schedule
Potential team Composition	Team leader, work selection specialist, turnaround specialist, inspection specialist, process supervisor	Team leader, experienced planner, turnaround specialist, process supervisor

Conclusion Everyone is looking to identify and implement practices that can help them achieve benchmark performance. We presented some practices that have been successfully used by a number of refineries in planning and executing turnarounds. Implementation of the practices has resulted in more effective and efficient turnarounds. Some of the practices discussed are also applicable to other management areas in a process plant.

The area everyone believes has the biggest impact on turnaround cost, and duration is work scope, but without effective planning and scheduling, the gains achieved by reduced work scope can be easily lost to inefficient planning and non-optimum scheduling. Poor contracting is the other area that can easily negate the gains of a reduced work scope. This can be due to the work split among contracts, not being wisely divided, or if contractors are required to perform work outside their capabilities (manpower, skill, etc.), or are hindered because of too many interfaces either with other contractors or with site personnel. To overcome these potential problems, an effective contracting strategy needs to be developed during early planning. The strategy should not only complement experience and the current contracting situation, but also include innovative approaches.

Implementing proven turnaround practices is not an easy task. One of the problems is that all the areas interact. Implementation of practices in one area will highlight the need to improve the practices of another area in order to achieve full benefit from that implementation. How one goes about implementing new practices and changing long existing paradigms is a challenging task and takes a great deal of continuous effort. Every plant is already using some practices that can be considered "Best Practices." The challenge is to identify and strengthen those practices while replacing practices that are not best or industry proven. One means to quickly identify the good and the bad is to use an outside team to conduct a cold-eye review of a major turnaround. The team will be able to quickly identify shortcomings and pass along new techniques for the poor and non-optimum practices.

The other approach is to undertake the task internally. Then, the first thing that needs to be done is an assessment of current turnaround performance. Is it benchmark or is it causing the maintenance budget to be out of line with competition? From this assessment and past turnaround critiques, can weakness be identified? The next step is to understand all the practices being used to plan, schedule, and execute a turnaround. Are they considered by the industry as proven or BPs? Combining these assessments should identify the areas that need improvement and the practices that need replacement. This is not an easy task and requires skilled and experienced people who have limited, or no, ownership of the current practices.

Hopefully, the practices that were presented here will be a start to improving your turnaround performance.

Machinery Maintenance Effectiveness Assessment: An Example

If we are involved in the IMR&O or turnaround activities of our machinery, we should perhaps take the time to step back and look at how our crew is doing. There is always room for "continuous" improvement.

The format shown in Table 16-6 was used to assess maintenance effectiveness on a couple of major pipeline stations in eastern Europe each containing several 25 MW gas turbine driven compression trains.

It served the purpose of getting everybody involved to focus on the issues that are important to the success of machinery inspections and overhauls. We were also interested at the time in gas-supply reliability promised by our east European pipeline operations partners. The assessment was to help us to understand how they were executing self-directed machinery turnarounds without the assistance of an OEM. By the way, the numbers above indicate the score or assessment number, 10 is the best. The total score of 5.33 was arrived at by adding up the individual score numbers and dividing the sum by 11.

We were particularly impressed by a machinery inspection activity our partners called "defectoscopy," something that we would consider closely related to non-destructive testing or NDT. This activity was supported by a group of highly qualified experts who examined every single machinery part that was disassembled, performed a failure analysis, and formally judged its serviceability. There can never be enough said about the need for such an activity because it is tied to the adage that every maintenance occasion should be considered an opportunity for improvement. No real improvement is possible if we do not understand the cause of our machinery component deterioration or failures.

Table 16-6
Machinery overhaul effectiveness assessment

Criteria	1	2	3	4	5	6	7	8	9	10
Historical record keeping Equipment cv (what, when, where why, how, who) including cost records									X	
Assembly data taking Fits and clearances, etc. Quality of inspection forms							X			
Maintenance instructions Quality of written procedures, i.e. books or manuals; general adherence to OEM recommended procedures									X	
Organizational effectiveness Role distribution, conflict resolution, etc.				X						
Safety practices Injury prevention, safety equipment				X						
Planning and preparation The T/A			X							
Spare parts organization Parts staging and availability, decision-making on repairability or re-use of questionable parts		X								
Work practices Quality and skills, quality of rigging practices, cleanliness, protection of dismantled components, etc.							X			
Tools and fixtures Condition of tools, supplies, fixtures, rigging, and lifting equipment					X					
Inspection practices NDT practices, defect assessment/"defectoscopy"								X		
Overall job co-ordination			X							

Total score: 5.33

As to the formal side of Table 16-6, somebody might say:

The order of your criteria is wrong! We would decidedly put Safety Practices – Injury Prevention and the Absence of Unsafe Conditions and Acts as our most important criterion. Then we would let perhaps Maintenance Instructions – Adherence to Written Procedures follow.

After that we would go on with ranking the remaining criteria from 9 to 11 – 11 being the criterion of the highest importance. We would then multiply our ranking number by the score number, put the product into a separate column, add this column up and make the sum our final score.

This will result in yet some more sophistication which would nevertheless be required if we wanted to arrive at objective assessment results – especially, if we wished to compare operating Unit A to Unit B, and so forth. However, we believe that for many maintenance professionals, Table 16-6 above is a sufficiently accurate snapshot of what this particular operation was like, namely not good and not bad. Would you be satisfied with such an assessment?

The reason for why we did not want to present Table 16-6 in a more sophisticated fashion is the fact that our opinions about ranking of its criteria may vary. For instance, there are people who contend, why make the safety issue all-important? If all the other criteria are addressed by training, practice, and reinforcement, the safety concern will eventually become less pervasive.

A good beginning would be to discuss the above criteria – and probably others contributing to the success of our turnarounds – in our periodic safety meetings instead of engaging in generalities relating to personnel injury prevention. You might want to try this approach and be surprised by the response and input from your people who will support the idea of "continuous improvement" when it comes to the organization and execution of their machinery turnarounds.

Machinery Uptime and Computerized Maintenance Management Systems*

Defining CMMS Versus EAM Versus the Other TLAs

First let us start with approximate definitions of what CMMS, Enterprise Asset Management Systems (EAM) and other similar systems are and what they are supposed to do. Contrary to the popular belief, CMMSs have been around for well over 20 years. They grew out of the recognition that just as accounting systems could increase the paper processing speed of payables, receivables, etc, so a work order system could speed the organizing and production of work orders and other core documents in the maintenance arena. As the systems became more powerful and more flexible, so the ideas about their functionality increased; the paper production focus gave way to improving processes – the transition of a work request through a work order to a purchase requisition, purchase

---

* Contributed by Ben Stevens of OMDEC, www.omdec.com

order, and eventual material receipt, for example. This was accelerated by the insertion of workflow capability into some of the more advanced systems, so that companies with different methodologies could advance at different speeds – even while using the same CMMS. Finally we are arriving at the fourth phase of CMMS – the use of the CMMS as a means of driving improvements in the quality of maintenance, the reliability of equipment, and the predictability of production. This evolution is summarized in Figure 16-9. Later in this chapter, we will speculate as to where Phase 5 will lead us.

Most CMMS users will recognize themselves as being somewhere in Phases 1 and 2, while most leading vendors are in Phases 2 and 3. Only a very small number of either are seriously investigating Phase 4.

Phase 3 saw the emergence of EAM as a subdivision of the CMMS marketplace. The cynics will argue this is really only a marketing ploy at market differentiation to separate the gurus from the wannabees, aided and abetted by the leading vendors to distance themselves from the pack. So EAM was born – to focus on the more asset intensive businesses and to more easily permit inter-plant comparisons. But in reality, only the dedicated insiders recognize much difference between the two approaches.

A major incentive from this differentiation, and the associated Best of Breed label, came as a result of the big ERP vendors launching or buying their own Asset Management capability. Hence Oracle's development of its Asset Management modules, SAP greatly enhancing their PM module's capability, and Peoplesoft's acquisition of JD Edwards. Hitherto, the ERP vendors emphasized the automatic integration of the modules with the accounting, HR, and manufacturing functions as key selling advantages. But as the flexibility and capability of the APIs have increased, this advantage has been eroded and has greatly diminished the core differences between the approaches.

However, the Best of Breed vendors* have continued to focus on improving the quality, functionality, and ease of use of their products,

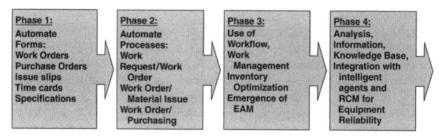

Figure 16-9. Evolution of computerized maintenance management systems (CMMS).

---

* Indus, MRO, Mincom

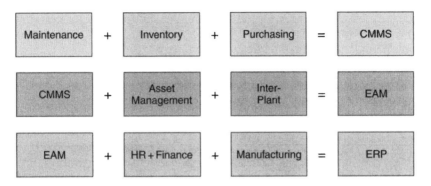

Figure 16-10. Evolution from maintenance management to enterprise resource planning (ERP).

especially in those asset-intensive industries where the cost of maintenance is a significant portion of the overall operating cost. The functionality race in particular has left the average user far behind; in most applications, only a fraction of the system's potential is being effectively used. Figure 16-10 summarizes at the highest level the differences among these three concepts.

Having laid this groundwork, for the balance of this chapter, we will use the CMMS and EAM terms interchangeably and will blithely ignore the difference between the Best of Breed and the ERP.

Purpose of CMMS/EAM

As we have noted earlier, the initial purpose of the CMMS was to streamline the production of the maintenance work orders. As the functionality has grown, so the opportunity to use the CMMS as the means to achieve far greater benefits for the maintenance function has become apparent.

In Phase 1 of the CMMS evolution, the benefits were measurable in terms of:

- reduced overhead and administration cost of preparing, issuing, and closing work orders;
- the same is true for purchase orders, materials issues, receiving, time cards, etc.
- improved timeliness and consistency of work orders.

Linking the individual procedures together in Phase 2 brought additional administrative and maintenance overhead time savings of a similar type. However, improvements in the quality and cost of maintenance and the reliability of equipment really came only as a secondary result of the maintenance staff having the option of trading admin work for organizing and analytical work.

Companies operating in Phase 3 should expect to see benefits emerging from:

- streamlined workflow;
- better planning of maintenance work and more focus on PMs;
- improved inventory numbers (both consumption and stores values);
- reduced overall maintenance costs.

To be effective in the Phase 4, companies have to not only build on their successes of the first three phases, but also embark on a new way of doing maintenance that emphasizes:

- tracking and enhancing reliability;
- advanced analysis of data;
- continuous improvement;
- the use of more sophisticated tools (including software);
- using the CMMS as a lever for best practices and profitability improvement.

In short, because the work order is so central to maintenance quality improvement, the CMMS should act as the essential link between corporate knowledge of maintenance and the employee's ability to translate this into action. A CMMS cannot become the central repository for reliability improvement knowledge. Its structure and purpose preclude it from accumulating the mass of condition-based monitoring data and the consequent analytical processes. However, it can become the means by which this knowledge is provided to the employee.

Maintaining and Enhancing Maintenance Knowledge

Given that this is so, then the essential question is "what are the requirements for maintaining and enhancing this knowledge?" These are several in number:

1. First, the clear understanding that maintenance knowledge is advancing very rapidly; organizations need a specific process for ensuring it is gathered, validated, synthesized, processed, retained, and updated. Too often the CMMS project is seen just as a one-time project rather than a continuous process of renewal and improvement.
2. Second, tying the CMMS into a performance management process is vital to avoid CMMS stagnation. The performance management process is more than just KPIs. To a large extent, KPIs have hijacked the entire performance management process; while they

are very important, they are by no means all or sufficient. Increasingly, CMMSs are including KPIs as part of their core functionality and increasingly they are being more attractively displayed. Nevertheless, the missing pieces are detracting from their value, but fortunately add-on software is helping in some cases.

3. Before KPIs are selected, the linkage between corporate strategy and the KPI needs to be made. After all, there are not much point in promoting maintenance trends that are contrary to the overall direction of the company. Again the CMMS is a key link in this chain as the output from the CMMS should act both as the scorekeeper showing the results of KPIs and also as the prompt to fine tune the objectives and strategy.

   Trends (both good and bad) need to prompt action to make something different happen. If tasks and activities do not change, then no improvement occurs. Measurements are only as good as the actions they prompt.

   Next, the KPIs have to drive lasting changes in behaviour by the staff responsible for the actions behind the measures – i.e. in our case, the maintenance staff. So understanding how these changes can be put into practice, and more importantly, the impact on the people involved becomes another critical link in the chain.

   In turn, so as to have maximum leverage, we need to measure the impact the changes will have on the KPIs – or, to adjust the phrase of the day, how do we decide what is the low hanging fruit. This cause–effect relationship between changed actions and KPIs is only dimly understood.

   Finally, the data behind the KPIs needs to continually refresh the maintenance knowledge base so as to ensure that we gain maximum benefit from the added experience.

4. Once the performance management system is in place, then we need to reverse engineer the data sources. We frequently hear the cry from CMMS users that the reports are not used because they do not trust the data. While the report format and the information it purports to show may be quite valid, the lack of data integrity undermines the confidence in and therefore the value of the report. Thus:

   – *Content*: We must examine the source data to ensure it is consistent, accurate, reliable, and needed:

   i. consistent in that a data description from plant 1 and department 1 represents the same base information as that from plant 2 and department 2;

   ii. accurate in that it is correctly collected, processed, and stored;

iii. reliable in that today's data parameters match that of yesterday's and tomorrow's;

iv. needed in that the data collection processes have far outstripped our capability for data analysis, and data not used wastes everyone's time and money. So the "need" is to link the data variable with the root cause.

– *Process*: If the data collection is not simple, straightforward, and quick then people being people, it will be resisted, fall into disrepute, and become worse than useless. Sporadic data or sloppy data collection leads to incorrect results from analysis and consequently the wrong maintenance response. Similarly, data that is consolidated and allocated while in process can also lead to incorrect conclusions from the analysis. The onset of automatic data collection (bar-coding, CBM, process control logs, etc.) is a mixed blessing. Too often the solution is seen to be "collect more data" rather than smarter analysis of data we already have.

5. Sitting between the data collection and the CMMS needs to be an array of analysis tools. Raw data is essential to the setting up of the CMMS – for the tombstone information in the asset, inventory, resource, and vendor files, for example. But once the data has been entered, then the focus should switch to a constant upgrading of the quality of the data. The tool kit needs to contain four major types of tools:

– Content tools – such as RCM – which will help to establish and improve the library of maintenance tasks.
– Analytical tools which are used to better understand the data that is being presented.
– Diagnostic tools which can be used to help to determine the root cause behind the symptom, and therefore prompt the appropriate action.
– Predictive tools which will provide the essential link between event data, the measurement readings of high impact variables and the probability of failure within a given time frame.

A key issue with each of these tools is the process by which their databases are updated with the new information – which in turn have to work their way through to changed maintenance practices as shown in the equipment maintenance plan. For example, make sure the RCM program is dynamic, revisit it to test what you are actually experiencing. Returning equipment to its original condition in the long run is impractical. Hence, for example, the long-term degradation curve must be recognized, and with it, a shorter maintenance interval.

6. Given that equipment reliability and maintenance performance improvement are key outputs from this process, then using the data derived from the above steps as the basis for helping to select the best equipment would seem to be natural outcome. However, experience shows that all too rarely is the maintenance –> design and select –> procure loop completed. Figure 16-11 shows a simplified data flow for this concept.

Keeping CMMS in Context

It is obvious from the above on the one hand, that a CMMS by itself has significant value. On the other hand, a CMMS supported and surrounded by other complementary tools can become the most powerful force in the maintenance arsenal for the improvement of quality in general and equipment reliability in particular. Hence the objectives of a CMMS are a far cry from the initial Phase 1 benefits; instead they should include:

- driving maintenance improvement and increase equipment reliability;
- acting as the basis for analysing, reporting, and recording an equipment's status and events;
- establishing the link between condition variables and the creation and issue of a work order;
- acting as the source library for improved practices;
- creating the transmission document for executing improved practices;

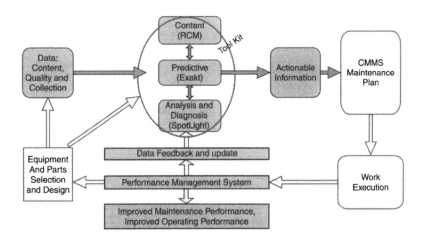

Figure 16-11. Intelligent equipment life cycle.

- tracking costs and resource consumption to equipment;
- being the basis for decision-making about priorities on expenditures about repairing and replacing.

Analyzing Failures Through CMMS

Maintenance effectiveness is impossible without good information. Good information is the synthesis of tombstone data, "as-found" and "as left" data, condition data, and operational and maintenance data (equipment events and minor maintenance). This data can reveal knowledge vital to optimal decisions about scheduled rework and condition-based tasks. There are several key issues:

1. Defining the probability of any failure (the age–reliability relationship) requires enough of the right data – including data about failures.
2. However, if the consequences of the failure are serious, then that fact by itself means that we will act conservatively to prevent the failure. Hence we lose key information about equipment behaviour at or close to the failure point.
3. On the other hand, where consequences of failure are small, lots of data exists – but of course, it is less valuable and is frequently expensive to collect.
4. The middle range between these extremes can pay huge dividends in terms of maintenance cost savings as this is where the bulk of resources are spent.

Figure 16-12 is adapted from the familiar PF curves and will illustrate both the dilemma and the objective.

Clearly, the diagram over-simplifies the situation as most equipment in today's industrial economy is more complex – i.e. it will experience failures based on a number of often unrelated component failures. The trick here is to separate the failure cycles of each of these critical components and record and analyze them.

Data from these component failures is the most fertile territory in which to conduct statistical analyses. One key point to make is that to achieve satisfactorily high confidence levels in statistical terms, the required data sample need not be numerically very large. Many maintenance analytical tools include procedures to compensate for small data sets.

The equipment's recorded history as collected in the CMMS thus can become the central data source for this analysis – but only if is it accessible, consistent, and reliable. One of the essential difficulties in

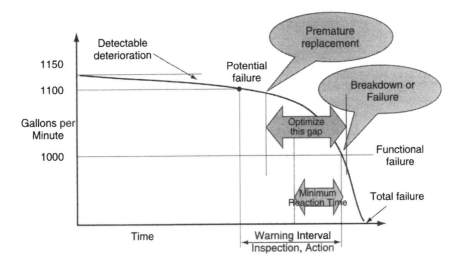

Figure 16-12. Mechanical equipment failure.

data analysis is that of reporting failures in a consistent manner. We are all familiar with the drop down lists of fault codes in the CMMSs and EAMs. And we all know that the top few on the list attract the most votes! In most cases, the tie-in between the RCM's consistent language of failure description and the CMMS's fault code is simply not made – let alone made accessible by a system interface.

What should be happening here is that the work order planner should prepare the task list by examining the RCM symptom analysis process, and the work order completer should have the same access to RCM failure causes through a symptom-related drop down list.

A second reason for paying close attention to CMMS data is that the RCM conclusions only remain valid if the operating context, performance requirements, failure modes, and effects remain essentially unchanged. If changes have taken place, then the RCM analysis needs to be revisited. Indeed in a strict RCM regime, a breakdown of an equipment indicates a breakdown in the RCM logic; both need to be analysed and repaired.

Implicit in the above discussion is that relative to the entire life of an item, the advanced warning period between the detection of deterioration and the actual failure is relatively short and is measurable. That means that gathering data to establish the age–reliability relationships using the potential failure rather than the functional failure helps to reduce the problem of lack of data for failures with serious consequences. This would then suggest changes in the inspection intervals and the times at

Combining Scheduled tasks with CBM

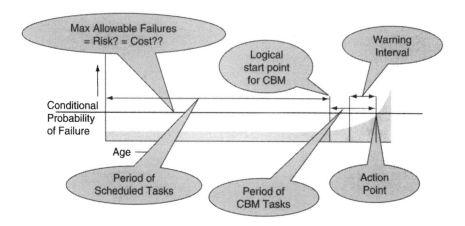

Figure 16-13. Changing patterns of inspection intervals.

which to intensify such inspections. Figure 16-13 uses a familiar curve to clarify this point.

Graphing the data will also quantify the effectiveness of the CBM program as measured by the gap between the functional failure and the potential failure curves of Nowlan and Heap's [4] diagram reproduced in Figure 16-14.

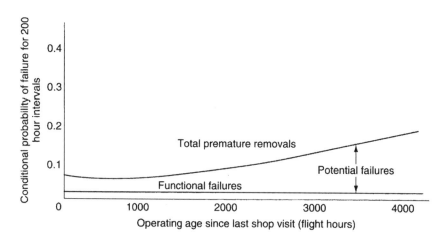

Figure 16-14. Frequency of functional failure versus potential failures.

Furthermore, graphs based on statistical analysis of this type will help to identify the dominant failure modes which themselves could be managed by some form of scheduled maintenance or redesign, improving overall reliability. Additionally, knowledge that the age–reliability relationship of an item is an exponential survival curve would point out whether current scheduled rework tasks are ineffective and in need of replacement by a more applicable form of maintenance.

Linking Maintenance Priorities to Executive Priorities

So where do we start? First, we have to close the gap between the maintenance understanding of what is important and what is the executive's understanding. The link has to be cost and finance. The report in Figure 16-15 can be readily adapted from the standard CMMS output reports.

Here we can summarize the breakdowns that have occurred over the past period. Common usage for this chart typically stops at number of breakdowns and breakdown hours. We need to take this several steps further. First, breakdown cost per hour is defined as the value of lost production; from this we can immediately see that only equipment critical to the production process should be included. This reduces the size of the report dramatically. There are two contentious areas to deal with:

1. How long did the breakdown last. There are many definitions of this – select one and stick to it.
2. Cost calculation is a tough one as it needs inputs and agreement from production and finance.

Breakdown report for the period May 2004						
Equipment	Breakdowns - #	Breakdown - Hrs	B/D cost per hour $	B/D cost	Emergency Repair $	Total B/D cost $
#5 winder	4	16	500	8,000	2,400	10,400
#4 winder	6	12	100	1,200	600	1,800
Slurry Pump	15	50	50	2,500	500	3,000
Turbine	2	4	6,000	24,000	18,000	36,000
Extruder	2	6	15,000	90,000	12,000	102,000

Figure 16-15. Example of a breakdown report.

Next, we have to identify the breakdown repair cost – not only the standard cost of the labour and materials, but also the cost of emergency buying, expediting, maintenance or contractor overtime and call-in costs, potential damage to the equipment from doing a rush job, and other problems overlooked for the same reason.

Finally (and not shown in the report summary above) is the root cause and the remedy for each significant item. And by "remedy," do not just show the fix for this breakdown, but what has to be done to prevent re-occurrence. Complete this loop by updating the PM program for this equipment in the CMMS.

A neat way of presenting the need for action is by a simple adaptation of the familiar Pareto chart (Fig. 16-16). In the chart in the left, showing the raw number of breakdowns, the slurry pump needs to be the focus for action. But by simply modifying the vertical axis to show the cost of breakdowns as opposed to the number of breakdowns, the slurry pump becomes irrelevant and the dominating issue rightly becomes the extruder. The value of these two charts shown together is that maintenance now has a compelling case to invest the company's resources in a modification program for the extruder.

Continuous Improvement Through the CMMS

We have argued earlier that the CMMS Work Order is the key connector between corporate maintenance knowledge and reliability improvement. This in turn demands a regular process for upgrading the work orders to include them with the best knowledge that we have available about the

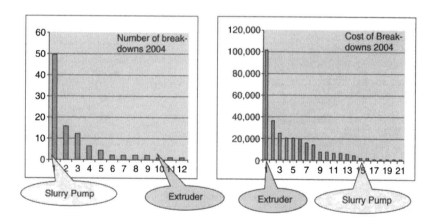

Figure 16-16. Pareto chart: Breakdowns.

equipment and its maintenance process. A few key steps will make this a standard part of the weekly maintenance department routine:

1. Set up a KPI to record the number of work order tasks that have been reviewed this month, track and post the results.
2. Create a PM-type work order to sequentially review all work orders – starting with those that are most frequently issued and cost the most to complete. The work orders should be reviewed for:

   – continuing need
   – frequency
   – task content
   – skills assigned
   – materials and tools required
   – data collected
   – follow-up action required.

3. By making them a PM-type work order that is issued each week, staff will be assigned along with a priority. This also means that incomplete reviews finish up on the backlog list for special attention.
4. Eliminate unnecessary steps, materials, etc. and match the frequency to the need. For time-based PMs and repairs, can they be changed to condition based?
5. Note that those difficult analytical tasks can also be treated in the same way; tagging them as a priority PM-type job ensures that staff and time are assigned and management attention is focused on them on a regular basis.

The role of CMMS in developing and maintaining BPs should by now be fairly obvious. But let us also realise that the label "best practice" is just that – a convenient label. In reality, BP – whether best maintenance practice or best EAM practice – is a continuously moving target and indeed a target that will vary according to the industry, its regulations, and the culture in which the business is transacted. For example, the intense focus on "safety above all" in the nuclear and the aircraft industries demands a different BP to that in sawmills or food processing. And the BPs that are the target in Scandinavia are quite different from those in the Middle East because of the different cultural demands. So "maintenance practice improvement" becomes a more sensible and pragmatic approach. Clearly, by focusing on the quality of the instructions in the work order, maintenance management has a very simple and effective means of improving quality and performance.

Phase 5: Extensions to CMMS

Despite their increasing functionality, CMMSs cannot and will never be the ultimate solution. Instead, they need to be combined with other practices and systems to enhance their value. In an earlier section, mention was made of content, analytical, diagnostic, and predictive systems – all can provide significant added value. These are becoming more and more sophisticated and can now be used to pinpoint solutions for specific problems. The most advanced CMMS vendors and users are clearly seeing the value of making these linkages. Before investing too heavily in this area, check on a couple of key factors:

1. Will the add-on specialist system do a high-quality job of providing results in your environment. For example, assumptions are sometimes built in about the number of samples that are needed or that the relationship between inputs and outputs taken from one industry are valid for others.
2. Make sure that the data collection and transfer process is easily accomplished. Difficult data collection routines do not last long.
3. Check that the results to be achieved do in fact provide sufficiently compelling proof that action becomes almost automatic. Low confidence levels on the result lead to low confidence in the tool.
4. Check that the analysis is based on sound statistical or mathematical analysis.

How to decide whether you are receiving value from your CMMS. Here is a quick value indicator that has proven itself many time over. Take the following self-test (Table 16-7). Give your CMMS implementation a five for a perfect fit, down to a zero for the reverse. Then total your score.

Companies scoring less than about 65 on this value test typically are spending too much scrambling from crisis to crisis and not enough on solving the important issues of quality and reliability. On the other hand, if you have a lot of "don't knows" in your answers, then that indicates a slightly different, but equally important problem.

Fortunately, the fix is simple and well known; unfortunately, it is tough to do among all the other priorities. But one thing is sure, if you do not fix it, then the other problems will not likely get fixed either.

Steps to increasing CMMS value. So what to do? Here is a simple 10-step approach.

Step 1: Decide what you want to get out of the system – one piece at a time.

This is the point at which senior managers may need to get involved. A key problem is the lack of common understanding (and even language!) between senior executives and maintenance

Table 16-7
CMMS value test

1. All critical equipments are in the system
2. All critical spares are in the system
3. All the data is accurate and reliable
4. All non-critical equipment and parts should be in the system
5. 70 + % of your work is done against PM WOs
6. All PMs are reviewed annually for relevance and accuracy
7. All corrective work is done from WOs
8. All breakdown work is recorded on templated WOs
9. WOs collect data – not only on material and hours used, but on the equipment condition and failure causes
10. All WOs make sense (according to the Mtce Tech on the job)
11. 95 + % of your work is scheduled
12. Parts pick-lists are prepared automatically from WOs
13. Parts replenishment is driven by automatic ordering
14. Overdue WOs are reviewed weekly
15. All breakdowns are scrutinized to update the PM program
16. The system prompts regular ABC counts
17. Reports are useful, timely, and accurate
18. Your CMMS ties into your performance management system
19. Your CMMS analyses equipment reliability and failure causes, and ties directly into RCM
20. Someone is specifically responsible for generating more value from the system

Total

management. This chasm can only be bridged by maintenance managers learning "Exec-speak." A good starting point is to make sure that proposals for spending company money are formalized in company proposal format, complete with summary, objectives, process, ROI targets, cash-flow projections, etc. To take a simple example, assume we wanted to link your CMMS with your RCM program and use it to increase equipment reliability.

Step 2: Confirm that the system can deliver what you want (it almost always can!)

The CMMS functionality is so strong these days that almost always it will out-perform the capability of the user. If only you could figure out how it is done? If you are in any doubt as to whether it will perform your requirements, then call the Help-line. In our current example, you probably have two different databases – one for the RCM work and one in the CMMS. And they probably never talk to each other. Which in turn means that when you collect event and failure data through the work order, it neither matches the symptom-root cause work you did in the RCM nor does it update the RCM database. Rarely do CMMSs do this well.

Step 3: Understand what you are currently doing with the system.

Many people will try to bypass this step. Do not. What it frequently does is to provide the basis for a simple modification rather than creating a whole new procedure. For our RCM link, you will probably have used the RCM process to create the library of tasks for critical equipment, but probably went no farther. Some companies are using the RCM fault codes list as the basis for the failure codes in the work order completion, but it is usually as a result of re-typing them into the CMMS drop down list. Which means they are not symptom related, rather they are equipment related. This in turn means there is a tendency to select the top one on the list because it is the easiest to access.

Step 4: Do a gap analysis between current and target.

Here we need to define what our change in procedure is going to produce for us – system linkage between RCM and work order so that the work order planning, execution, data collection, and closure routines are based on the equipment's symptoms as built into the RCM analysis. Plus the data collected from the work order needs to be analyzed to see if there is a change prompted in the RCM analysis and the future work orders.

Step 5: Itemize tasks to be done to close the gap.

First decide on how the RCM–CMMS linkage can be made to happen – through an external utility, through an extension to the CMMS screens and forms, or by requiring the CMMS vendor to buy into the upgrading of their capability. Next, list the changes in the CMMS procedures for work order planning, release, execution, and closing. Then, identify the changes in the CMMS output reports and in analysis of the results from the CMMS report. Next, show the methodology for updating the RCM database and the work management routines. And please do not forget the training that needs to go along with it, otherwise it will not work.

Step 6: Prioritize tasks based on both payback and cost and difficulty of implementing.

In our simple example, we should consider all the steps as essential, so prioritising them is not really an issue. However, we should do a financial evaluation to assure ourselves (as well as the senior executives) that we are spending company money wisely. The payback will be measured in terms of reduced frequency of breakdowns; so we need to know the current cost of breakdowns (repair cost plus lost production value), plus a realistic target for improvement.

Step 7: Plan the tasks in detail – what has to be done, by when, and by whom.

This step should proceed only if we expect a positive return on investment. In more complex projects, and especially if you have a specific deadline, you will want to use a project management package like MS Project. In our example, this is likely not needed, so an action planning work sheet will cover it off nicely. Identify in detail what the tasks are, what is the sequence of activities, who is to do them (whether internal or external), what is the expected outcome, and who is to do it. Allow for approvals where needed and also lead times for any purchased items.

Step 8: Use the CMMS work order process to issue and execute the tasks

Most companies reserve the CMMS work order process strictly for maintenance work. No reason it cannot be used for these types of mini-projects. The work order can be used for task definition, resource allocation, materials required, deadlines, cost collection, and reporting. Plus any late tasks will appear on the overdue work order list at the next maintenance department meeting. It also has the advantage of letting other maintenance staff know what is happening.

Step 9: Install a simple progress tracking process – both for the tasks and the resulting improvements.

Tying CMMS into performance management is very useful exercise – especially if the performance management process is broader than just KPIs. Set up the KPIs to track the improvements you are forecasting and post them so that we can all see the progress. A good way of presenting the data is through Pareto charts – except you should track not the frequency of failure, but the cost of failure (again the repair cost plus the lost production value).

Step 10: Get on with it!

As the results start appearing, do not be concerned if the turnaround is not immediate. It frequently takes time to work through the system. However, the analysis you do should be prompting you to take specific action with regard to individual equipments or types of equipment. Make sure the action plan targets the root causes of the failure, the revised work plan is actually being executed by the technician on the job, and the required data is being accurately and consistently recorded.

A complete overhaul to bring you up to date could take up to a year depending on the resources you can put on the job. If you prefer to use an outside person or company to help you, then they should concentrate on:

- diagnosing the problem;
- helping to plan the solution;
- guiding the upgrade;
- providing technical help.

Make sure your own people do the work; that way the knowledge stays in-house and the costs stay reasonable.

Maintenance Effectiveness Audits

Periodic auditing of the state of your reliability management practices is needed if your plant wants to be assured of not being overtaken by the competition. One effective way to conduct such audits is to first provide key members of your organization with detailed checklists that highlight how BOC companies:

- address reliability issues (the "team model");
- arrange for input at the specification phase;
- organize follow-up after issuing purchase documents;
- carry out RCFA;
- perform equipment alignment tasks;
- implement rotor balancing procedures;
- perform vibration monitoring and analysis;
- define and execute equipment growing work;
- define, implement, and carry out machinery repairs.

These key staff members might represent such job functions as equipment engineers, maintenance supervisors, equipment technicians, operators, stores personnel, inspectors, and others. They would be asked to review available plant records prior to meeting with the specialist performing the audit. Their input would then form the basis for a 27-point summary matrix (Table 16-8).

As an example, the overall impression of the survey results at the "XYZ Specialty Gas Company" may illustrate the audit approach. This plant excelled in the way major machinery was selected, installed, monitored, and repaired. Assuming a maintenance cost versus asset replacement value of 2.0%, this company ranks near the very top of the best performers.

Challenges Remained

The auditor pointed out that the reliability/maintenance organization should, in some measure, disengage itself from the obvious priority of

Table 16-8
Maintenance Effectiveness and Reliability Organization Survey

#	Description/item	Observation/explanation	Ranking
01	Reliability group acts as a *service* (on-demand) organization rather than a fully proactive *support* organization		0
02	Recognition and implementation of component/equipment *surveillability* and *supportability* in specification, bid evaluation, and procurement practices		0
03	Assignment of operators to surveillance tasks, i.e. optimization of reliability group personnel for analysis of deviation from norm, pursuit of improvement measures rather than routine data collection		0
04	Input in project development and execution sequence; advisory and justification input		0
05	Definition and compilation of maintenance and reliability-related documentation, including installation guidelines, repair procedures, troubleshooting checklist, etc.		0
06	Organized to view every maintenance event as an opportunity to evaluate upgrade options		0
07	Performing true and effective RCFA wherever appropriate and following up in most effective fashion		0
08	Resourcefulness in capturing vendor/OEMs prior or available work product. For example, mechanical seal application statistics		0
09	Utilization of appropriately detailed, rapidly retrievable failure histories from accessible computers		0
10	Networking, access to mentors, effective, and continuous training and expansion of knowledge base; detailed training plans		0
11	Understanding of reliability impact of special tooling, e.g. torque wrenches, coupling hub pullers, bearing heaters, etc.		0
12	Taking lead role in identification and verification of critical spare parts, including decision-making process for OEM versus non-OEM spare parts procurement		0
13	Development and utilization of component specification, e.g. rolling element bearing code letters identifying acceptable internal clearances, cage materials, etc.		0
14	Utilization of cost-effective condition monitoring methods, including temperature, vibration, velocity, spike energy, lube oil analysis, ferrography, particle count, etc.		0
15	Optimization of synthetic lubricant usage, cost justification for on-stream or batch purification of lubricants		0
16	Grease lubrications methods, including compatibility and frequency, sealed versus re-lubricated bearings, avoidance of single-point device		0
17	Recognition of sound foundation, baseplate, grouting, and leveling criteria. Disallowing (!) installation of mounted pump and driver sets		0
18	Applying two-hand rule to piping installations, piping "away" from equipment casings, two PTFE plates at sliding supports		0
19	Understanding of standardization and its limitations, participation in decision-making process		0
20	Vendor stocking programs, *intelligent* alliances, and partnerships		0

Table 16-8
Maintenance Effectiveness and Reliability Organization
Survey–cont'd

#	Description/item	Observation/explanation	Ranking
21	Process (operations) interface, e.g. involvement in definition of cavitation and pump internal recirculation susceptibilities, etc.		0
22	Review-type participation in PSM document development		0
23	Well-defined role statements disseminated to other plant functions; status summaries and accomplishments routinely documented; activities given proper visibility		0
24	Work scope specifications developed and consistently utilized		0
25	Contractor interface and experience upgrading efforts delineated and pursued		0
26	Repair versus replace decisions based on solid financial analysis and economics		0
27	Awareness and pursuit of equipment reliability upgrade and maintenance/operations cost reduction opportunities (e.g., surge control, lube oil purification, etc.)		0

Note: Rankings are either numerical (10 = organization applying best available practices, 1 = Organization applying least desirable/unacceptable practices) or letter based (A = excellent, F = failing).

ascertaining availability and reliability of major machinery at this facility. Money could still be made in failure avoidance of pumps and other general-purpose equipment.

Plant management was advised that:

- There were clear opportunities in such areas as providing guidance and direction in assembling better repair and failure data. XYZ must know its pump MTBF and maintain a running log of "bad actor" pumps.
- The company employed above-average quality-reliability technicians and engineers. However, some of these employees were spending time on routine tasks such as collecting general vibration data. Instead, they should be redirected to concentrate more on the value-added task of analyzing excursion data collected by others and developing permanent remedies for equipment at risk.
- Although XYZ had a small number of reliability professionals, for a company its size, it was asked to defer adding more personnel until their demonstrated abilities were fully utilized. This can be done by emphasizing value-added assignments and encouraging them to be resourceful.
- Resourcefulness has many facets. For instance, letting vendors and suppliers provide data on bearing designations, the reliability professional would efficiently restructure these into important component

specifications. The latter would enable the purchasing department to buy least-risk or lowest lifecycle cost bearings, mechanical seals, etc.

- Root-cause failure identification is key to failure elimination. Repeat equipment failures at XYZ indicated that the true causes of equipment distress had not been identified. The obvious recommendation was to arrange for refresher courses. Problem-solving exercises should use XYZ's own examples. Thereafter, RCFA should be institutionalized.

- XYZ was asked to adopt the mindset that repeat failures are as unacceptable as safety incidents. Every unanticipated maintenance event should be seen as an opportunity to upgrade the equipment. Providing answers to the questions – Is an upgrade possible? and, if yes, is it economically justified? – should be the primary goal of reliability professionals.

References

1. Mitchell, John S., *Physical Asset Management Handbook*, ISBN 0–9717945–1–0.
2. Modified from H. Grothus, *Die Total Vorbeugende Instandhaltung*, Grothus Verlag, Dorsten, Germany, 1974.
3. T. Idhammar, Maintenance management legends, *Plant Engineering*, September 2004, pp. 31–34.
4. Nowlan, F. S. and Heap, H. *Reliability-centered Maintenance*. Springfield VA: National Technical Information Service. US Deparment of Commerce, 1978.

Chapter 17
Continuous improvement

Needless to say, optimizing machinery uptime is a continuous improvement task. This chapter now shows how the leading companies, the Best-of-Class practitioners, try to interrogate their available data in an effort to "predict the future." What at first seems a step up in complexity is not really that difficult and competent teacher-consultants can be enlisted for these analyses.

Predict Future Failures From Your Maintenance Records*

What are CROW/AMSAA Reliability Plots?

They are cumulative failures plotted against cumulative time on log-log graphs form Crow/AMSAA reliability growth plots. The plots can handle data from single failure modes or multiple failure modes. Slope of the trend line is an important statistic telling if failures are increasing, decreasing, or the failure rate is unchanged. The method is simple and visual.

The challenge of every reliability engineer is to make reliability improvements to avoid failures. Improvements, with longer times between failures, will put a cusp on the trend lines. The cusp will demonstrate a real change has occurred by substantially stretching the time until the next failure. The longer intervals to the next failure will cause localized trend lines to appear with flatter slopes. When the former trend line is extrapolated to longer times, improvements must demonstrate measurable, vertical gaps which measure the cumulative failures avoided by the

---

* Contributed by H. Paul Barringer, P.E., Barringer & Associates, Inc.

improvements. Thus improvements are visual and quantifiable likewise deteriorating conditions produce steep slopes, and situations of no change are identifiable.

The view from your office may be spectacular, but can you see the future failures and make your information visual to the organization? You need a vision for forecasts of future-expected failures along with their costs and alternatives for reducing the costs. The tools for gaining this vision are your maintenance failure data and Crow/AMSAA plots. The view for reliability growth plots comes from the simplicity of straight lines on log-log plots.

Today, log-log plots are emerging from unusual studies. The straight-line plots make explanations easy and understandable. Web crawler robot studies on the Internet find a "power law distribution" relating incoming links on Web pages and outgoing links to Web pages. Studies of computer networks spell out straight-line relationships on log-log plots. Scientists fail to see straight-line relationships on log-log plots because they have not looked for them [1]. Barabási's unique exponents for his network equations have negative values, over a limited range of values, whereas reliability growth curves have positive exponents, again over a narrow range. The log-log plots describe natural laws at work.

Why do CROW/AMSAA Growth Plots Make Straight Lines?
Why do Crow/AMSAA Plots of Cumulative Failures Versus Cumulative Time Produce Straight Lines on Log-log Plots?

The forerunner of the C/A concept has parallel roots in manufacturing with exhaustive demonstration as log-log phenomena. It is a natural occurrence of learning and improving. Consider the following parallel which began before Crow/AMSAA plots.

T.P. Wright [2] pioneered an idea in 1936 that improvements in man-hours to manufacture an airplane could be described mathematically – a very helpful concept for management production planning. Wright's findings showed that, as the quantity of airplanes produced in sequence, the direct labor input per airplane decreased in a mathematical pattern that forms a straight line when plotted on log-log paper. If the rate of improvement is 20% (the learning percentage is 80%) and thus when large processes and complicated operations production quantity is doubled, the time required for completing the effort is 20% less. Thus, a unit of production will decrease by a constant percentage each time the production quantity doubles. Figure 17-1 illustrates the concept [3].

Wright's method in the 1940s was a helpful concept for the USA War Production Board in estimating the number of airplanes produced for a given complement of men and machines. After World War II (WWII),

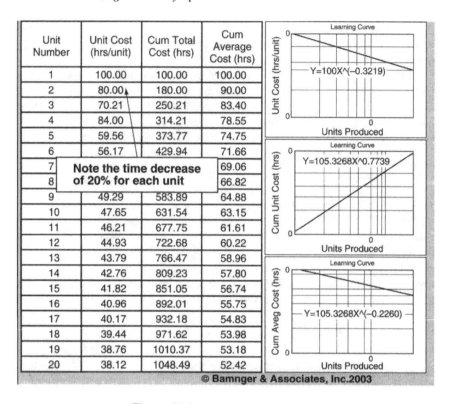

Unit Number	Unit Cost (hrs/unit)	Cum Total Cost (hrs)	Cum Average Cost (hrs)
1	100.00	100.00	100.00
2	80.00	180.00	90.00
3	70.21	250.21	83.40
4	84.00	314.21	78.55
5	59.56	373.77	74.75
6	56.17	429.94	71.66
7	Note the time decrease		69.06
8	of 20% for each unit		66.82
9	49.29	583.89	64.88
10	47.65	631.54	63.15
11	46.21	677.75	61.61
12	44.93	722.68	60.22
13	43.79	766.47	58.96
14	42.76	809.23	57.80
15	41.82	851.05	56.74
16	40.96	892.01	55.75
17	40.17	932.18	54.83
18	39.44	971.62	53.98
19	38.76	1010.37	53.18
20	38.12	1048.49	52.42

© Bamnger & Associates, Inc.2003

Figure 17-1. Learning curve concept.

the US government employed the Stanford Research Institute (SRI) to validate improvement curve concepts. Stanford Research Institute studied all USA airframe WWII production data to validate the concept and SRI developed a slightly different version than the simple case offered by Wright (DOD 2003) which also plotted on a log-log plot as a straight line. Today we know the log-log concept as learning curves when involved with production units and time/cost. Other names are cost improvement curves, or progress function, or Crawford curves (J. R. Crawford was on the SRI validation team – Crawford's model is considered less technical than Wright's model), or Boeing curves, or Northrop curves, and so forth to represent the findings of each manufacturer of airframes. Each manufacturer developed a variation on T. P. Wright's simple equation.

The simple improvement curve was $Y = AXB$. This curve will produce a straight line on log-log paper. Y is the unit cost (hours/unit or $'s/unit), X is the unit number, A is a theoretical cost of the first unit (hours or $'s) and B is a line slope constant that is related to the rate of improvement – B is literally equal to ln(learning percent)/ln(2) where the learning percent = 100 − (rate of improvement). For example, if the first unit took 100 hr to complete (A = 100) and if we had an improvement rate of 20%, the learning percentage would be 80%, so that B = ln(1.00 − 0.20)/ln(2) and B = −0.32193. Thus we would expect production of the second item would require 80 hr and the fourth item produced would require 64 hr , and so forth, as the production quantity doubles we save 20% from the production time.

Some typical learning curve slopes are described at the NASA Cost Estimating Website (NASA 2003) and the learning percent varies from a low of 96% for raw materials to a high of 75% for repetitive electrical operations with most values around 80–90%. The plots can have three different formats:

1. hours/unit or $/unit versus cumulative production;
2. cumulative (hours or $'s) versus cumulative production;
3. cumulative average (hours or $'s) versus cumulative production.

General Electric (GE) Company made extensive use of learning curves in their manufacturing operations. A GE reliability engineer (James Duane) made log-log plots of cumulative MTBF versus cumulative time which gave a straight line for reliability issues (Duane 1964). Duane argued for the use of failure data on complex electromechanical systems. He recommended the Y-axis should be $Y =$ (cumulative failures)/(cumulative time) $= KT − \alpha$ where the value K is a constant which is dependent upon equipment complexity, design margins, and design objectives for reliability. Duane said the value for $\alpha \approx 0.5$ with the expectations that some designs would be better (meaning $\alpha > 0.5$) and some would be less (meaning $\alpha < 0.5$) and T is cumulative time. Duane drew his conclusions from studying five different data sets and found remarkable similarity in patterns for the curves (meaning the line slopes were about the same). Duane also rearranged his equations and showed cumulative failures $F = KT(1 − \alpha)$ so the formula allowed forecasting future failures based on past results. James Duane had a deterministic postulate for monitoring failures and failure rates of a complex system over time using a log-log plot with straight lines.

At the US Army Material Systems Analysis Activity (AMSAA) during the mid-1970s Larry Crow converted Duanes postulate into a mathematical and statistical proof via Weibull statistics. MILHDBK-189 (DOD 1981) describes the details. The military handbook addressed:

- *Reliability growth*: The positive improvement in a reliability parameter over a period of time due to changes in product design or the manufacturing process.

- *Reliability growth management*: The systematic planning for reliability achievement as a function of time and other resources, and controlling the ongoing rate of achievement by reallocation of resources based on comparisons between planned and assessed reliability values.

The ultimate goal of the improvement program was to make reliability grow to meet the system reliability and performance requirements by managing the development program. The management effort required making reliability

1. visible, and
2. a manageable characteristic.

Reliability growth programs required goals and forecast of progress. The failure data usually produced straight-line segments on log-log plots with $N(t) = \lambda t\beta$ where N is the expected number of failures, λ is the failure rate at time $t = 1$, t is cumulative time, and β is the line slope for cumulative failures versus cumulative time (and $\beta = 1 - \alpha$ from Duane's equation). Scientific principles determine that failure data fit $N(t) = \lambda t\beta$ and thus failure data trends can produce a straight line on log-log paper.

Data from maintenance failure databases plotted on a log-log plot will build a Crow/AMSAA relationship for finding the Y-axis intercept at $t = 1$ to identify λ and the slope of the line will define β changes in the programs. Thus, future failures can be forecasted and cusps on the data trends will tell if the system is improving (failures are coming more slowly, $\beta < 1$), deteriorating (failures are coming more quickly $\beta > 1$), or if the system is without improvement/deterioration (failures rates are unchanged, $\beta \approx 1$).

Recently AMSAA updated the information from the USA Military Handbook MIL-HDBK-189 and produced the AMSAA Reliability Growth Guide TR-652 [4].

Examples

Example 1

Actual pump maintenance interventions are reported from a Brazilian chemical plant (Barringer 1997) based on data shown in Table 17-1. The Crow/AMSAA plot is shown in Figure 17-1 using reliability software (Fulton 2003) and Crow/AMSAA reliability technology [5].

The cumulative failures versus cumulative time produce two straight lines. The trend line before starting a TPM program [6] shows slight improvement ($\beta = 0.947$). After introduction of a TPM program operators were taught a few fundamental things they could do to reduce failures. Notice how the failure trend line shows a distinct cusp in Figure 17-2. The improvement curve shows a slope $\beta = 0.529$ which is almost as predicted by Duane at $\alpha = 1 - \beta = 1 - 0.529 = 0.471$.

Using the data in Table 17-1 and Figure 17-2 the savings from the TPM program at time $t = 36$ months (29 months into the TPM effort) have been Nbefore $= 34.65(36)^\wedge 0.947 = 1032$ interventions, Nafter $= 77.49(36)^\wedge 0.529 = 516$ interventions which is an avoidance of 516 interventions in 29 months or 18 interventions/month. Assume each intervention has an average cost of US\$1000, the savings from the TPM program has been (516 interventions)*(1000\$/intervention) $= \$5,16,000$ over the last 29 months. The net savings for the TPM program will be amount saved less amount spent for introducing the TPM effort. In most cases, you can easily justify a TPM program based on this scorecard data. Every maintenance program requires factual justification of costs and benefits, and Crow/AMSAA plots organize the facts into straight lines.

Table 17-2 is a forecast of failures for the next 12 months using the trend line after implementation of the TPM program in Figure 17-2. This monthly forecast of failures will be for months 37 through 60 to cover a 2-year forecast interval.

Table 17-1
Maintenance interventions

Month	1995	1996	1997
January	35	12	8
February	32	13	3
March	28	12	15
April	30	11	5
May	41	11	10
June	30	11	9
July	16	15	8
August	18	9	7
September	21	8	7
October	14	8	9
November	12	10	7
December	11	10	8
Total =	288	130	96
TPM began in August 1995			

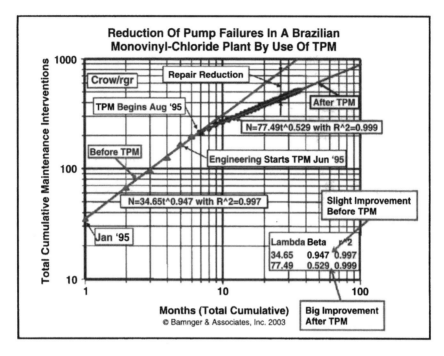

Figure 17-2. Reduction of pump failures.

Table 17-2
Maintenance interventions: Forecast

Month	1995	1996	1997	1998	1999
January	35	12	8	8	7
February	32	13	3	7	7
March	28	12	15	7	6
April	30	11	5	7	6
May	41	11	10	7	6
June	30	11	9	7	6
July	16	15	8	7	6
August	18	9	7	7	6
September	21	8	7	7	6
October	14	8	9	7	6
November	12	10	7	7	6
December	11	10	8	7	6
Total =	288	130	96	85	74

TPM began in August 1995

Major improvements for Example 1 were achieved by operating pumps at the best efficiency point (BEP) and introducing a pump maintenance training program [6]. This required cooperative efforts between operations and maintenance. A Pareto distribution was established prior to the kick-off of the TPM program to identify 'bad actors' and build a Pareto priority list for action by the team – in most cases, the pumps required trimming of the impellers using the laws of affinity along with correction of net positive suction pressures. Pumps operate on their BEP by decisive action. Pumps operate off their BEP by benign neglect and errors. Insufficient net positive suction head and off-BEP causes vibration, cavitation, and other harmful actions which drive-up the need for maintenance interventions.

Example 2

Failures strongly influence most total maintenance department expenditures. The "failure data" is simply maintenance cost (as cost is a precursor for failures). A TPM was initiated in January 2002 (but not advertised), operator involvement began in February 2002, and hand-held computers went active in July 2002 (advertised as commencing a new program). Maintenance costs are for a specific area of a petroleum refinery operation. The improvements involved use of mobile, hand-held data logging equipment to verify touching the equipment and proper equipment monitoring so operators take responsibility for both equipment and the process.

In January 2003, an assessment occurred to find the improvement savings. The data is not very clean as shown in Table 17-3. Note the data in Table 17-3 is not monotonically increasing in maintenance costs (i.e., a credit was received for maintenance costs overcharges representing 2 year-end corrections and 1 mid-year correction). Three italicized cost values show the specific data points not used in the calculation of trend lines (although the cumulative maintenance costs are included). Thus Table 17-3 represents dirty data with imperfections.

The Y-axis of Figure 17-3 is US \$ (not failures). It shows savings began almost as soon as operators were involved in the improvement effort. Furthermore, the trend line of Figure 17-3 includes the data points to the left of the cusp. Notice the trend line slope, $\beta > 1$, tells that maintenance costs (a precursor for failures) are accelerating with time.

Figure 17-4 zooms in on the plotted data points in the upper right-hand corner, so that the cusp is clearer. The trend line for most of the data is based on years 1999 through 2001 plus 1 month of 2002. The trend line after the cusp is comprised of the last 11 data points in Table 17-3 and the cusp is literally computed as 1151 days. February 2002 was decided based on good engineering judgment along with a few trial-and-error

Table 17-3
Failures represented by monthly maintenance costs

Petroleum Refinery Department Maintenance Cost History For One Area

	1999		2000		2001		2002	
	Cum. days	Cum. $'s	Cum. days	Cum. $'s	Cum. days	Cum. $'s	Cum. days	Cum $'s
Jan	31	210,097	396	4,146,017	762	8,805,297	1127	13,627,145
Feb	59	456,441	425	4,450,893	790	9,077,531	1155	14,076,446
Mar	90	756,350	456	4,846,968	821	9,435,355	1186	14,275,526
Apr	120	1,028,044	486	5,129,931	851	9,746,244	1216	14,537,284
May	151	1,262,368	517	5,673,580	882	10,135,413	1247	14,937,865
Jun	181	1,540,101	547	6,147,311	912	10,674,844	1277	14,732,077
Jul	212	1,815,380	578	6,896,160	943	10,957,464	1308	15,075,166
Aug	243	2,121,788	609	7,537,645	974	11,420,963	1339	15,310,813
Sep	273	2,769,953	639	7,856,635	1004	11,932,656	1369	15,589,596
Oct	304	3,047,065	670	8,254,432	1035	12,857,704	1400	15,826,120
Nov	334	3,360,486	700	8,716,149	1065	13,402,128	1430	15,944,082
Dec	365	3,748,406	731	8,440,050	1096	13,214,697	1461	16,275,941

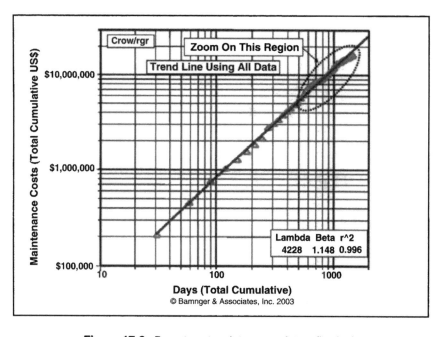

Figure 17-3. Department maintenance data – first look.

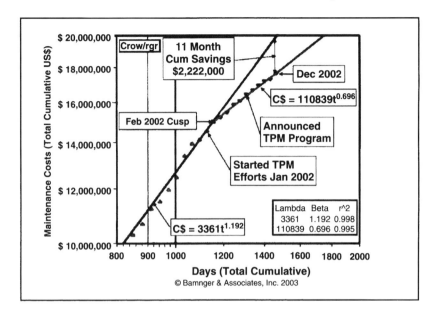

Figure 17-4. Department maintenance data – second look.

selections of the data points in each set. Figure 17-4 quantifies savings during the year 2002 from the improvement program.

In Figure 17-4 notice how much better behaved (lower variability) the data is on the plot following operator involvement in the maintenance programs. The trend line slope, β, after the cusp tells that costs are growing more slowly with time. Trend line savings at the end of year 2002 was $3361(1461)1.192 - 110839(1461)0.696 = $ US \$2.222 million as the gap between the two trend lines at month 36 = 1461 days. Since the trend lines were diverging, the savings for 2003 was larger than for 2002 – does this remind you of the adage "the rich get richer and the poor get poorer!"

For the 2003 fearless forecast, the cumulative savings by the end of year 2003 $(1461 + 365 = 1826$ days) was $3361(1826)1.192 - 110839(1826)0.696 = $ US \$5.313 million. The savings for the year 2003 alone was (US \$5.313 – US \$2.222) = US \$3.09 million. No tree grows to reach the heavens and no improvement program continues indefinitely. It is reasonable to consider the line slope for the improvement curve will begin to swing toward a slope of $\beta = 1$ in 3–5 years from the start of the program.

All TPM programs require selling (not telling) and persuading (not forcing) the workforce to "make a change to get a change" in performance. Most TPM programs require relinquishing control of maintenance decisions to the operators [7]. All TPM programs require training of the operators in fundamental information about the equipment and how the process can affect the equipment all in the quest for reducing costs. Think of the capital expenditure and instruments required to achieve the information easily acquired by the operators with an assist from hand-held data logging equipment and the five senses of the operator on a mutual quest for making improvements. Suppose you do not like the TPM concept, just find another smart way to make the improvements and then use your data to predict future failures – do not wait time flies [8].

Example 3

A chemical plant, with a fairly stable level of employment, has recorded the following reportable safety incidences over a long time as shown in Table 17-4 for a 9-year time period. Each safety incidence represents a

Table 17-4
Safety record: Major chemical plant incidents

Cum. days	Cum. incidents	Cum. days	Cum. incidents	Cum. days	Cum. incidents	Cum. days	Cum. incidents
1	1	367	26	1046	53	2622	88
8	2	368	27	1096	54	2742	89
23	3	429	28	1184	55	2754	90
47	4	526	29	1195	56	2825	92
53	5	553	30	1291	57	2846	93
58	6	585	31	1345	58	2851	94
65	7	598	32	1397	59	2888	95
67	8	599	33	1565	60	2922	96
72	9	600	34	1591	61	2969	97
78	10	632	36	1598	62	2984	99
94	12	635	37	1624	63	3099	100
105	13	660	39	1626	74	3106	101
106	14	677	40	1634	75		
108	15	690	41	1655	76		
124	16	719	42	1670	77		
149	17	759	44	1692	78		
226	18	773	45	1711	79		
228	19	830	46	1753	81		
248	20	878	47	1759	82		
285	21	1009	48	1990	83		
288	22	1018	49	2186	84		
289	23	1031	50	2430	85		
310	24	1040	51	2472	86		
312	25	1044	52	2509	87		

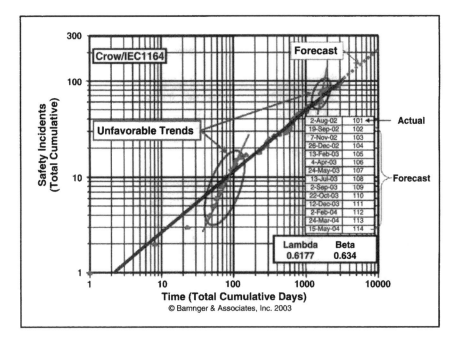

Figure 17-5. Safety incidents (January 1994–August 2002).

failure. The bold horizontal lines separate data by year. Is the plant safety program improving? How long until the next failure incident? Table 17-4 data produces the Crow/AMSAA plot in Figure 17-5.

Figure 17-5 shows a long-term improvement in the safety records at this plant – incidents are declining as reflected in the line slope with $\beta < 1$. A forecast of when safety incidents (failures) can be expected are shown as an inset in Figure 17-5; the next failure is expected in 49 days. In Figure 17-5, notice the steep upward trends that highlight troublesome periods with a return to the trend line.

Safety failures (incidents) occur in an insidious manner. You need trend lines (preferable straight lines as sales tools) to show the team how safety programs are progressing. The long-term safety incident graph in Crow/AMSAA format shows two interesting line slopes. The "unlearning" trend lines display steeper slopes for degradation than the improvement trend line. Clearly, safety improvements are a learning process and likewise deterioration in safety is an unlearning process where humans can impact the records.

The important task in safety programs is to put cusps on the data to make the trend line turn sideways toward more shallow slopes where incidents occur over increasing long-time periods. The goal is to have a safety incident-free environment.

Safety failures are occurring over increasingly longer periods of time as shown in Figure 17-5 as inferred by $\beta < 1$. This plant is operating with roughly ~50 days per incident. Is this good enough for a safety record? – Never! Compared to other chemical plants, this facility has a good record. Yet, it can still be improved.

Example 4

Table 17-5 shows failure records for environmental spills. A double line separates the new improvement initiative from the old practice.

Spills are Failures. Spills incur clean-up expenses. Spills incur governmental non-compliances. Clean-up for spills is hundreds of times the cost of lost fluids from the spills. Spills should never be taken lightly. Figure 17-6 shows the Crow/AMSAA graph of the actual data along with a projection of failures reduced from the new initiative.

The gap between old practice and new practice is easily observed. Spill reduction is calculated from the simple equation $N(t) = \lambda t^{\wedge}\beta 3$ for the statistics defining the trend line of failures. The calculated failures saved from the improvement initiative is the delta between the improvement trend line and the old method trend line.

Table 17-5
Fearless forecast of spills

Raw data			Crow/AMSAA data		Forecasts	
Spill date	Days between spill	Spill events	Cum. days	Cum. spills	Failures predicted by old method	New method savings
11/18/1995	35	1	35	1		
1/31/1996	74	1	109	2		
5/8/1996	98	2	207	4		
5/22/1996	14	1	221	5		
7/29/1996	68	1	289	6		
8/23/1996	25	1	314	7		
8/25/1996	2	1	316	8		
6/20/1997	299	1	615	9	18	9
2/22/1998	247	1	862	10	27	17
2/10/1999	353	1	1215	11	41	30

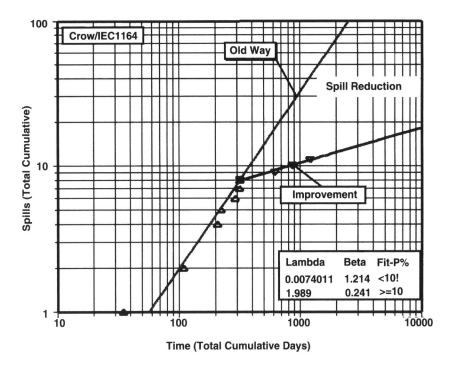

Figure 17-6. Spills – improvements.

When processes are pushed for improvement, they often require continued nursing to maintain the improved conditions otherwise they have relapses. Unfortunately, for this case, the new track is maintained only for a short interval (three failures in 899 days), then attention shifts to other issues and set backs occur. Many organizations accept deterioration without objection and resume the previous bad behavior unless they have clear signals for re-initiating improvements. This is illustrated in Table 17-6 and shown in Figure 17-7 where the relapse data shows 13 failures in 1283 days! Without visual clues, too many organizations fail to correct the bad behavior resulting in significant retrenchment from good performance [9].

The missed opportunity column represents the delta between the improved trend line and the relapse line. You can argue that even with the relapse we have a savings and this is true. However, the relapse from better performance shows ever-growing missed opportunities from not carefully "tending the farm." The relapse line slope is $\beta \sim 1$. The slope tells we are neither making improvements nor suffering from deterioration.

Table 17-6
Relapse data

Raw data			Crow/AMSAA data		Forecasted failures		
Spill date	Days between spill	Spill events	Cum. days	Cum. spills	Failures predicted by old method	New method savings	Missed opportunities from relapse
11/18/1995	35	1	35	1			
1/31/1996	74	1	109	2			
5/8/1996	98	2	207	4			
5/22/1996	14	1	221	5			
7/29/1996	68	1	289	6			
8/23/1996	25	1	314	7			
8/25/1996	2	1	316	8			
6/20/1997	299	1	615	9	18	9	
2/22/1998	247	1	862	10	27	17	
2/10/1999	353	1	1215	11	41	30	
8/16/1999	187	1	1402	12			1
11/7/1999	83	1	1485	13			2
2/12/2000	97	1	1582	14			2
4/29/2000	77	1	1659	15			3
11/16/2000	201	1	1860	16			4
12/25/2000	39	1	1899	17			5
3/25/2001	90	1	1989	18			5
8/1/2001	129	1	2118	19			6
10/28/2001	88	1	2206	20			7
7/10/2002	255	1	2461	21			9
7/25/2002	15	1	2476	22			9
9/6/2002	43	1	2519	23			9
2/18/2003	165	1	2684	24			11

Generally speaking, processes either improve or deteriorate and the status quo rarely continues for very long. Experience says this process will deteriorate and failure will grow unless corrective action is applied to significantly reduce the number of spills. Unfortunately, the action from many management groups is to declare the improvement changes are of no value and to trash the good work that achieved two spills in 18 months instead of correcting the problems associated with the relapse conditions. Here is where the Crow/AMSAA plots are of great use in providing the effective sales tools to show changes and sell the organization in getting back on track for the improvement curve.

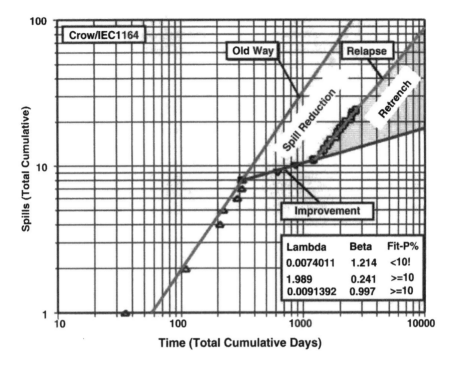

Figure 17-7. Spills – relapse.

Example 5

Chemical plants and refineries around the world are adding co-generation facilities expecting sale of their excess power into the national power grid to pay for the cost of the capital installation. Co-generating units produce electric power and steam for manufacturing processes and they function at high efficiency for the combined plants to get the highest return for capital expended.

Co-generating units have many different operating modes. Most plants supplement power supplied from the national grid (outages are not critical). A few other co-generating plants function as islands to carry the full demand load as any power outage has huge costs of unreliability for the manufacturing operations – but basic greed causes many companies to consider this for low cost power (island outages are extremely critical and highly reliable systems are required). Others function as islands of supply with backup power available from the national grid to provide uninterrupted electrical service – of course, this backup source has a fixed fee for the life-line to the grid (island outages are mitigated for a price paid by the life-line attachment to the national power grid).

Table 17-7 shows the failure record for a co-generating system. Data commences with the commission date and reflects 31 forced outages in 1432 days or some 46 days per forced outage. The typical thought process is "We are moving through the new problems and soon we will be all right."

Figure 17-8 uses data from the two right-hand columns in Table 17-7 for the Crow/AMSAA plot. The failure data generates a good straight line on the log-log paper with $\beta = 0.996$. The line slope infers a system functioning without improvement or deterioration. Figure 17-8 tells we are not working our way through the problems (as if we were correcting infant mortality problems)! We are in a static condition of failures that respond as if the forced outages occur from chance events.

Use the λ and β statistics to predict seven failures expected during 2003 for failures 32–38 where $t = (N/\lambda)^{\wedge}(1/\beta)$, where t is the cumulative future time and N represents cumulative future failures for the "fearless failure forecast" number/date:

32. February 23, 2003
33. April 11, 2003
34. May 27, 2003
35. July 13, 2003
36. August 28, 2003
37. October 14, 2003, and
38. November 29, 2003.

Make Fearless Forecasts. Alert the organization to the high cost of expected failures. Take preventive action to avoid the future failures. Make this co-generating system more durable to avoid outages and prevent failures from occurring by funding the improvements from the pool of expected cost of unreliability. Do you suppose the design criteria for this system would have allowed "We expect this system will fail every 47 days"? – One could make a substantial bet that the system was assumed to fail maybe once every 5 years; so, there is a huge reliability gap between expectations and reality!

Summary. Five actual examples of industrial failures show typical straight-line patterns of failures when plotting cumulative failures against cumulative time on log-log plots. The slope of the line (β) tells if failures are increasing, decreasing, or resulting in no changes in failure rates. Statistics for the straight-line (λ and β) plots of cumulative failures versus cumulative time allow forecast of future failures if the system proceeds on the same course since stable processes produce straight lines on log-log paper [10] [11] [12] [13].

Table 17-7
Failures in a co-generating system

Date	Event outage	Event description	Days between event	All outages Cum. days	Cum. failures	Forced outages Cum. days	Cum. failures
2/1/1999	Planned	Tie in	0	0	0	0	0
2/20/1999	Planned	Tie in	19	19	1		
2/24/1999	Forced	Gas line outage	4	23	2	23	1
5/22/1999	Forced	Animal contact	87	110	3	110	2
7/9/1999	Planned	Interconnect energized	48	158	4		
8/9/1999	Forced	Switching error	31	189	5	189	3
9/13/1999	Forced	Tie wrap failure	35	224	6	224	4
10/13/1999	Forced	Lightning strike	30	254	7	254	5
11/3/1999	Forced	Static wire short	21	275	8	275	6
11/6/1999	Forced	Switch failed	3	278	9	278	7
11/10/1999	Forced	Not logged	4	282	10	282	8
1/3/2000	Forced	Cable bond fault	54	336	11	336	9
6/12/2000	Forced	Underground cable fault	161	497	12	497	10
6/21/2000	Forced	Bird contact	9	506	13	506	11
9/11/2000	Forced	Lightning strike	82	588	14	588	12
11/7/2000	Forced	Animal contact	57	645	15	645	13
12/2/2000	Forced	Animal contact	25	670	16	670	14
12/12/2000	Forced	High winds	10	680	17	680	15
4/11/2001	Forced	Not logged	120	800	18	800	16
4/12/2001	Forced	Not logged	1	801	19	801	17
4/19/2001	Planned	Tie in	7	808	20		
6/7/2001	Forced	Not logged	49	857	21	857	18
8/22/2001	Forced	Pole damage	76	933	22	933	19
9/13/2001	Forced	Interconnect opened	22	955	23	955	20
9/16/2001	Forced	Supplemental power out	3	958	24	958	21
10/6/2001	Forced	Power dip	20	978	25	978	22
10/12/2001	Forced	Control tripped	6	984	26	984	23
10/31/2001	Forced	Power dip	19	1003	27	1003	24
12/1/2001	Forced	Power dip	31	1034	28	1034	25
1/1/2002	Forced	Steam outage	31	1065	29	1065	26
4/15/2002	Forced	Switching error	104	1169	30	1169	27
4/18/2002	Forced	Load shedding error	3	1172	31	1172	28
9/27/2002	Forced	Water in switch gear	162	1334	32	1334	29
12/6/2002	Forced	Generator air intake frozen	70	1404	33	1404	30
1/3/2003	Forced	UPS failure	28	1432	34	1432	31

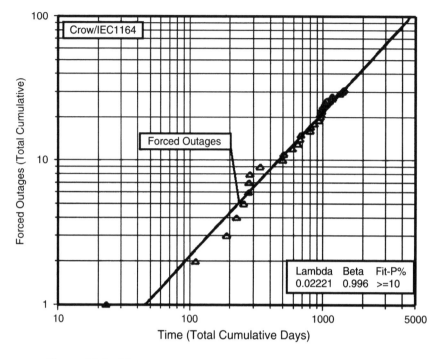

Figure 17-8. C/A plot of outages – combined cycle co-generating plant.

The purpose of "fearless future failure forecast" is to sound the alarm. Tell the organization about impending problems. Take corrective action for preventing future failures and thus avoid high cost of failures. Proactive involvement can prevent future failures. Passive involvement encourages failures.

Use failure data from your maintenance records to predict future failures. Set up a system to defeat the forecasted failures. Ignorance of future failures is not bliss and you cannot afford the failures!

Laying Hands On

Example 1: Retrofitting Oil Mist Eliminators

Modern compressor installations have a good many pieces of equipment, which one would call "ancillary." These are items that do not contribute directly to the availability of a compression train but are needed to maintain long-term unit integrity. One such item was an oil mist separator furnished originally by the manufacturers (OEM) of a large gas turbine

Figure 17-9. Vent oil mist eliminator with integrated filter and blower [14].

driven gas pipeline booster train. It was now not meeting today's strict pollution control requirements. A retrofit seemed in order, and looking around for suitable replacements fit for purpose, we selected a blower-powered oil demister or vent mist eliminator to be located on top of the lube and seal oil tank (Fig. 17-9).

What is the Purpose of an Oil Demister? Oil demisters prevent oil mist from entering the compressor building interior and ultimately the atmosphere. Oil mist issuing from the breathers of turbo train gearboxes, for instance, has been known to interfere with gas and fire detector instrumentation. Oil demisters further allow lube oil to be recovered that would otherwise go to waste. Finally, they can preclude the formation of explosive mixtures in lube or seal oil tanks.

In the not too distant past oil mist carried by air or gas from lube and seal oil tanks was routinely vented above the roofline into the atmosphere. Occasionally it was being ducted into gas turbine exhaust stacks thus becoming "invisible" but contributing to even more severe environmental pollution through its combustion products.

How is Oil Mist Formed? Gases and air accumulate over the surface of the oil being stilled and "degassed" in the reservoir. In gas turbines, the source of air in the oil are massive amounts of sealing and cooling air mixing with the oil as it drains from the bearings into the lube oil tank. In the case of pipeline compressors equipped with oil seals, degassers and traps have the job of separating the gas from the oil. These devices return the "degassed" oil into the lube oil tank but it usually still contains a considerable amount of gas.

As air or gas disengage from the oil, it is making its way to the reservoir vent all the while forming a mixture commonly referred to as oil mist. This oil mist or aerosol consists of oil droplets in fine distribution with an average particle size of 0.1 μm similar to smoke and fumes.

Frequently, there are baffles mounted in oil reservoir vent lines to reduce oil mist emission and to recover some of the oil. In spite of this provision, some pipeline operators have found that concentrations of oil in their roof vents were unacceptable.

To overcome the problems of the past, the market offers high-performance coalescing filters with integrated suction blowers. The blowers, usually regenerative type machines, generate a small vacuum inside the oil reservoirs. Oil mist cannot enter the area surrounding the tanks.

Figure 17-10 illustrates a tailor-made approach to oil mist vent elimination. The oil containing air is being continually sucked from the reservoir into the filter pot. Here it moves through a special filter element from the inside to the outside. The small oil droplets remain hanging on the filter's glass fibers to grow into larger drops and finally run down the outer surface of the filter element into the bottom collection space. From here the oil is being returned to the system – consequently there is no loss. Dirt particles contained in the air are caught by the filter. The exiting air is therefore not only oil free but also clean and can be introduced into the work environment.

Manufacturers of oil demisters claim that 99.9% of oil droplets are being eliminated. The same applies to the capture rate of dirt particles.

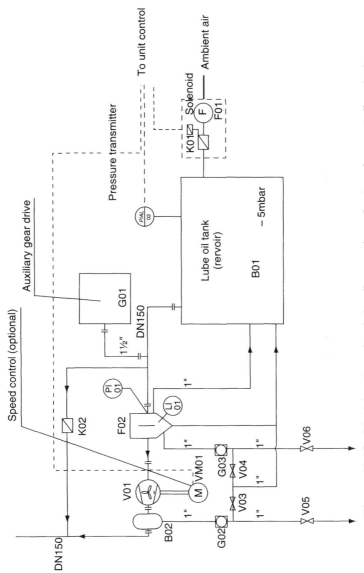

Figure 17-10. Principle schematic – Lube oil reservoir venting to reduce vent oil content.

Vent oil content reduction has been quoted from levels of $1500\,mg/m^3$ without mist elimination to $\leq 20\,mg/m^3$ or less with mist elimination [15].

We believe in one important advantage of the described system as it applied to the operation of pipeline compressors equipped with oil seals: When a shutdown or trip of the unit occurs gas continues to disengage from the lubrication oil inside the reservoir and an ignitable gas/air mixture may result. Some operators have therefore decided to keep their vent blower running. The blower is started 5 min before unit startup and shutdown some 72 hr after lube and seal oil pumps have stopped. One pipeline company even activates the blower for 5 min every 9 hr while the compressor is on depressurized standby. Further, to assure no inappropriate pressure rise in the reservoir occurs, its pressure level is monitored continuously.

Instrumenting the oil tank as mentioned above will also help in maintenance because frequently ancillary equipment does not get attention until indicated: a high pressure alarm will prompt the required maintenance action.

Example 2: Retrofitting Gas Seals

One day in the fall of 1988, my manager said to me: "You must have felt last night like one of the people launching the first space rocket – not knowing whether or not it would work." He was referring to our self-directed retrofit of gas-lubricated shaft seals and a successful startup on two critical motor-driven 3000-hp ethylene refrigeration turbocompressors. Our achievement was a first for our large multi-national petrochemical company, even though, admittedly, gas transmission companies in North America were applying gas seals in the early 1980s well ahead of the HC process industries.

Things have changed since then. It seemed the ice was broken in those days and many compressor operators followed suit by replacing their cumbersome oil-lubricated shaft seals with gas seals. Today, gas seals have become to be known as a reliable and cost-effective alternative to oil-type seals. Their key characteristics are low leakage rates, wear-free operation, and an extremely low level of power consumption.

Gas seals are non-contacting, dry running mechanical seals. Figure 17-11 shows the principal design features of such a seal. They are usually furnished as a cartridge containing a spring-loaded stationary seal face or sliding ring [1] sealed by an O-ring and a rotating seat or mating ring [2]. The sealing faces slide over each other without contact. This results in almost no wear and a long seal life. The seal face of the rotating mating ring is divided into a grooved area at the high pressure side and a dam-area at the low pressure side (Fig. 17-12). The stationary

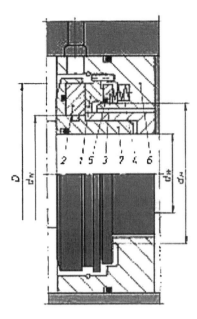

Figure 17-11. Cross section of a gas seal. 1 Stationary seal face (sliding ring); 2 Rotating seat (mating ring); 3 Thrust ring; 4 Compression spring; 5 O-ring; 6 Housing; 7 Shaft sleeve with cupped retainer (Burgmann DGS) [16].

Figure 17-12. Mating ring V-grooves, U-grooves. (Note: Arrows indicate sense of rotation).

sliding ring is pressed axially against the mating ring by spring forces and sealing pressure.

The sealing gap is located between the mating ring and the sliding ring. For proper non-contacting operation, these two rings have to be separated by a gas film acting against the closing forces in the sealing gap. The gas film is achieved by the pumping action of the grooves and the throttling effect of the sealing dam. Groove geometry is critical for trouble-free operation of the seal. Current general operating limits of gas seals are shown in Table 17-8.

Can we Always Justify Retrofitting Gas Seals? It was easy to justify our 1988 gas-seal conversion. At the time we arrived at a very favorable payback well worth taking the risk of a retrofit, eventhough the original floating ring-type oil seals on the refrigeration compressors had generally worked satisfactorily. However, when they failed they had invariably caused expensive clean-up work, as lost seal oil would permeate the entire refrigeration system and reduce process yield for a long time.

On pipeline compressors equipped with oil-type seals, I would look once more at seal oil losses. We met operators who would not admit to malfunctioning oil-type seals indicated by excessive oil losses while routinely emptying oil pots and low point drains of compressor discharge lines. Table 17-9 shows a comparison of seal leakage rates. With natural gas prices rapidly moving past the $5.00/MSCF mark, it might be time again to look at gas leakage rates in order to justify a conversion to gas lubricated seals.

Finally, we suggest to study the failure statistics of your oil type seals and compare them with this: Current quotes of failure rates of gas

Table 17-8
General operating limits for compressor gas seals [16]

Criterion	Limits customary units	Limits SI units	Remarks
Nominal shaft diameter	≤10.0 in	≤254 mm	
Pressure, absolute	≤1450.0 psi	≤10 MPa	Elastomeric seals,
temperature	≤3626.0 psi	≤25 MPa	non-elastomeric seals
Temperature	−4 °F ... + 392 °F	−20 °C ... + 200 °C	Elastomeric versions,
	−70 °F ... + 482 °F	−55 °C ... + 250 °C	non-elastomeric versions
Sliding velocity referred to outer diameter of mating ring	≤656 ft/sec	≤200 m/s	

Table 17-9
Comparison of leakage rates from various compressor shaft seals

Seal type	Geometry*	Normal recoverable oil leakage gal/h (l/h)	Gas consumption** CFH (m³/h)
Floating ring oil seal	2 rings, 0.787 in. (20 mm) each, rad.cl. = 0.002 in. (0.05 mm)	gas side 1.9 (7.2)	S.O. trap vent 636 (18)
Mechanical seal	gap = 0.04 mil (1 μm)	gas side 0.019 (0.072)	S.O. trap vent 636 (18)
Gas seal	gap = 0.20 mil (5 μm)		12.7 (0.36)

* Nominal shaft diameter = 5.5 in. (140 mm); 5000 rpm; gas pressure p1 = 87 psi (600 kPa); buffer (seal) pressure p3 = 109 psi (750 kPa); gas: air; buffer fluid: oil.
** Frequently vented to atmosphere and not recovered due to the unavailability of containment at a lower pressure.

seals are in the neighborhood of 0.175 failures/year meaning that we could expect a problem every 6 years or so. One seal manufacturer bases recommended maintenance intervals around gas seals on limits set by the elastomer aging process. He suggests the following maintenance routine after 60 months of operation:

- replace all elastomers;
- replace the springs;
- replace all seal faces and seats;
- carry out a static and dynamic test run on a test rig.

We cannot entirely agree because we tend to adhere to an old law originating from the Canadian Navy that goes: "Leave well enough alone!"

Example 3: Retrofitting MJTs on a Combustion Gas Turbine*

Introduction. Global competition has been moving the petrochemical industry to ever-higher levels of efficiency and production rates. This has led to an increase demand on both the equipment and the manpower to higher reliability and shorter turnaround times. One major problem faced when dealing with high temperature and pressure equipment is their bolting and unbolting. This is due to the very large torque required to properly seal the joint and the difficulty of its application especially in the field.

* Courtesy of Husain Al-Mohssen, ARAMCO, Saudi Arabia.

Bolting is one of the most commonly used methods of holding different parts together in modern plants and machinery. The reliability of a bolted joint is absolutely essential for the safety and reliability of the equipment and plant.

Figure 17-13 shows a sketch of a typical bolted connection. When the bolting is tightened, the main body of the bolt (or the stud if there is a stud and two nuts) is elongated to a new length. The increase in the length of the bolts introduces tensional stresses into the bolt or stud. At the same time, the distance between the nuts has decreased by the same amount. This introduces compressive stresses in the flange faces, which holds the flange together and prevents leaks.

The tensional stress in the stud is produced by the rotation of the nut or the bolt. Essentially the bolt is an element that converts torque to stress which resides in itself, the nut and the joint components. This stress is also called preload and varies due to many factors. This includes the friction coefficient between the nut and the bolt, the cross section of the bolts, as well as many other factors. By far the biggest contributor to the required torquing value is the cross-sectional area of the bolt or stud. Table 17-10 and Figure 17-14 show the different values recommended from one gas turbine manufacturer for casing bolts. These torquing values are representative of most other manufacturers as well as many local and industrial standards.

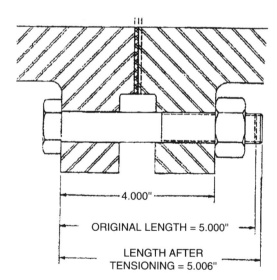

Figure 17-13. A typical bolted connection with a stud and a nut. As the nut is tightened the stud elongates.

Table 17-10
Torque values

Torque Value to reach 45,000-psi preload

Nominal diameter (in.)	Torque (ft lbs)
0.5	43
0.625	90
0.75	150
0.875	240
1	369
1.125	534
1.25	750
1.375	1019
1.5	1200
1.625	1650
1.75	2250
1.875	3000
2	3313

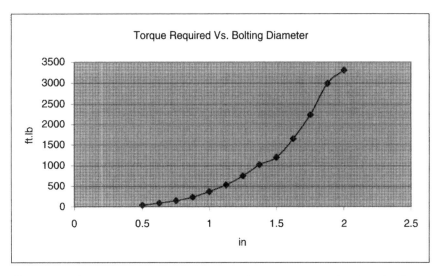

Figure 17-14. Manufacturer recommended bolting torque as a function of different bolt diameters.

As can be seen from the table and the plot, the torquing value required to set a bolt to its proper preload is a non-linear function of the bolt diameter. While it only takes 369 ft-lb to seat a bolt to 45,000-psi preload when its diameter is 1 in. It takes 3313 ft-lb – 8.9 times as much! – to set a 2-in. diameter bolt to the same preload. In fact, it can be shown that the required torque is a polynomial of degree 3 as a function of the bolt diameter if other parameters are kept constant.

Hand tools can be used to apply the required torque with relative ease and accuracy for the smaller diameter bolting. However, as the bolting diameter – and consequently the torquing requirement – gets higher, methods for applying the required torque get more and more extreme. For example, it is quite common to use large sledgehammers and field-made "tools" to bolt and unbolt large joints. In fact, some times so much force is required that cranes and/or other heavy equipment have to be used to tighten up large nuts or to free up seized bolts. Apart from being totally inaccurate preloading to the joint and possibly causing joint failure, these extreme methods have major safety concerns especially when the bolting is confined to awkward areas.

In fact, there are many reported cases in the industry of serious injuries happening to workers as a consequence of trying to bolt or unbolt a large size bolt or stud. Apart from this the inaccurate, uneven and often insufficient bolting stress will sometimes lead to serious joint leaks that will be a major problem of its own.

There have been a number of methods proposed and used to help in solving or easing the problem of large diameter bolting. Some of the most common ones are as follows.

Bolt Heating. In this method, studs that have a specially designed heating cavity in their center are used. The principle is to heat up the bolt or stud using special heating rods and tighten it while it is hot. The rods are later removed and the preload will gradually develop as bolt gradually cools down to the temperature of the casing. This method although helpful in many cases, still does require special studs and heating elements and has a large torque requirement. In addition, the torque accuracy can be limited in many cases though it is generally better than standard bolting practices.

Hydraulic Stud Tensioning. This method of tightening the bolts relies on stretching the stud of the bolt a certain amount to induce the proper preload in it. After that the bolt or nut is tightened by hand until it touches the flange face. The bolting is complete when the hydraulic tension is removed and the nut holds the preload in the bolt. On the one hand, this method can be used for more than one bolt at the same time, which can be very advantageous especially when bolting gasketed flanges. On the other hand, this method can be inaccurate (especially for short studs) and it cannot be used for all types of bolting. In addition, a special hydraulic tensioning devices and associated pumps and auxiliaries have to purchased and maintained.

Hydraulic Bolting. Is identical to normal bolting except that the normal torque is applied using a special hydraulic mechanism. Again this method works in many applications but can not always be used especially when there are space limitations. Apart from not having a high accuracy of preload they suffer from some of the same limitations of normal bolting including thread galling. And like hydraulic stud tensioning, there is an upfront cost of the hydraulic mechanism that may be significant in situations.

Multi-Jackbolt Tensioners. Multi-Jackbolt tensioners are a relatively new method of trying to deal with the problems described in the last section. Multi-Jackbolt tensioners (known commercially as Superbolts or Supernuts™) are special patented design fasteners that replace existing bolts or nuts. The main idea behind MJTs is very simple: to tighten a number of smaller jacks instead of one large bolt or nut. Figure 17-15 shows a cross section of a MJT which will help to visualize how it works.

In the figure, we see a cross section of a Supernut™, which is just one of the many forms MJTs come in. The main parts are common to all MJTs and they are [17]:

- A number of high compressive strength jacking bolts that are tightened when the bolts are installed. These jacking bolts are embedded in the body of the MJT and push against the bottom washer.
- A special hardened washer that sits between the bolt and the body of the flange. The jacking bolts push against this washer when they are tightened creating a gap between the bolt body and the washer.
- In the case of Supernuts™, an internal threading that holds the stud just as in the case of a standard nut. When the jacking bolts are tightened the nut is pushed away from the body of the flange that causes the stud to be pulled to the proper preload of the bolt.

The advantage of MJTs lies in the fact that the torque required to achieve the target preload using the jacking bolts is much less than what is required to preload the big nut or bolt [18]. Table 17-11 shows the jacking bolt torque required versus the original bolt required for one particular application. In addition, the last column shows the torque advantage, which varies from 26:1 to 273:1 for larger diameter bolting!

Retrofitting MJTs to Combustor Casing of a Gas Turbine. Industrial combustion gas turbines are well known for being especially hard to bolt and unbolt. Not only are the sizes of the bolts and nuts very large, they are not forgiving and will leak when not preloaded properly since the flanges are metal to metal with no gaskets. The problem can be exasperated when

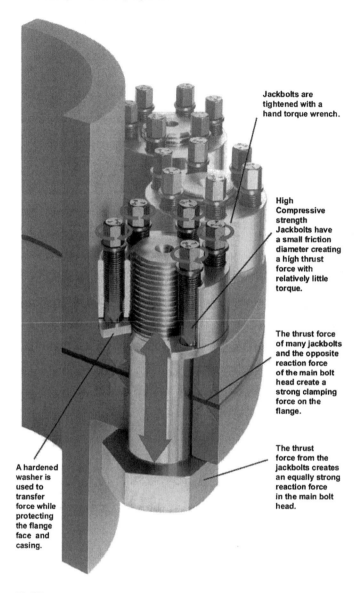

Jackbolts are tightened with a hand torque wrench.

High Compressive strength Jackbolts have a small friction diameter creating a high thrust force with relatively little torque.

The thrust force of many jackbolts and the opposite reaction force of the main bolt head create a strong clamping force on the flange.

The thrust force from the jackbolts creates an equally strong reaction force in the main bolt head.

A hardened washer is used to transfer force while protecting the flange face and casing.

Figure 17-15. A cross section of a multijackbolt tensioning assembly (Supernut™).

the bolts are in a hard to access area or lack heating holes for thermal tightening. One particular unit in a refinery suffered from all of these problems with a large number of its bolts. Some bolts literally took hours to bolt and unbolt especially when they have not been opened in a long

Table 17-11
Torque required for both the standard bolts as well as MJT
of the same size. Last column shows the ratio between them

Size (in.)	Torque required for standard Hex Nut	Torque required for MJT	Advantage (regular torque/MJT torque)
1	955	36	26.53
1.5	2,890	65	44.46
2	6,880	152	45.26
3	25,200	310	81.29
4	50,500	310	162.90
5	99,000	520	190.38
6	1,42,000	520	273.08

time. Sometimes bolts could not be freed up and would have to be cut with a torch to be removed. On the top of all of this the unit was notorious for developing hot air leaks after it has been overhauled and consequently would have to be shutdown for bolt re-tightening. Even then the leak would not completely disappear as the casing faces are distorted due to previous leaks.

All of these problems made this unit a very good candidate to replace the original turbine bolting with MJTs. The first step in this was to decide which bolts to replace with MJTs during the next overhaul of the unit. The unit has many bolts and it would be uneconomical to replace all the bolts, instead efforts were concentrated on selecting a few large diameter bolts in inaccessible and critical area to be replaced. Figure 17-16 shows a cross section of the gas turbine casing with arrows pointing to the bolts that were selected to be changed. As can be seen from Figure 17-16, these bolts are situated at the corners of the casings and are among the largest in diameter (2 in.). Furthermore, the original bolts had a large hexagonal socket cap, which meant that a special socket had to be used to torque it. It was felt that the replacement of these bolts not only would take care of the most problematic bolts in the area but will also have the most positive effect on the reduction of leaks.

Once it was decided which bolts would be retrofitted with MJTs, a custom design would have to be developed with help from the manufacturer. The bolt would have to fit in the recessed area intended for the original bolt, it would also have to allow enough clearance above it to access the jacking bolts. Since detailed drawings for the bolts and the casing were not available, detailed measurements of some spare bolts along with the casing had to be taken. Once these details were collected

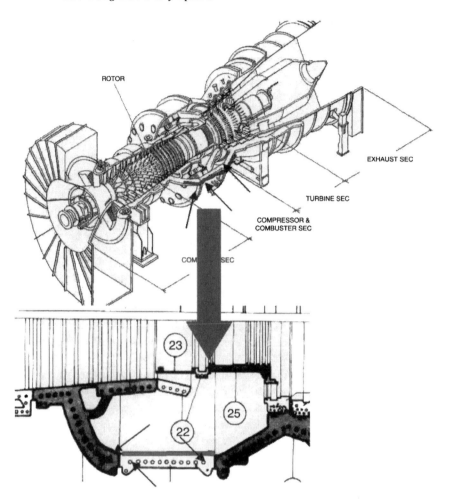

Figure 17-16. Cross section of the combustion gas turbine. The red line shows the location of the leaks while the blue arrows show the location of the bolts to be replaced.

they were sent to the manufacturer who came up with a design similar to what is shown in Figures 17-17 and 17-18.

The next step was to come up with an appropriate material for this application taking into account both the designed preload of the bolting and the temperature that the bolts will be having to withstand. After some background search as well as consultation with different turbine manufacturers, a preload close to the original design was selected that would help in minimizing the leak problem. The actual operating temperature of the casing was measured by taking a thermal image of the unit when

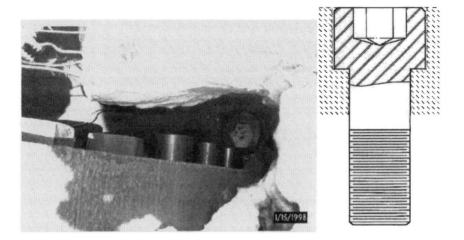

Figure 17-17. Top view of the original bolts. A sketch of it is visible on the right.

it was running (Fig. 17-19). Consequently, an estimate of the maximum temperatures the bolting will be seeing over its life was reached. Samples of the original bolting material were analyzed to try to guess its composition and the design constraints the original bolting designer must have had. A final selection of bolting materials was reached with help from the manufacturers of the MJT (Superbolt Inc.) as well as in-house company engineering expertise. Finally, the designed bolts were manufactured and installed in a major overhaul of the unit.

An Assessment of MJTs. In our experience we have found that they were more expensive than the original OEM bolts (about 40% more expensive in this particular application). In addition, most machinist do not know how to properly install MJTs so some time has to be spent explaining the proper sequence of installation. Having said that, we feel that MJTs have performed very well in our particular application. The time required to install bolts has shrunk from about 30–60 min to less than 5 min per bolt. Moreover, all the problems that we were facing with the original bolts have now disappeared. No heavy hammering to set the bolts or heating with torches is required to set these bolts. In our application, the required torque has decreased from the original 3313 to 65 ft-lb (an advantage of more than 50:1). This new torque can be easily and accurately applied by hand. This is achieved with confidence that the bolts were tightened to the target preload of the bolts.

As has been mentioned earlier, this turbine has a long history of developing leaks in its horizontal and vertical flanges. In the past, the

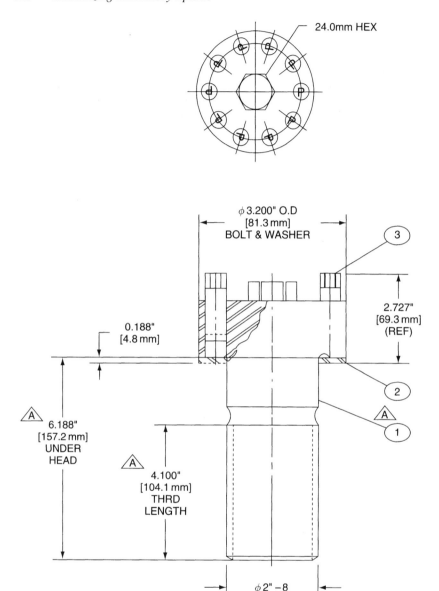

Figure 17-18. Cross section of a MJT designed to replace the bolts in Figure 17-16 and 17-17.

leaks were so severe that they have caused unevenness in the casing of the turbine (Fig. 17-20). After MJTs were installed and the unit was run, some leaks did develop but only in the areas where the new bolts were not installed on the casing. This shows that these bolts are as good or

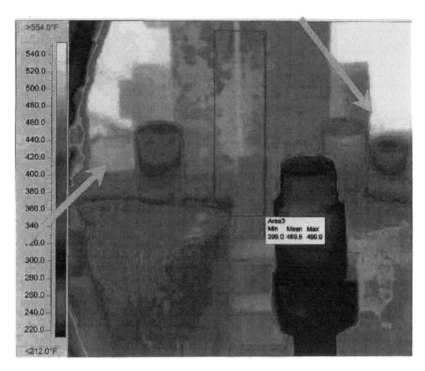

Figure 17-19. A thermal image of the casing while the unit was running. The arrows points at the area of bolt replacement.

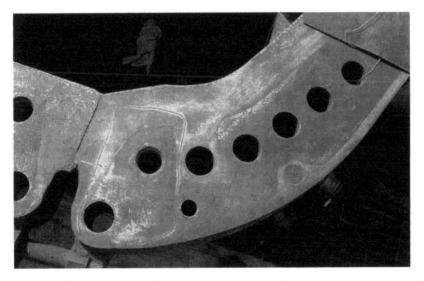

Figure 17-20. Picture of the flange face of the casing. Note the unevenness caused by past hot air leaks. The deep narrow groove is caused by previous attempts to stop the leak by crushing a wire between the two flanges.

most likely better than the original bolts in applying the proper preload and preventing hot air leaks on the casing.

Conclusion. We have attempted to give an introduction to some of the basics of bolting. In addition, common problems associated with achieving proper joint preloads were described as well as some existing methods of overcoming these problems. Finally, a fairly detailed description of how MJTs work was given. Additionally, we described the details of how some were custom designed to be retrofitted in a large industrial gas turbine generator working in a refinery environment.

Although they are more expensive than standard bolts, in our experience MJTs have accomplished what they were designed to do. They are easier, safer, and much faster to install than standard bolts and although the evidence is not yet conclusive they do seem to have helped to minimize leaks from our casing.

The decision to use MJTs should not be automatic; instead, each bolting problem should be evaluated separately to decide if they can be used and if they are worth the engineering effort to retrofit them to solve the problem. There are other bolting methods that are very useful and effective in many applications that should always be considered. It is our opinion, however, that MJTs should be at least considered for large or especially critical bolting applications to improve safety and reliability. They should be considered for new installations that requires large bolts or high torquing values and preloads.

Example 4: Redesigning Coupling Guards for a Major Compressor Train*

The K-002 sales gas compressors have several historical problems with their OEM coupling guards. The major deficiencies are due to the horizontal split line design of these guards, which utilize unreliable RTV sealant that has an excessive curing time and the numerous oil leakage paths through the flanges or the flange bolts themselves. In addition, at the connection to the compressor/gearbox, the OEM guard does not utilize a standard O-ring sealing design, such as a rectangular groove, which leads to more oil leaks. This coupling guard has considerable design oversights causing constant problems in service.

Based upon operations request to resolve these issues, we obtained quotations for new, modern coupling guards from the OEM. Unfortunately,

---

*Courtesy of Abdulaziz Al-Saeed and Abdulrahman Al-Khowaiter, ARAMCO, Saudi Arabia.

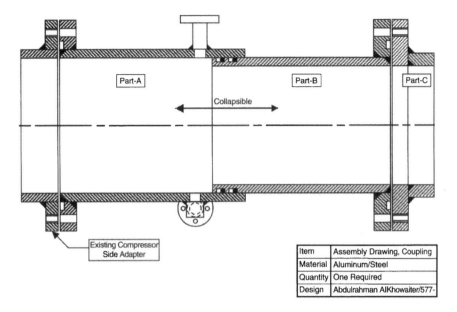

| | Part-A | Collapsible | Part-B | Part-C |
| | | | | |

Existing Compressor
Side Adapter

Item	Assembly Drawing, Coupling
Material	Aluminum/Steel
Quantity	One Required
Design	Abdulrahman AlKhowaiter/577-

Figure 17-21. Coupling guard – assembly drawing.

these guards as quoted were expensive and still retained many of the design weaknesses of the original guards, with only marginal improvement.

Later, we developed a new heavy-duty Barrel type construction coupling guard, which utilizes two removable cylindrical parts which slide together, using O-ring radial seals, and one static component – see assembly drawing, Figure 17-21. The goal of this design was to develop the simplest possible configuration of coupling guard, which will have a minimum of parts, minimum size, cost, weight, and maximum sealing reliability. In addition, simplicity of fabrication at local vendors and ease of installation were also important design factors. The average installation time for the new guard will not exceed 60 min and there are only 32 bolts required as compared to 62 in the past. No RTV is necessary as all sealing is by O-rings.

Many sources have assisted in analyzing and reviewing the new design, including staff engineers and the unit senior machinists. Their comments have been helpful toward ensuring the best possible design solution.

The drawings shown in Figures 17-22 and 17-23 as well as the scope of work have sufficient detail to cover all aspects of fabrication by local vendors (through the mechanical services shops division) and installation by unit maintenance. In addition, all required materials are readily available with qualified fabrication shops.

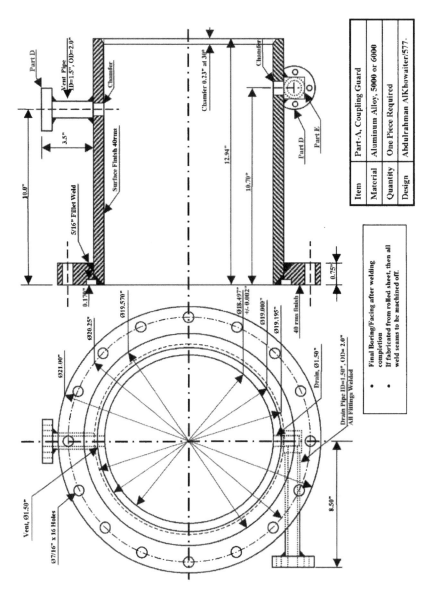

Figure 17-22. Coupling guard – detail 1.

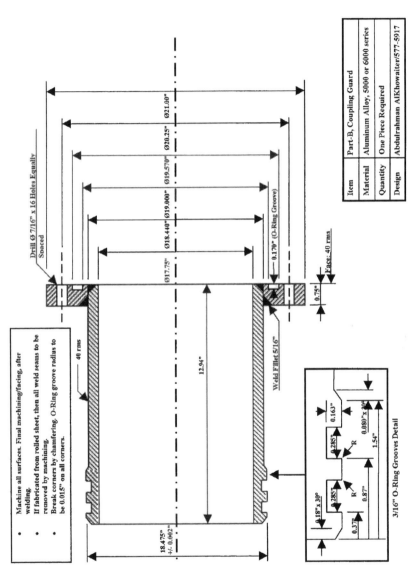

Figure 17-23. Coupling guard – detail 2.

We propose that, after installation, when the new coupling guard passes a 60-day field trial test, the package can be reissued for the fabrication of two additional coupling guards for the remaining compressors.

Example 5: Pump Protection to Increase Uptime: One of Many Ways [19]

There is a large body of evidence that centrifugal process pumps do not perform well at flows away from their best efficiency point. At anywhere between 20 and 35% of their best efficiency flow point, they begin to show signs of distress. This distress has many symptoms. They run the gamut from operational difficulties, indicated by noise and vibration and attendant discharge pressure fluctuation to frequent repair outages and total wrecks. Things become even more difficult when pumping fluids near their boiling point. Here, in liquid petroleum gas (LPG) plants and tankage pump stations, we often experience failures that are not only caused by pumps running below their minimum continuous safe flow (MCSF) but also by pumps gassing up due to insufficient venting and subsequent dry running. These problems occur mainly during startup and pump restart, and happen frequently even with operating personnel in attendance. In many cases, such pumps are being remotely controlled and started up from a central control room. It is therefore, not always possible to ensure that the pumps are sufficiently vented and filled with liquid prior to startup. The consequence can be considerable damage and outages of critical impact.

With fluids near their boiling point or with LPGs at their boiling point and the pump just shut off or idle, only a small increase in temperature can result in a change from liquid to gas form. The gas pushes the remaining liquid out of the pump into the suction line with the result that the pump is partially or completely filled with gas. The cause for a pump "gassing up" is heat transmission from ambient as well as residual heat retention of the machine immediately after it is shutdown. Depending from pump design, it will be rendered completely "dry" or it will be filled with gas so the impeller will not be able to generate any liquid head on startup. The pump, for all practical purposes, will run dry and within seconds after startup it will experience a failure due to internal rubbing followed by seizure – an event that can lead to a wreck and considerable other plant damage.

So we just came across this interesting fitting – an automatic recirculation valve (ARV) as it has been used for many years in the power generating industry for protection against low flow on boiler-feed water pumps. This multifunctional valve – see Figure 17-24 – ensures automatic

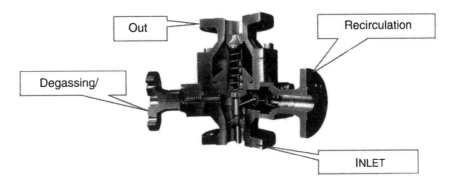

Figure 17-24. Multi-function automatic recirculation valve (ARV).

degassing of an idle or spare pump in order to maintain the liquid inside. It cannot be repeated enough.

The complete filling of a centrifugal pump with liquid is a prerequisite for a proper startup, or restart, so there can be an instant building of head and movement of the pumpage. It is a condition for any type of warrantee for process or pipeline operations by contractors as well as pump manufacturers.

How does the Valve Work? It is installed near the discharge flange of the centrifugal pump similar to a standard check valve. Due to the valve's elevated installation, a high point on the pump discharge side is established just below the check valve cone seat. When the pump is idle, accumulating gas starts collecting in the high point. The degassing mechanism of the multi-functional valve will be automatically held open ensuring continuous venting to vapor space of the suction vessel. This ensures that the pump stays liquid filled. The same would be the case for multiple parallel pump sets. A typical installation schematic is shown in Figure 17-25.

With liquefied gas pumps at low temperatures, the machines are continuously kept cold, ready to be started or restarted. Immediately after startup the pump generates the necessary differential pressure and the auto-degassing mechanism of the valve closes the vent passage tightly. If the pump is shutdown, decreasing differential pressure causes the degassing mechanism to open automatically so nascent gas – generated by residual heat – can be immediately and efficiently vented. The pump stays filled by liquid ready for the next startup or automatic restart. The main area of application of this valve is in the process and movement of liquid gases, especially low temperature gas processing, tankage, and

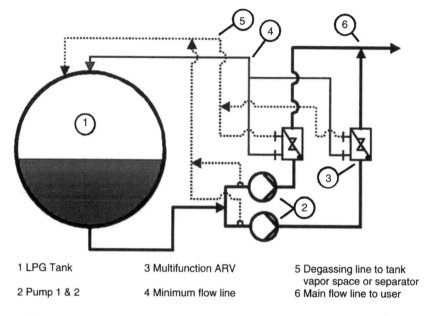

1 LPG Tank   3 Multifunction ARV   5 Degassing line to tank vapor space or separator

2 Pump 1 & 2   4 Minimum flow line   6 Main flow line to user

Figure 17-25. Application example: LPG Tankage for NH_3, C_2H_4, C_3H_6, etc.

loading of liquid gases. It is easy to see how this multi-functional valve can find its application in all pump installations that handle fluids:

- near their boiling points;
- containing gas;
- consisting of two-phase mixtures;
- where gas is injected through pump sealing provisions.

The valve is particularly suited for the protection against dry running of magnetic drive and canned motor pumps.

We have presented this example to show to our readers how frequently these type of armatures will not be provided as part of a capital project because process designers and owners do not consider it essential. It will become a typical retrofit later in the life cycle of the project or the plant, again, in the constant effort to increase uptime of the facility.

References

1. Barabási, Albert-László, *Linked-The New Science of Networks*, Perseus Publishing, Cambridge, MA 2002, pp. 67–72.
2. Wright, T. P., Factors affecting the cost of airplanes, *Journal of Aeronautical Sciences*, February 1936.

3. Teplitz, Charles J., The Learning Curve Deskbook, Ouorum Books, Westport, CT, USA, 1991.
4. Department of Defense, AMSAA Reliability Growth Guide TR-652, September 2000, downloadable from http://www.barringerl.com/nov02prb.htm (2000).
5. Abernethy, Robert B., The New Weibull Handbook, 4th ed., published by the author, North Palm Beach, FL (2002).
6. Torres da Silva, Jairo, Lúcio Antonio Moreira Ivo, Paulo Soares de Oliveira, Sérgio Lins Pellegrino, *Bombas Centrifugas Passo a Passo* (Centrifugal Pumps Step by Step). Turbotech Engenharia LTDA, Salvador. Bahia, BRASIL, http://www.turbotech.com.br (1999).
7. Suzuki, Tokutarō, *TPM in Process Industries*, Productivity Press, Portland, Oregon, 1992.
8. Barringer, H. Paul, Problem of the month November 1997 – Total Productive Maintenance Results, http://www.barringer I.com/nov97prb.htm (1997).
9. Bloch, Heinz P. and Fred K. Geitner, *Machinery Failure Analysis and Troubleshooting*, 3rd ed., Gulf Publishing Company, Houston, Texas, 1998.
10. Department of Defense, Reliability Growth Management. MIL-HDBK-189, February 13, 1981.
11. Duane, J. T., Learning curve approach to reliability monitoring, *IEEE Transactions on Aerospace*, **2**(2), April 1964.
12. Fulton, Wes, WinSMITH Visual Software, Version 4.0T, Fulton Findings, Torrence, CA, 2003.
13. NASA Learning Curve Calculator. http://www.jsc.nasa.gov/bu2/learn.html (2003).
14. Technical literature from Contec Industrieausrüstungen, Bad Honnef, Germany.
15. Hilliard Corporation, Elmira, NY, USA, Sales Literature.
16. BURGMANN, *Gas-Lubricated Mechanical Seals*, ISBN 3–929682–18–4, Burgmann, Munich, Germany, 1997.
17. Steinbock, R., "Multi-bolt mechanical tensioners," ASME PVP – vol. 158, *Advances in Bolted Joint Technology*, 1989.
18. Bickford, J. H., *An Introduction to the Design and Behavior of Bolted Joints*, Marcel Dekker New York, 1981.
19. H. Schroeder and Co., Pumpenschutzarmaturen, Gummersbach, D-51647 Germany. (s.welker@schroeder-elbach.com).

Chapter 18

Review of mechanical structures and piping for machinery

Overview, Design, and Case Studies*: Introduction

Proper design of mechanical structures and piping is essential to machinery equipment reliability. The installation and support of the piping also have a significant impact on the reliability of the connected machinery equipment. The engineer or technician involved in machinery uptime assessment must therefore have more than a superficial knowledge of the basic design parameters governing piping and machinery support elements.

In this chapter we present some fundamental principles and procedures for the analysis, design, and installation of mechanical structures and piping. We will also examine the effect of these components on the reliability of the connected machinery.

Design of a structure consists of the determination of shape, size, and material of the structure, connecting elements and support system. Design of the associated piping system consists of the determination of size and material of the pipe, flanges, bolting, gaskets, valves, and fittings. Items such as expansion joints may also be present. Piping design also includes the determination of the number, type, and size of the required pipe support elements. In this chapter, more emphasis is placed on the analysis and design of piping systems than of structures.

* Contributed by Bill Moustakakis, Ph.D., P.E. Mr. Moustakakis is the Lead Mechanical Engineer of ExxonMobil, Houston, TX. He has advised numerous hydrocarbon processing plants in the United States and Europe on reliability improvement, failure avoidance, and troubleshooting issues relating to piping and mechanical structures.

The design of any structural or piping system generally includes the following steps:

- Establishment of design criteria.
- Selection of materials.
- Establishment of acceptable layout.
- Calculation of the required sizes, dimensions, thicknesses, etc.
- Establishment of acceptable support configuration.
- Stress analysis to establish compliance with design criteria.

Stress analysis methods are used to determine displacements, stresses, and internal forces and moments in structures and piping, and to determine the fitness of these systems for the intended use. Of particular interest to the machinery engineer is the determination of forces and moments imposed by the piping to the connected machinery. Analysis methods range from simple engineering judgment and comparison to similar units already in operation, to the most rigorous applied mechanics methods using the most modern computing equipment. The choice of analysis is generally dictated by considerations of safety, operability, and economics.

Design Criteria

Design criteria for piping and structures are usually set by the various codes, regulations of the several governmental bodies, industry standards, and the owner's standards and specifications. The designer must consider all loading combinations that might result during fabrication, erection, testing, and operation. The highest possible loading that the structure or piping system is expected to experience will govern the design.

The various codes, developed over the years, specify the minimum requirements for safe construction in order to provide public protection. They include the Uniform Building Code (UBC) and the American Institute of Steel Construction (AISC) Code for buildings and structures, the ASME Boiler and Pressure Vessel Code for vessels, and the ANSI/ASME piping codes for piping.

The ASME Boiler and Pressure Vessel Code includes the following sections:

- I   Power Boilers
- V   Non-Destructive Examination
- VIII   Pressure Vessels, Divisions 1 and 2
- IX   Welding and Brazing Qualifications

The ANSI/ASME piping codes include:

- B31.1   Power Piping
- B31.3   Chemical Plant and Petroleum Refinery Piping
- B31.4   Liquid Transportation Systems for Hydrocarbons, Liquid Petroleum Gas, Anhydrous Ammonia and Alcohols
- B31.5   Refrigeration Piping
- B31.8   Gas Transportation and Distribution Piping Systems
- B31.11   Slurry Transportation Piping Systems

Of primary importance to the machinery engineer are the industry standards dealing with pumps, compressors, and turbines. These standards give, among other things, the allowable loads on machinery nozzles. They include:

- API 610   Centrifugal Pumps for General Refinery Service
- API 617   Centrifugal Compressors for General Refinery Service
- NEMA SM23   Steam Turbines for Mechanical Drive Service
- API 560   Fired Heaters for General Refinery Service
- API 661   Air-Cooled Heat Exchangers for General Refinery Service

Design Loads

The various codes make provisions for the minimum loads that must be considered in the design of structures and piping. Loads include the dead loads, live loads, and other externally imposed loading.

Dead loads include the weight of the structure, piping, platforms, insulation, fireproofing, vessel or pipe internals, and the like. A contingency of the order of 5% is usually added to the dead load.

Live loads are variable loads such as the weight of the operating or test fluids, the weight of people that might be present on a platform or building floor, snow or ice on a building roof, and the like.

Externally imposed loads include applied pressure, wind or earthquake loads, impact or surge loads, flow-, wind-, or machinery-induced vibration loads, construction equipment loads, etc. In addition, externally imposed loads include thermal expansion effects. Thermal expansion effects must be considered under normal operating temperatures, startup, shutdown, and emergency or process upset conditions. This requires that the piping engineer have adequate knowledge of the process.

Normal operating temperatures, as the name implies, are associated with the planned everyday plant operation. Stable production conditions are expected to occur most of the time that the plant is in operation.

Startup and shutdown conditions must be examined in detail to establish actual flexibility conditions. For example, if a tower is heated in selected sections during startup, while other sections remain cold, this situation must be identified and the piping flexibility must be examined for adequacy under such condition. Piping serving a dual purpose, e.g. regeneration in powerformers or decoking in furnaces, in addition to normal operation, must be analyzed separately to have adequate flexibility for each purpose and for switching from one service to the other. Different operating conditions for the same piping, as for example when the piping connects spare equipment and may be hot in one operating case and cold in another, must be considered in the design.

Emergency and upset conditions are unplanned events occurring as a result of equipment malfunction or other circumstances that result in predictable or unpredictable temperature fluctuations. These temperature fluctuations must generally be considered in the design. Emergency and upset conditions can include loss of cooling medium flow, loss of process flow while heating facilities continue to operate, etc. Clearly, then, this would be an area which the machinery reliability professional may wish to question or investigate. More than once has the piping designer overlooked the upset potential of certain machines!

The piping code requires that both the maximum and the minimum temperatures expected during the operation of a line be used to determine the amount of thermal expansion that must be accommodated. This means that the temperature range to be used in the flexibility analysis must be equal to the maximum metal or ambient temperature minus the minimum metal or ambient temperature, whichever number is greater.

Classification of Computed Stresses

The loads applied to a structure or piping system result in stresses that are generally separated into primary, secondary, and localized stresses. The codes provide for different allowable stresses for each category.

A primary stress is the direct stress generated by the imposed loading that is necessary to satisfy the laws of equilibrium. Primary stresses are not self-limiting, and if they exceed the yield strength of the material, they will result in excessive distortion or even structural failure. Examples of primary stress are the circumferential or longitudinal membrane stress due to internal pressure, pipe or beam bending stresses due to gravity loading, etc.

Secondary stresses, usually attributed to temperature effects, are self-equilibrating stresses necessary to satisfy compatibility. Unlike the primary stresses, secondary stresses are not the cause of direct structural failure in ductile materials with a few applications of the load. Secondary stresses exceeding the material yield strength will merely result in local yielding and stress redistribution. If the applied loads are of cyclic nature however, secondary stresses constitute a potential source of fatigue failure. Examples of secondary stresses are bending stresses due to thermal expansion effects.

Localized stresses are those found near discontinuities, which diminish rapidly with distance away from their source. Examples of localized stresses are peak stresses developed at elbows, miter bends, tee intersections, and other geometric discontinuities.

Evaluation of Computed Stresses

The codes provide for different allowable values for primary, secondary, and localized stresses. Generally, a basic allowable stress value is determined for the material at the temperature, and then the allowable stress values for the different types of stresses are given as a function of this basic allowable value. As an example, the basic allowable stress value S for a material other than bolting, cast iron, or malleable iron is given by the ANSI/ASME B31.3 Piping Code as the lowest of the following strength values:

- One-third of the specified minimum tensile strength at room temperature
- One-third of the specified minimum tensile strength at design temperature
- Two-thirds of the specified minimum yield strength at room temperature
- Two-thirds of the specified minimum yield strength at design temperature
- The average stress required to cause 1% creep in 100,000 hr
- Two-thirds of the average stress required to cause rupture in 100,000 hr
- Four-fifths of the minimum stress that could cause rupture in 100,000 hr.

For austenitic stainless steels and nickel alloys, 90% rather than two-thirds of the specified minimum yield strength at temperature may be used, but this higher value is not recommended in deformation sensitive applications (e.g., flanges, where the extra deformation can cause leakage). Allowable stresses for structural grade materials are limited by the code to 92% of these values.

These basic allowable stresses are typically tabulated in code appendices for each material and for various temperatures in 50° increments.

The material may not be used at temperatures higher than the highest temperature for which a basic allowable stress is given.

The circumferential pressure stress in the piping is limited by the code to ES, where E is the casting quality factor, E_c (0.85 to 1, depending on the level of casting examination) or the weld joint quality factor, E_j (0.60 to 1, depending on weld type and the level of weld examination). The combined longitudinal stresses due to pressure, weight, and other sustained loads are limited by the code to the basic allowable value of S. In calculating these stresses, the corroded pipe thickness, $t - c$ must be used.

The range of secondary stresses due to thermal expansion, S_e, must not exceed the allowable stress range

$$S_a = f(1.25\,S_c + 0.25\,S_h) \tag{18.1}$$

where S_c is the basic allowable stress for the material at the minimum (cold) metal temperature expected during the displacement cycle, S_h is the basic allowable stress for the material at the maximum (hot) metal temperature expected during the cycle, and f is the stress range reduction factor for cyclic conditions, depending on the number of full temperature cycles expected over the life of the piping. The value of f is equal to 1 for piping expected to undergo not more than 7000 cycles of loading and decreases to 0.5 in accordance with the following table:

- Over 7000 up to 14,000 cycles, $f = 0.9$
- Over 14,000 up to 22,000 cycles, $f = 0.8$
- Over 22,000 up to 45,000 cycles, $f = 0.7$
- Over 45,000 up to 100,000 cycles, $f = 0.6$
- Over 100,000 cycles of loading, $f = 0.5$

For castings, the hot and cold stress allowable values, S_h and S_c, must be multiplied by the casting quality factor, E_c. When the hot allowable, S_h, is greater than the combined longitudinal stress, S_L, the difference between them is added to the 0.25 S_h term in Equation 18.1 for the allowable stress range. In that case, the equation becomes

$$S_a = f[1.25(S_c + S_h) - S_L] \tag{18.2}$$

The ASME Boiler and Pressure Vessel Code also contains appendices with basic allowable stresses for each material covered by the code and for a range of temperatures. The allowable stresses are based on criteria similar to those outlined in the piping code. Division 2 provides for higher allowable stresses than Division 1, but requires more stringent examination and excludes certain construction details that are permissible in Division 1 construction.

Selection of Materials

Material selection is made on the basis of design conditions, most notably temperature and pressure. Additional guidelines for material selection include economics, availability, properties, code rules, and corrosive properties of the contents. These guidelines may dictate the use of alternate materials if the first choice is not available at a reasonable cost, etc.

ASTM specifications give the physical and chemical properties of the materials. Commonly used materials are carbon steel, carbon-molybdenum steel, chromium-molybdenum steel, and stainless steel. Depending on the conditions of temperature and pressure, the materials may have to be impact-tested to ensure sufficient toughness.

Carbon steel is the most frequently used material of construction in chemical plants and refineries. Carbon steel piping is usually in accordance with ASTM specification A-106 or A-53. The two materials are chemically identical, but A-106 undergoes more rigorous testing. Both are made in grades A and B. Grade A has greater ductility, and B has higher strength. For that reason only grade A is permitted for cold bending. Carbon steel piping should generally not be used at temperatures beyond 775 °F.

Carbon steel vessels at temperatures of 60–650 °F and API storage tanks at temperatures of 60–200 °F can be constructed of A-283-C with flanges and forgings made of A-105 steel. Code vessels at temperatures up to 750 °F can be built of A-285-C and A-516-70 steels.

Carbon-molybdenum steels can be used in a range of temperatures not exceeding 1000 °F. At temperatures beyond 800 °F, however, it should be used with caution. Piping specifications include A-204 (electric fusion welded) and A-335 (seamless). For vessels, A-204-B or C is one material of choice, used with A-182-F1 flanges and forgings.

Chromium-molybdenum steel can be used at temperatures up to 1100 °F. Piping specifications include:

- ½Cr ½Mo A-335 Grade P2 for temperatures up to 950 °F
- 1Cr ½Mo A-335 Grade P12 for temperatures up to 1000 °F
- 1-¼Cr ½Mo A-335 Grade P11 for temperatures up to 1050 °F
- 2-¼Cr ½Mo A-335 Grade P22 for temperatures up to 1100 °F

In cases of high temperature combined with corrosive action, 5Cr ½Mo A-335 Grade 5 can be used. For vessels, A-387-11 C12 is one material of choice.

Stainless steel material is used at temperatures below −20 and above 1100 °F. For piping, A-213 Grade TP321 and A-213 Grade TP347 are two examples. For vessels, A-240-304 and for forgings A-182-F304 are choices. For nozzle necks and welding fittings A-312-TP304 is often used. Bolting of A-320-B8 at low temperatures and A-193-B8 at high temperatures and nuts of A-194-8 at both ends of the temperature range are the choice.

Layout Considerations

The plant layout designer, with input from the design engineers, determines the piping layout using common sense, his or her knowledge of how the plant operates and the way the equipment is maintained, and certain general principles to arrive at an optimum configuration that meets the client's requirements, standards, and specifications. The objective of the layout designer is to create a safe, functional, and cost-effective layout.

Input to the layout process is obtained from engineers of all disciplines, including process, civil, structural, vessel, project, mechanical, furnace, exchanger, rotating equipment, instruments, electrical, inspection, and construction. It is at this juncture that the machinery reliability professional will again address such issues as accessibility, maintainability, and surveillability. Surely, an inappropriate equipment layout could present a serious impediment to achieving one or more of these key ingredients of a reliable machinery installation.

As a first step in-plant layout, the location of all equipment on the plot plan is determined. The questions of sequence of construction, handling of large pieces of equipment, operability, maintenance, and economics need to be addressed. As an example, a very large vessel may need to be manufactured and installed in two pieces, welded together in the field, and the weld pressure-tested. Parts of the surrounding structure may be erected last to leave room for the installation of the large vessel.

The locations of all nozzles required for process, utility, and instruments are then determined on the plan, and finally safety and miscellaneous items are located before the piping layout begins. The piping layout is best done on the basis of treating the unit as a whole, rather than locating one line at a time.

The inlet line of a centrifugal compressor is designed with a minimum of three pipe diameters of straight run between the inlet nozzle and the first elbow. Preferably, the horizontal run is parallel to the compressor shaft. Strainers are installed between the block valve and the inlet nozzle. All lines that must be removed for maintenance are flanged. All operating valves must be accessible.

Line design should be simple and close to the ground for easier support. Supports can be on individual foundations, separate from the compressor foundation, to minimize the transmission of piping vibrations. This may not be desirable however where soil conditions make it difficult to control differential support settlement. In that case, the support should be put on the same foundation as the compressor, because differential support settlement can be more detrimental to the piping.

Pump piping may include large expansion loops for needed flexibility. Pump nozzle allowable loads are very low, and care must be taken to avoid overstressing the pumps. Overstressing will not only void the pump manufacturer's warranty, but may lead to internal misalignment and high failure rates in mechanical seals and bearings. This is an important issue which will be addressed in more detail later.

To keep pump nozzle loads within manufacturer's allowable, the piping must be properly supported. The need to remove the pumps for maintenance must be taken into account. The piping configuration is often duplicated for various groups of pumps of the same size under similar service conditions, i.e. a standard pump layout is used. Multiple pump piping arrangements should be such as to minimize the support requirements. Refer to Figures 18-1 and 18-2 for typical pump suction and discharge piping arrangements, respectively.

Figures 18-3 and 18-4 illustrate additional piping-related requirements that must be verified on pump layouts. Because turbulent flow through valves may adversely affect pump reliability, prudence requires valves to be located a sufficient distance from the suction nozzle. This is especially important where double-flow pumps are concerned (Fig. 18-3). Similarly, elbows should be located at least five pipe diameters away from the pump suction nozzle. The effect of not having at least five diameters of straight run between elbow and nozzle is shown in Figure 18-4.

Steam turbine piping is laid out with the steam trap at the low point of the system in order to avoid the introduction of steam condensate into the turbine case and the resulting blade damage.

Support Configuration

Proper supports and restraints must be used to ensure reliable operation of a structure or piping system. Support and restraint location and type are selected so as to minimize the resulting stresses and control the direction of thermal expansion. In addition, the support system must be such as to prevent joint leakage, excessive forces and moments on connecting equipment, excessive sag or distortion in the piping, unintentional disengagement of the pipe from the support, resonance and excessive vibration, etc.

It is important to note that unimpeded sliding in the direction of thermal expansion is best achieved by placing *two* plates or blocks of polymeric

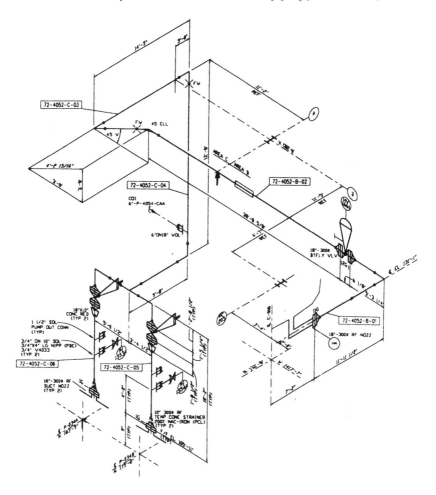

Figure 18-1. Typical pump suction piping arrangement.

material under the pipe shoe, as illustrated in Figure 18-5. Using two instead of only one of these items allows the sides in contact with rough, or perhaps rusty steel, to stick to the metal. With the slippery polymer thus touching its equally slippery mirror-image part, the coefficient of sliding friction will remain suitably low to assure piping movement as desired.

Rigid anchors can divide a piping system into more than one section. Not only does division of the system into smaller systems enable the engineer to independently analyze each of the smaller systems, but the division also results in better isolation and control of systems subjected to vibration.

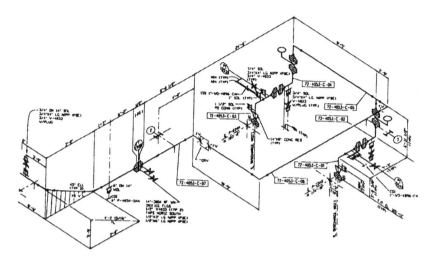

Figure 18-2. Typical pump discharge piping arrangement.

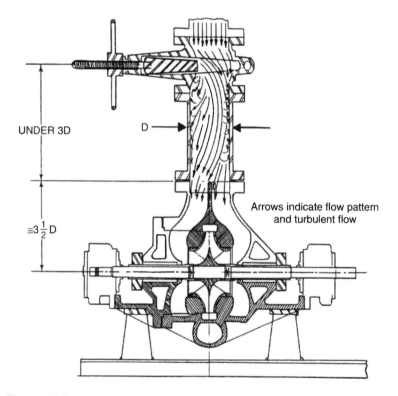

Figure 18-3. Effect of turbulent flow through a valve on double suction pump.

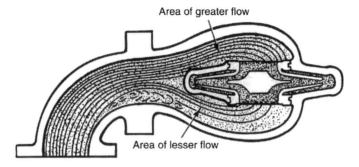

Figure 18-4. Effect of elbow in horizontal plane on suction flow to a double suction pump.

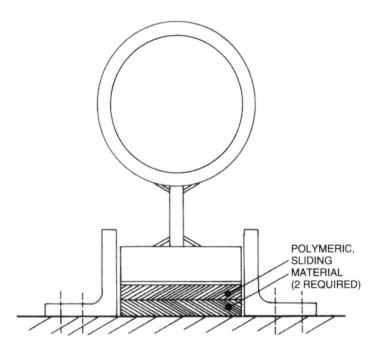

Figure 18-5. To allow free sliding in the direction of the pipe, it is necessary to use two, not just one, polymeric slide plates (or blocks).

Selection of supports is often a balancing act because the proper support system must provide for sufficient stiffness for proper operation under dynamic loading, without resulting in excessive stresses due to the restriction of thermal expansion.

Other rules of good practice in pipe support design include the locating of the pipe supports on the pipe rather than fittings, valves, expansion joints, etc. on straight runs rather than the already highly stressed bends, and as close as possible to heavy load concentrations, such as vertical runs, heavy valves, etc. Piping attached to connections high in a tall vessel should be supported from that vessel to minimize the effects of thermal expansion.

In addition to minimizing stresses in the piping system, stresses in the supporting structures also need to be minimized. The same principles apply. An effort should be made to apply the piping loads near a supporting column rather than at the mid-span of a supporting beam to minimize bending. Unnecessary torsion and compression of slender members should also be avoided.

Types of Support

Several types of pipe supports and restraints are available for the designer. Their definitions are given here:

- An *anchor* provides a completely fixed support. No translation or rotation is permitted in any direction.
- A *support* is used to resist gravity loads.
- A *restraint* is used to resist thermal expansion (Fig. 18-6).
- A *stop* prevents translation in one direction (single-acting or unidirectional stop) or in two directions (double-acting or bi-directional stop).
- A *limit stop* (gap) permits translation up to a specified displacement and then engages to prevent further movement. A limit stop can be single- or double-acting.
- A *guide* prevents rotation about one or more axes.

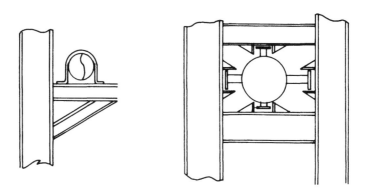

Figure 18-6. Typical rigid piping restraints.

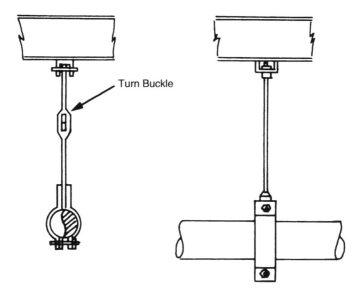

Figure 18-7. Typical hanging rod arrangements.

- A *sliding support* prevents translation in the support direction, but permits translation in the perpendicular directions except as limited by friction.
- A *hanger* suspends the piping from a structure. Figure 18-7 shows typical arrangements.
- A *variable force spring* exerts on the piping a force proportional to the displacement at the point. The variability is needed because of different displacements of the piping during shutdown and under the various operating conditions (see Fig. 18-8).
- A *constant force spring* (Fig. 18-9) exerts a constant force on the piping, balancing the pipe weight.
- A *snubber* permits the gradually occurring thermal displacements as if no restraint were present, but locks, acting as a rigid support when dynamic loading, such as earthquake, wind gust, or fluid slug force causes the piping to move rapidly. Snubbers can be mechanical, using a system of gears, or hydraulic, using viscous fluid. Figure 18-10 shows a mechanical snubber.

The most common method of supporting the pipe, in the absence of vibration loads, is to simply have it rest on the support. A hanging rod can also be used. Turnbuckles are used when adjustable hanging rods are required. As illustrated in Figure 18-11, hanging rods should be kept as close to the vertical position as possible, forming an angle of not more than 4° with the vertical. If necessary, the rod can be offset from the

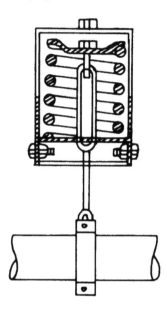

Figure 18-8. A variable force spring hanger.

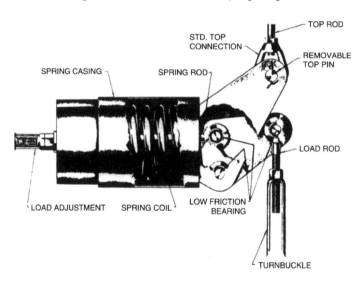

Figure 18-9. Constant force spring.

vertical in the cold position, so as to limit the angle in the hot position. If it is not possible to keep the angle within the 4° limit, then sliding supports or roller-type assemblies are used. Vertical thermal movements are small.

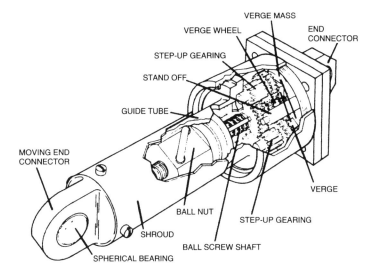

Figure 18-10. Mechanical snubber.

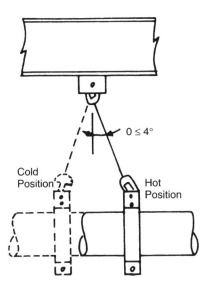

Figure 18-11. Rod hanger angle limitations.

When vertical thermal movements are large, rigid supports are not practical as they cause excessive thermal stresses in the piping. Excessive gravity stresses can also occur if the thermal movement lifts the pipe off the support thus redistributing the load to other supports that are still engaged. The problem is addressed by using variable force spring supports to carry the piping weight. To avoid large, unbalanced forces, the spring support load in the cold position must not differ greatly from the load in the hot position. The spring variability is usually limited to 10–25%, depending on the installation.

When thermal movement is excessive (usually over 2 in.) and use of variable force spring is not feasible, constant force supports are used. Constant force supports may be provided through a system of pulleys and weights or levers and weights, but constant force springs are preferred for compactness and reliability. They use the principle of mechanical advantage to produce a nearly uniform load throughout the travel range of the spring.

Rigid restraints need to be provided in locations where thermal movements are small enough to permit their use in supporting gravity, wind, earthquake, and vibration loads. In cases of high temperature piping, it may not be possible to identify points of sufficiently small thermal movement for the location of rigid pipe supports, but rigid supports may still be needed to resist earthquake or other occasional shock loading. In that case, snubbers may be employed to permit the slow, gradual motion resulting from thermal expansion, but prevent the sudden motion of the pipe under shock loading.

ASME Code Calculation for Shells Under Internal Pressure

The ASME Boiler and Pressure Vessel Code gives formulas for determining vessel thicknesses. For cylindrical shells under internal pressure the formulas are

$$t = PR/(SE - 0.6P) \text{ or } P = SEt/(R + 0.6t) \tag{18.3}$$

where R is the inside radius, S is the maximum allowable stress, P is the internal design pressure, t is the minimum required thickness, and E is the joint efficiency factor. For spherical shells, the code requires that

$$t = PR/(2SE - 0.2P) \text{ or } P = 2SEt/(R + 0.2t) \tag{18.4}$$

For ellipsoidal heads, the governing formulas are given as

$$t = PD/(2SE - 0.2P) \text{ or } P = 2SEt/(D + 0.2t) \tag{18.5}$$

where D is the inside diameter of the head skirt.

For torispherical heads, the code provides the formulas

$$t = 0.885PL/(SE - 0.1P) \text{ or } P - SEt/(0.885L + 0.1t) \tag{18.6}$$

where L is the inside spherical or crown radius. For conical shells,

$$t = PD/2\cos\alpha(SE - 0.6P) \text{ or } P = 2SEt\cos\alpha/(D + 1.2t\cos\alpha) \quad (18.7)$$

where α is half the included (apex) angle, and D is the inside diameter at the point under consideration.

ASME Code Calculation for Shells Under External Pressure

For shells and heads under external pressure, a more elaborate, iterative procedure is given by the code. Typically, a value is assumed for the required minimum thickness, and a chart is entered using the appropriate L/DO and DO/t to determine the value of a factor A. This factor is then used with another chart and is entered to determine the value of a factor B. The code then gives formulas for the allowable external pressure in terms of factor B. If this allowable external pressure is smaller than the given external pressure, the assumed thickness is increased and the calculation repeated.

Piping Code Calculation for Pipe Under Internal Pressure

The ASME/ANSI B31.3 gives formulas for determining the thickness of pipe under internal pressure. The code requires that the minimum thickness of straight pipe sections, considering the manufacturer's tolerance of 12.5%, shall not be less than

$$t_m = t + c \quad (18.8)$$

where c is the sum of all mechanical plus corrosion and erosion allowances and t is the pressure design thickness. For piping with $t < D/6$, the required pressure design thickness is given by

$$t = PD/2(SE + PY) \quad (18.9)$$

or by the simpler formula

$$t = PD/2SE \quad (18.10)$$

Here, Y is a factor given in the code ranging from 0.4 to 0.7 depending on the material and temperature, and E is the quality factor, E_c or E_j. Alternate formulas that may be used to determine the pipe thickness are

$$t = D/2\left[1 - \sqrt{(SE - P)/(SE + P)}\right] \quad (18.11)$$

and

$$t = P(D + 2c)/[2(SE - P(1 - Y)] \qquad (18.12)$$

For $t > D/6$ or $P/SE > 0.385$, the code requires special consideration of factors such as theory of failure and effects of fatigue and thermal stresses.

While it would rarely be within the scope of the machinery reliability engineer's assessment to perform all of these calculations, he should request the piping designer's cooperation in ascertaining that associated piping and vessels incorporate the design conservatism inherent in these formulas. In addition, the engineer may indeed wish to perform some of the calculations on a spot-check basis.

Branch Connection Reinforcement

In areas where the pipe is cut to insert a branch connection, the pipe is locally weakened and, unless the wall thickness is sufficiently in excess of that required for pressure, added reinforcement is needed to restore the pipe strength locally. Essentially, the idea is to replace the amount of metal that was lost in creating the opening, if that was needed to resist the applied pressure.

When multiple branch connections are closely spaced so that their reinforcement zones overlap, the distance between centers of openings should be at least 1.5 times their average diameter. The reinforcement between the openings should be at least half of the total reinforcement. Each opening should have sufficient reinforcement, and no part of the reinforcement can be accounted for more than once.

Stress Analysis/Flexibility Analysis

Stress analysis is used to determine the magnitude of stresses, forces, and displacements in a structure or piping system. Modern methods of stress analysis involve the use of personal computers in association with finite element analysis. Software is available that makes analysis of even the most complex structures and systems possible. These methods include use of structural models for finite element analysis. The modeling process uses discrete finite elements to approximate the continuous structures. As an example, a number of discrete beams, columns, and elbows are used to collectively approximate a piping system, a number of discrete beams and columns are used to approximate a building frame, a number of discrete triangular and quadrilateral plate elements are used to approximate a plate or shell structure, etc.

Piping Flexibility Analysis

Not all piping is analyzed for stresses. The piping code provides that no formal analysis is required if a system duplicates or replaces another system operating with a successful record, or can readily be judged adequate by comparison to previously analyzed systems. In addition, analysis is not required if the system is of uniform size, has no more than 2 points of fixation, no intermediate restraints, is not subjected to severe cyclic conditions, and the geometry is such that

$$DY/(L-U)^2 \le K_1 \qquad (18.13)$$

where D is the outside diameter in inches (mm), Y is the total displacement to be absorbed, in inches (mm), L is the developed length of pipe between anchors in feet (m), U is the straight line distance between the two anchors, and K_1 is 0.03 for English units and 208.3 for SI units.

This criterion is not applicable to systems under severe cyclic conditions. The stress range reduction factor f should be taken into account in such cases. The criterion can be taken as a rule of thumb, and may not always be conservative. Therefore, care must be taken not to apply it to unusual piping configurations, such as unequal length U bends with one side longer than 2.5 times the other, large diameter thin wall pipe, etc. If this criterion is not met, then the piping system must be analyzed, either by simplified, approximate methods or by comprehensive methods to ensure that sufficient flexibility is provided by the layout.

Once the decision has been made to analyze the flexibility of the piping system, the choice of calculation method must be considered. It is usually more expedient to use computer methods to analyze the piping. In the case of relatively simple, two-anchor systems however, approximate solutions exist that can assist the designer in the evaluation of the piping flexibility. These approximate solutions are typically presented in chart or table form. Methods of solution, charts, and tables are given in many older textbooks on piping, including the M. W. Kellogg book, where the guided cantilever method of analysis is given, and the ITT-Grinnel book. Both are listed at the end of this chapter.

With the advent of modern piping flexibility analysis programs on personal computers, these methods are used less and less for piping analysis. But even though these methods are not used for a final check on piping flexibility, they are useful in the approximate or preliminary assessment of piping flexibility, for assistance in arriving at a suitable layout before detailed analysis, and to establish the location of restraints.

There are several commercially available piping flexibility analysis programs. These programs can accurately analyze virtually any piping configuration of virtually unlimited size using finite element and matrix methods of structural analysis, which are explained briefly in this chapter. The piping system may be composed of any of the commonly used elements, e.g. straight members, elbows, miter bends, tee intersections, rigid links, expansion joints. The elements may have any orientation in space, and can contain closed loops. Loading may result from thermal expansion, anchor movements, dead and live loading, or external restraints involving applied forces and moments or imposed displacements and rotations. In addition, partially restrained anchors or free ends can be accommodated.

It is generally difficult to determine whether a particular piping system should be analyzed by rigorous methods, because this decision depends on the type of service, severity of service conditions, and actual layout and size. The following guidelines are offered to help determine when detailed flexibility analysis should at least be considered to verify the acceptability of the piping system. It is recommended that the following lines be subjected to detailed analysis:

- All lines 4 in. and larger with maximum differential temperature exceeding 400 °F
- All lines 8 in. and larger with maximum differential temperature exceeding 300 °F
- All lines 12 in. and larger with maximum differential temperature exceeding 200 °F
- All lines 20 in. and larger at any temperature
- All lines 2 in. and larger connected to rotating equipment
- All lines 4 in. and larger connected to air fin exchangers
- All lines 12 in. and larger connected to tankage.

These are only guidelines intended to identify in principle the lines that need to be analyzed. The final decision should be made considering the complexity of the specific layout and considering the reliability philosophy of the equipment owners.

Flexibility Characteristics

When curved piping is used to change the direction of the piping system, its cross section assumes an elliptical shape subjected to bending, and its flexibility increases. In addition, higher stresses are present in the bends than would be indicated by the elementary theory of bending. These characteristics of curved pipe are recognized in piping system flexibility calculations by the use of flexibility factors and stress intensification factors, which are simply the ratios of actual flexibility and stress to those predicted by the elementary bending theory.

The ANSI/ASME Piping Code gives the flexibility factors, k, and in-plane and out-of-plane stress intensification factors, i_i and i_o, for pipe bends, miters, and tees, in terms of a geometric parameter called the flexibility characteristic, h.

The code provides that pipe stresses due to thermal expansion should be combined in accordance with the equation

$$SE = \sqrt{S_b{}^2 + 4S_t{}^2} \qquad (18.14)$$

where SE is the equivalent flexibility stress to be compared to the allowable stress range S_a, S_b is the resultant longitudinal bending stress, and S_t is the resultant torsional shear stress.

The resultant bending stress, S_b, is given in terms of the in-plane and out-of-plane bending moments, M_i and M_o, and stress intensification factors by the formula

$$S_b = \sqrt{\frac{(i_iM_i)^2 + (i_oM_o)^2}{Z}} \qquad (18.15)$$

where Z is the section modulus of the pipe.

The resultant torsion shear stress, S_t, is given by

$$S_t = M_t/2Z \qquad (18.16)$$

where M_t is the torsional moment in the pipe.

When the analysis is completed, the results are examined to verify the acceptability of the piping. The calculated flexibility stresses, SE, are limited by the code to the allowable stress range, S_a. This requirement is aimed at preventing failure of the piping anchors.

Excessive forces and moments on the nozzles of connected equipment must be avoided, as they may cause malfunction or mechanical failure of the equipment. Forces and moments on the nozzles of sensitive equipment, including pumps, compressors, steam turbines, and air-cooled heat exchangers should be limited to the allowable values given in the corresponding industry standards listed earlier. These allowable values are examined in more detail in the following sections.

Centrifugal Pump Piping Systems

Centrifugal pump piping, especially for high temperature service, may present a significant problem to the piping designer because of the low allowable nozzle forces and moments. The criteria for allowable forces and moments set forth in API 610 are stringent and may need to be satisfied for more than one loading conditions, as for example in the case of piping attached to a spare pump that may be cold when the rest of the system is hot. The governing criteria must be satisfied for all possible spare pump operating combinations. Not to be overlooked is the possibility of one pump being removed entirely while undergoing shop repair.

The standards set forth in API 610 specify that acceptable piping configurations should not cause excessive shaft misalignment or casing distortion. The allowable values, given in the standard, would limit shaft displacement to less than 0.007 in. and casing distortion to half the distortion that would impair the operation of the pump or seal, typically about 0.002 in. The sign convention adopted by API 610 is shown in Figures 18-12 through 18-16.

API 610 allows that the values given in Table 18-1 may be exceeded provided that no individual component force or moment acting on a pump nozzle exceeds the specified table range by a factor of more than two,

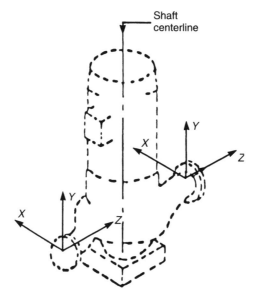

Figure 18-12. Coordinate system for the forces and moments in Table I: vertical in-line pumps.

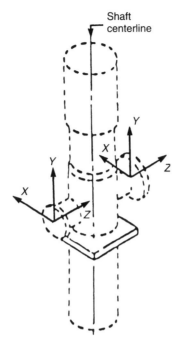

Figure 18-13. Coordinate system for the forces and moments in Table I: vertical double-casing pumps.

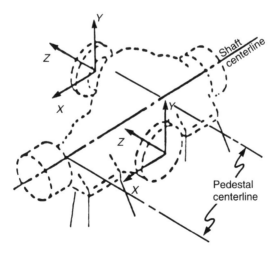

Figure 18-14. Coordinate system for the forces and moments in Table I: horizontal pumps with side suction and side discharge nozzles.

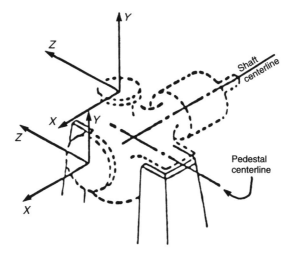

Figure 18-15. Coordinate system for the forces and moments in Table I: horizontal pumps with end suction and top discharge nozzles.

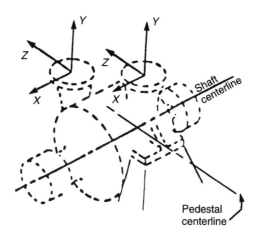

Figure 18-16. Coordinate system for the forces and moments in Table I: horizontal pumps with top nozzles.

the resultant applied forces and moments on each pump nozzle satisfy the appropriate interaction formula, and the magnitude of the resultant applied force and resultant applied moment are held to prescribed limits. Specifically, using the subscripts A for applied, R for resultant, S for suction, D for discharge and T to indicate the values given in Table I of the standard, the resultant force, F_R, and resultant moment, M_R, on the

Table 18-1
Nozzle loadings

Force/Moment*	Nominal size of nozzle flange (in.)								
	2	3	4	6	8	10	12	14	16
Each top nozzle									
FX	160	240	320	560	850	1200	1500	1600	1900
FY	200	300	400	700	1100	1500	1800	2000	2300
FZ	130	200	260	460	700	1000	1200	1300	1500
FR	290	430	570	1010	1560	2200	2600	2900	3300
Each side nozzle									
FX	160	240	320	560	850	1200	1500	1600	1900
FY	130	200	260	460	700	1000	1200	1300	1500
FZ	200	300	400	700	1100	1500	1800	2000	2300
FR	290	430	570	1010	1560	2200	2600	2900	3300
Each end nozzle									
FX	200	300	400	700	1100	1500	1800	2000	2300
FY	130	200	260	460	700	1000	1200	1300	1500
FZ	160	240	320	560	850	1200	1500	1600	1900
FR	290	430	570	1010	1560	2200	2600	2900	3300
Each nozzle									
MX	340	700	980	1700	2600	3700	4500	4700	5400
MY	260	530	740	1300	1900	2800	3400	3500	4000
MZ	170	350	500	870	1300	1800	2200	2300	2700
MR	460	950	1330	2310	3500	5000	6100	6300	7200

Note: Each value shown above indicates a range from minus that value to plus that value; for example, 160 indicates a range from -160 to $+160$.
*F = force, in pounds; M = moment, in foot-pounds; R = resultant.

suction nozzle must satisfy the interaction formula

$$\frac{F_{RSA}}{1.5F_{RST}} + \frac{M_{RSA}}{1.5M_{RST}} \leq 2 \qquad (18.17)$$

and similarly for the discharge nozzle

$$\frac{F_{RDA}}{1.5F_{RDT}} + \frac{M_{RDA}}{1.5M_{RDT}} \leq 2 \qquad (18.18)$$

The resultant of all applied nozzle forces (F_{RCA}) translated to the center of the pump is limited to

$$F_{RCA} < 1.5(F_{RST} + F_{RDT}) \qquad (18.19)$$

and similarly the resultant of all applied moments (M_{RCA}) translated to the center of the pump is limited to

$$M_{RCA} < 1.5(M_{RST} + M_{RDT}) \tag{18.20}$$

and the Z component of the resultant moment is limited to

$$M_{ZCA} < 2(M_{ZST} + M_{ZDT}) \tag{18.21}$$

Vertical in-line pumps that are supported by the attached piping may be subjected to piping loads in excess of the table values, provided these loads do not cause principal stresses greater than 6000 psi. The section properties of the pump nozzles may be based on schedule 40 pipe of nominal size equal to that of the pump nozzle.

API 610 gives the equations that must be satisfied in that case as

$$S = (\sigma/2) + (\sigma^2/4 + \tau^2)^{1/2} < 6000 \tag{18.22}$$
$$\sigma = [1.27FZ/(D_o^2 - D_i^2)] + [122D_o(MX^2 + MY^2)^{1/2}]/(D_o^4 - D_i^4) \tag{18.23}$$
$$\tau = [61D_oMZ/(D_o^4 - D_i^4)] + [1.27(FX^2 + FY^2)^{1/2}]/(D_o^2 - D_i^2) \tag{18.24}$$

Finally, it should be pointed out that the European ISO Standard for centrifugal pumps uses somewhat less conservative values for allowable forces and moments than API 610.

Just how serious excessive pipe forces and moments can be is evident from Figure 18-17(a) and (b). Both photographs depict mechanical seal components that failed when the pump shaft contacted the bore of these stationary seal faces. After numerous seal replacements, allowable versus actual nozzle loads were investigated. The results, Table 18-2, are rather startling. With the moment overload in the M_x plane exceeding allowable values by a factor of 29, the pump casing was so severely twisted that 0.040 in. of radial clearance between the gland plate and the shaft vanished to the point of rubbing. With the excessive pipe load, the pump typically ran well for a few days, at most. When the piping was reconfigured, the pump ran for 18 months before another downtime event was recorded.

A second example of pump piping is shown in Figure 18-18. The sum total of all forces imposed on the pump is given in Table 18-3. A balance has been achieved between the loads imposed in the sustained (weight) case and the operating case. One of the moments (Mz) is slightly over the vendor allowable. The spring hangers labeled SH-1 and SH-2 were set at 1050 lb and 760 lb respectively.

Shaft misalignment in this case is approximately 0.010 in., just above the API allowable. It can be seen that if one of the springs were set

(a)

(b)

Figure 18-17. Failed mechanical seal.

Table 18-2
Nozzle loads on an ISO standard centrifugal pump

Allowable	F_x	F_y	F_z	M_x	M_y	M_z
Allowable values, lb/ft-lb	325	300	250	1500	1000	750
Actual values, lb/ft-lb	68	784	8770	43500	97	518

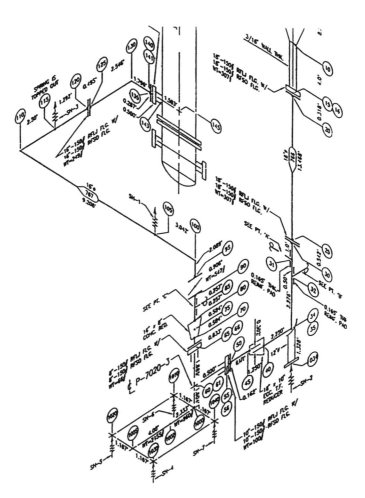

Figure 18-18. Example of pump suction and discharge piping.

Table 18-3
Loading imposed on the pump by piping

	F_x	F_y	F_z	M_x	M_y	M_z	$F_r + M_r/3$ Suction	$F_r + M_r/3$ Discharge
Sustained case	41	−492	−21	−167	−259	−2153	1138	1717
Operating case	−51	−7	317	1022	−404	−1277	871	2054
Vendor allowable	670	1670	1335	3350	1670	1670	1870	1730

100 lb higher or lower, the moments M_z and M_r would be 300–500 ft-lb higher, and the loading imposed on the pump would not be acceptable. An additional force of 300 lb would approximately double the moments imposed on the pump nozzle, increasing shaft misalignment to 0.020 in., which is unacceptable.

Steam Turbines

Allowable forces and moments on steam turbine inlet, extraction, and exhaust nozzles are given by the National Electrical Manufacturer's Association (NEMA) standard SM 23.

The total resultant force and total resulting moment imposed on any nozzle should not exceed the values per limit 1:

$$3F_R + M_R \leq 500 D_e \qquad (18.25)$$

where F_R is the resultant force in pounds,

$$FR = \sqrt{F_x^2 + F_y^2 + F_z^2} \qquad (18.26)$$

and M_R is the resultant moment in foot-pounds,

$$M_R = \sqrt{M_x^2 + M_y^2 + M_z^2} \qquad (18.27)$$

For connections up to 8 in. in diameter, D_e is the nominal pipe diameter in inches. For larger sizes, D_e is defined as $D_e = (16 + \text{nominal diameter})/3$.

The combined resultants of the forces and moments on the inlet, extraction, and exhaust nozzles should not exceed the values per limit 2:

$$2F_c + M_c \leq 250 D_c \qquad (18.28)$$

where F_c is the combined resultant force on the inlet, extraction, and exhaust nozzles in pounds. M_c is the combined resultant of inlet, extraction, and exhaust moments and moments due to the forces in foot-pounds. D_c is the diameter of an opening of total area equal to the sum of the areas of the inlet, extraction, and exhaust openings up to a value of 9 in. For values of equivalent diameter larger than 9 in., $D_c = (18 + \text{equivalent diameter})/3$.

In addition, the components of these resultants must be limited to

$$F_x = 50D_c \qquad\qquad M_x = 250D_c$$
$$F_y = 125D_c \qquad\qquad M_y = 125D_c$$
$$F_z = 100D_c \qquad\qquad M_z = 125D_c$$

The coordinate system used with these criteria has the x-axis parallel to the turbine shaft and the y-axis vertically upward. The right-hand rule applies.

Centrifugal Compressor Piping

API standard 617 on centrifugal compressors establishes the requirements for compressor piping. This standard provides that the compressor shall be designed to withstand external forces and moments at least equal to 1.85 times the values calculated using NEMA SM 23. Whenever possible, these allowable loads should be increased, after consideration of the size and location of compressor supports, nozzle length, degree of nozzle reinforcement, and casing configuration and thickness.

The standard also provides that the compressor casing and supports shall be designed to have sufficient strength and rigidity to limit coupling misalignment caused by imposing the allowable forces and moments to 0.002 in. (50 μm).

Air-Cooled Heat Exchanger Piping

Piping connected to air-cooled heat exchangers is covered in API 661. The allowable loads on nozzles given by the standard are shown in Table 3 of that document. The nozzles must be able to withstand these forces and moments in the corroded condition.

The design must be such as to withstand the simultaneous application of all nozzle loadings on each header without damage. In addition, the sum of all nozzle loads on a single header must not exceed the values given in Table 3.

Fired Heater Piping

Fired heater piping is covered in API Standard 560. The heaters must be designed to support the forces and moments shown in Table 7 of the standard.

Tank Piping

Tank designers are concerned with the magnitude of the loads connecting piping transmits to the shell of large diameter tanks. These loads are the result of shell radial movement and nozzle rotation during filling and emptying, thermal expansion of the piping, tank foundation differential settlement, and weight of piping, valves, and contents. In addition, all possible combinations of these loading cases should be considered.

The supports of large valves and pipe loops near the tanks must be designed to allow unrestrained movement of the pipe during filling and emptying of the tank. Variable force springs and flexible piping are the preferred combination. This combination will minimize the part of the piping weight supported by the tank nozzle.

For locations where soil conditions are such that large settlements are anticipated, special care must be taken in the design of the spring supports. Generally, spring supports are located close to the tank because of the heavy weight of the valve, pipe, and liquid. These supports frequently incorporate shims or other means of adjustment, in order to accommodate large settlements. Frequent removal of shims is not required as the piping is generally designed to withstand significant amounts of differential settlement.

Pitfalls and Misunderstandings*

As was demonstrated earlier in this chapter, the load and stress imposed by a connecting piping system can greatly affect equipment reliability. We have seen that these loads, either from expansion of a pipe or from other sources, can cause shaft misalignment as well as shell deformation interfering with the internal moving parts. Therefore, it is important to design a piping system imposing as little stress as possible on the equipment. Ideally, having no piping stress imposed on equipment is preferred, but that is impossible. The usual practice is for the equipment manufacturer

---

* Contributed by L. C. Peng, P. E. Mr. Peng is a consulting engineer in Houston, Texas. He originally wrote this segment for the 2nd International HP Equipment Reliability Conference organized by Gulf Publishing Company in Houston, Texas, November 1993.

to specify a reasonable allowable piping load and for the piping designer to design the piping system to suit the allowables. The allowable piping loads given these days are generally determined solely by the equipment manufacturers without any participation from the piping engineering community. The values so determined are often too low to be practical.

In fact, designing for extremely low allowable piping loads may well result in a machine that has little or no margin of safety. This not only means mechanical vulnerability, but also complicates the layout of the piping system in meeting the allowable. Unusual configurations and restraining systems are often used to make the calculated piping load satisfy the given allowable. However, all these efforts are very often just exercises in computer technology. The main reliability problem has not been solved. Better-designed equipment with some common-sense piping arrangement is the basis for improved reliability.

Allowable Load Revisited

Process equipment, and especially rotating machinery, generally has a very low allowable piping load. Piping engineers often think the manufacturers give low allowables to protect their own interests. This notion is not necessarily true because some equipment indeed cannot take too high a load. The problem is that a weak link exists that is often overlooked in the design of equipment. Figure 18-19 shows a typical pump installation that can be divided into three main parts: the pump body, the foundation, and the pedestal/base plate. Without input (or threat) from piping or equipment engineers, the routine design of the pump assembly can have different significance to different parts of the pump. The pump body is designed to be as strong, if not stronger, than the piping so that the body

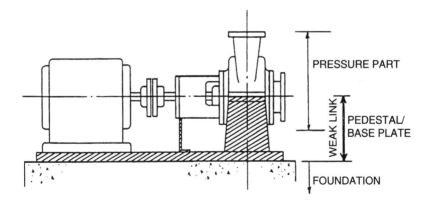

Figure 18-19. The weak link in machine mounting arrangements.

can resist the same internal design pressure as the piping. The foundation, normally designed with the combined pump/motor assembly weight, is also massive and stiff due to limitations of the soil bearing capacity. However, the pedestal/baseplate is a different story. Without considering the taking of any piping load, the pedestal/baseplate is generally designed to support only the pump weight. This design basis creates a very weak pedestal/baseplate that can take very little load from the piping, hence, the famous story of the vendor who claimed his equipment cannot take any piping load. Currently, most vendors have more sense than to claim such a thing, but the allowable piping load is still not large enough to be desirable. The weak link, of course, is the pedestal/baseplate assembly.

By understanding the situation, the problem can actually be rectified very easily. Improvement has already been seen in pump applications. Pump application engineers who for long realized the low allowable piping load problem customarily specified double-strength (2X) or triple-strength (3X) base plates to increase the allowable piping load by two or three times respectively. Most engineers were surprised to learn the cost of a 2X or 3X pump was only marginally more than that of a regular pump. Actually, this should not have been the least bit surprising, because the only thing a vendor has to do to make a pump 2X or 3X is to provide a couple of braces or stiffeners. Recognizing the popular demand for the 2X or 3X baseplate, the API formally adopted it to its pump standards. Since the sixth edition of API Std-610, the allowable has been increased to a level that makes the 2X and 3X specification no longer necessary. In other words, the strength of the whole pump assembly has become fairly uniform, and no additional allowable can be squeezed out without adding substantial cost. Unfortunately, at present, this philosophy has not been shared by other manufacturers. For example, the 1956 NEMA turbine allowable load is probably the most unreasonable of its kind. API Std-617 for centrifugal compressors and ASME/ANSI B73.1 pump are not far behind. API Std-617 uses 1.85 times the NEMA allowable, and the ANSI B73.1 vendors often use 1.30 times the NEMA values for the allowables. Figure 18-20 shows the comparison of the pipe strength, the allowable API Std-610 piping load, and the NEMA allowable piping load. The pipe strength curve is based on a 7500 psi bending stress. It should be noted that the allowable pipe stress against thermal expansion can be as much as three times higher than 7500 psi.

Figure 18-20 shows the piping load that can be applied to equipment is much smaller than the strength of the pipe itself. Therefore, in designing the piping connected to an equipment item, the equipment allowable load is the controlling factor. For low allowable stress casings, such as a large size steam turbine, an extensive expansion loop and a restraining system are generally required. This fact should be understood by all parties concerned.

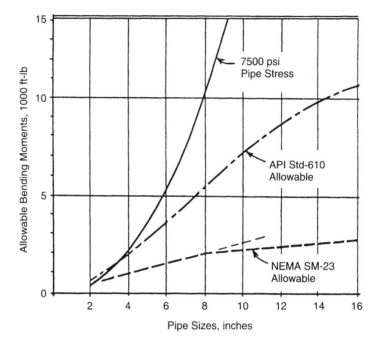

Figure 18-20. Allowable piping loads.

Because of the elaborate design of the piping system attached to sensitive equipment, engineers may sometimes get trapped in the computer maze and overlook engineering fundamentals. Typical examples that can cause unreliable operation are discussed next.

Excessive Flexibility is not Desirable

Adequate piping flexibility near the equipment is required to reduce the piping load to acceptable values. However, a good design should realistically consider support flexibility and the proper use of protective restraints. Without properly located restraints, a piping system, no matter how flexible it is, has difficulty meeting the allowable load imposed by the equipment. Figure 18-21 shows a pump piping system that was designed without any restraints installed. This is a common mistake made by inexperienced engineers who think that a restraint can only increase the stiffness, thus increasing the load. It is true that a restraint will tend to decrease the flexibility of the system as a whole and will increase the maximum stress and force in the system. However, a properly designed

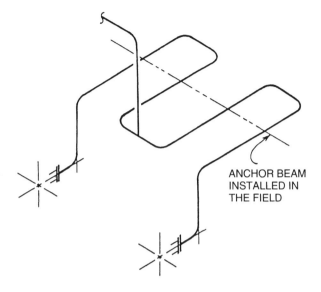

Figure 18-21. Too much flexibility in a piping system.

restraint can shift the stress from the portion of piping near the equipment to a portion further away from the equipment.

Although extensive loops are used in the piping shown in Figure 18-21, the piping load still may not meet the equipment allowable due to the lack of a restraining system. Excessive flexibility makes the system prone to vibration because it is easily excited by small disturbing fluid forces. In addition, the piping loops enhance the internal fluid disturbance by creating cavities and other flow discontinuities due to excessive pressure drops. A system similar to that shown in Figure 18-21 experienced very severe vibrations in one petrochemical plant. The operating engineer had to install a large cross beam to anchor all the loops in the field to suppress the vibration to a manageable level. This shows that the function of the original loops was lost by the anchoring system. The piping still experiences larger than normal vibrations due to flow disturbance caused by the loop which is structurally fixed, but hydraulically still open to many directional changes.

How Theoretical Restraints Often Differ from Actual Ones

A properly designed piping system generally has some restraints to control movements and to protect sensitive equipment. However, there are also restraints that are placed in desperation by piping engineers trying to

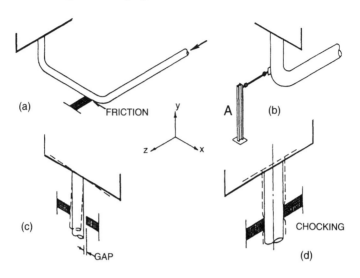

Figure 18-22. Problems with theoretical restraints.

meet the allowable load of the equipment. These so-called computer restraints give a very good computer analysis result on paper, but are often very ineffective and sometimes even harmful. Figure 18-22 shows some typical situations that work on the computer, but do not work on a real piping system. These pitfalls are caused by the differences between the real system and the computer model. Some important discrepancies are described next.

Friction is important in the design of the restraint system near the equipment. Figure 18-22(a) shows a typical stop placed against a long Z-direction line to protect the equipment. In the design calculations, if the friction is ignored, the calculated reaction at the equipment is often very small. However, in reality, the friction at the stop surface will prevent the pipe from expanding to the positive X-direction. This friction effect can cause a high X-direction reaction to the equipment. A calculation including the friction will predict this problem beforehand. A proper type of restraint, such as a low friction plate or a strut, would then be used.

An *ineffective support member* is another problem often encountered in the protective restraints. Figure 18-22(b) shows a popular arrangement to protect the equipment. The engineer's direct instinct is to always put the fix at the problem location. For instance, if the computer shows that the Z-direction reaction is too high, the natural fix is to place a Z-direction stop near the nozzle connection. This may be correct on the computer, but in reality it is very ineffective. For the support to be effective, stiffness of member A has to be at least one order of magnitude higher than the stiffness of the pipe. In this case the pipe is very stiff due to its relatively short distance from the nozzle.

A *gap* is generally required in the actual installation of a stop. Therefore, if a stop is placed too close to the nozzle connection, its effectiveness is questionable due to the inherent gap. As shown in Figure 18-22(c), because of the gap, the pipe has to be bent or moved a distance equal to the gap before the stop becomes active. Due to the closeness of the stop to the equipment, this is almost the same as bending the equipment that much before the pipe reaches the stop. This is not acceptable, because the equipment generally can only tolerate a much smaller deformation than the construction gap of the stop.

Choking is another problem related to the gap at the stop. Some engineers are aware of the consequences of the gap at the stop mentioned above and try to solve it by specifying that no gap be allowed at the stop. This gives the appearance of solving the problem, but another problem is actually waiting to occur. As shown in Figure 18-22(d), when the gap is not provided, the pipe will be choked by the stop as soon as the pipe temperature starts to rise. We all remember to pay attention to the longitudinal or axial expansion of the pipe, but we often forget that the pipe expands radially as well. When the temperature rises to a point when the radial expansion is completely choked by the support, the pipe can no longer slide along the stop surface. The axial expansion will then move upward, exerting an upward force on the equipment.

Special Concern: Expansion Joints

An alternative solution to meet the allowable pipe loading for machinery and other equipment is the use of bellows expansion joints (Fig. 18-23). Regardless of the constant objection from plant engineers, the bellows expansion joint is very popular in the exhaust system of a steam turbine drive that has an extremely low allowable pipe load for pipes 8 in. and above. Bellows joints are also often used for fitting the large multi-unit assemblies as shown in Figure 18-23(b). Although a properly installed and maintained bellows expansion joint should have the same reliability as components such as flanges and valves, it is often found to be very undesirable due to vulnerabilities in the event of fire, and due to the difficulty in maintenance. For instance, when covered with insulation, the expansion joint looks just like a pile of blanketed scraps. Nobody knows exactly what is going on inside the mixed layers of covering. Because of the blindness anxiety, many installers have resorted to an uninsulated arrangement. This not only creates an occupational safety concern, but it can also cause cracks due to thermal shock from the environment and/or

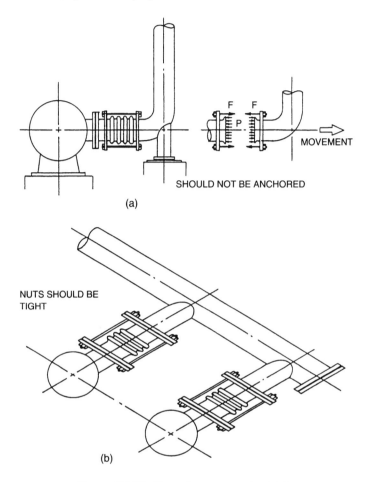

SHOULD NOT BE ANCHORED

(a)

NUTS SHOULD BE
TIGHT

(b)

Figure 18-23. Tie-rods on expansion joints.

weather changes, and to fire events and the possibility of water hitting
red-hot surfaces.

Another important factor often overlooked by engineers in the installation of a bellows expansion joint is the pressure thrust force inside the
pipe. The bellows is flexible axially. Therefore, the bellows is not able to
transmit or absorb the axial internal pressure end force. This pressure end
force has to be resisted either by the anchor at the equipment or by the
tie-rod straddling the bellows. With the exception of very low pressure
applicators, such as the pipe connected to a storage tank, most equipment items are not strong enough to resist the pressure end force equal
to the pressure times the bellows cross-section area. The pressure thrust

force has to be taken by the tie-rod. Somehow this idea is not obvious to many engineers, resulting in some operational problems. Figure 18-23 shows two actual problems. Figure 18-23(a) shows one of many steam turbine exhaust pipes installed at a petrochemical plant. The expansion joint layout scheme appears to be sound, but the construction was not done properly. The actual installation had a sliding base elbow anchored with four bolts. This problem often escapes the eyes of even experienced engineers. When the base elbow is anchored, the tie-rod loses its function as soon as the pipe starts to expand. In this case, the pipe expands from the anchor toward the bellows joint, making the tie-rod loose and ineffective. The large pressure thrust force pushes the turbine, causing shaft misalignment and severe vibrations. Figure 18-23(b) illustrates a similar situation. The bellows expansion joints were used solely for fitting up the connections. The tie-rods were supposed to be locked. However, before startup operation, an engineer had loosened the tie-rod nuts, apparently thinking the tie-rods defeat the purpose of the expansion joint. The startup was very shaky and had to be quickly halted. It took quite some time before anyone discovered that the problem was caused by the loose tie-rods. When the nuts are loose, the pressure end force simply pushes the equipment way out of alignment.

Other Practical Considerations

As discussed above, the reduction of pipe stress is not at all straightforward. Especially when dealing with the low allowable of some equipment, the technique becomes tricky and very often only works on paper. Other practical approaches may be explored to further improve overall reliability. One very important resource often ignored is the experience found in operating plants. We often see a good, simple working layout changed to a complicated, shaky layout only because a computer liked it that way. Undoubtedly, computers are important tools, but they are only as good as the information we give them. Because there are so many things, like friction, anchor flexibility, etc. that cannot be given accurately, computer results need to be interpreted carefully. It is time to realize that if something works well in a plant day in and day out, it should be considered good, regardless of whether the computer predicted it to be good. The process of evolution is very important in designing a good, reliable plant.

Other ideas, such as the use of sliding supports, spring supports, and more compact in-line arrangements as shown in Figure 18-24, can also be seriously considered. It is understood that engineers do not feel too confident on the movable assembly, but it is important to note the difference between movement of the whole assembly and movement of only the

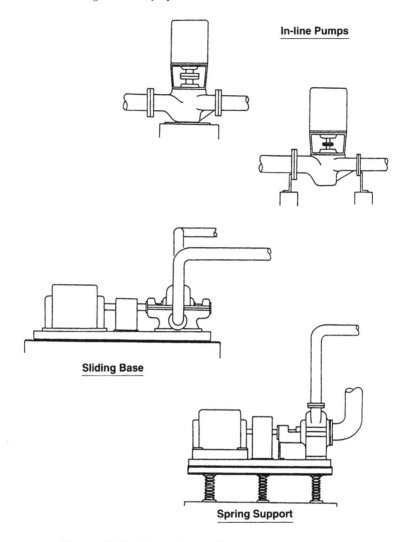

Figure 18-24. Alternative machine support arrangements.

pump or turbine. When the whole assembly moves, the shaft alignment can still be maintained, if distortion of the equipment is not excessive, i.e. if the piping load is still within the allowable. It should be noted, however, that these movable assemblies are just potential alternatives. One should not be oversold on the idea and blindly use it in a plant. To make the sliding base or the spring support scheme workable, an extra strong baseplate is required. Then again, if we have that strong a baseplate in the first place, the allowable piping load would have increased substantially.

What is it Worth to do it Right?

From the above, we hope to have shown that there is genuine merit in verifying the adequacy of equipment alignment and connected piping. Better yet, there are statistics to make the same point.

A North American chemical company operates a facility with over 20,000 pumps. Of these, an estimated 12,000 are centrifugal pumps in all sizes and age categories. In 1989, the company began using computerized failure tracking to identify problem pumps and to graphically display pump experience.

Some problem locations were identified quickly, and among these was one whose 11 pumps registered an average meantime between failures (MTBF) of 1 month. During the years of operation, many different types and brands of mechanical seals had been evaluated. Some had better success than others, but obviously none had given the desired 2 year plus MTBF. The pump installations were more closely examined and two changes recommended:

1. Improve the mounting by adding mass and stiffness. Implementing this recommendation in some cases required upgrading stilt-mounted pump bases (Fig. 18-25) to the considerably sturdier, conventional foundation arrangement found in major refineries.
2. Improve the piping by reducing pipe strain and ascertain that five straight pipe diameters existed before the suction nozzle and after the discharge nozzle.

The recommended changes were made during the annual shutdown for that area of the plant. The average MTBF for those 11 pumps is now 15 months plus.

Figure 18-25. Stilt-mounted pump set. This arrangement is not usually conducive to reaching optimum MTBF.

After analyzing and comparing a population of 700 pumps, it was established that foundation-mounted pumps with properly installed, fully grout-supported baseplates, and acceptable connected piping had a *five times better* MTBF rate than pumps with inadequate mounting support, piping deficiencies, or both. Not unexpectedly, vibration severity and failure frequency were often related, and equipment with overall levels in the "unsatisfactory" or "unacceptable" ranges given in Figure 18-26 was found to merit more frequent attention to shaft, baseplate, and piping alignment.

A correlation between seal failure rate and vibration severity is given in Figure 18-27. Collected from process plants in Great Britain and encompassing approximately 500 centrifugal pumps, these data again show that high vibration and frequent failures go together.

In conclusion, the reliability professional would be well advised to look over the shoulders of the piping and installation specialists. The procedure used to mount and pipe pumps definitely affects MTBF. A pump that has sufficient support and correct piping will have an MTBF of up to five times less than a pump that does not have sufficient base support mass and correct piping.

Unless the piping specialist has field experience, his design may incorporate pitfalls and oversights that will show up as decreased equipment reliability. More often than not, these will elude belated troubleshooting

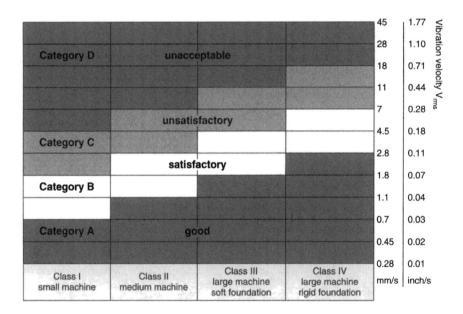

Figure 18-26. Vibration classification chart.

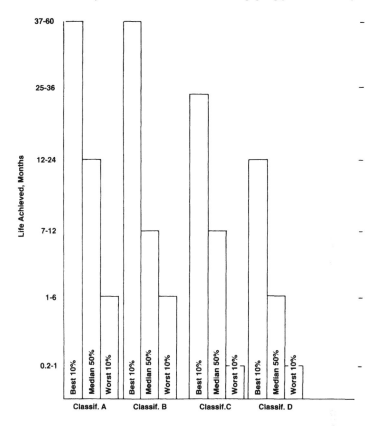

Figure 18-27. Seal life experience versus vibration severity.

efforts and can impose serious burdens on future equipment availabilities and maintenance budgets. We strongly recommend that machinery reliability assessment efforts include an appropriate review of the matters covered in this chapter.

Bibliography

1. Bausbacher, E. and Hunt, R., *Process Plant Layout and Piping Design*. Auerbach Publishers.
2. Pullman Power Products, *Design of Piping Systems*. New York: John Wiley and Sons, 1956.
3. Smith, P. R. and Van Laan, T. J., *Piping and Pipe Support Systems Design and Engineering*. New York: McGraw-Hill, 1987.
4. *ITT-Grinnell Pipe Hanger Design and Engineering*, Revised 1979 ITT-Grinnell Corp., Providence, RI.

5. Marks, L. S., Ed., *Standard Handbook for Mechanical Engineers*. New York: McGraw-Hill.

6. *API Standard 610, 6th Edition*. American Petroleum Institute, Washington, DC, 1981.

7. *API Standard 610, 7th Edition*, Centrifugal pumps for general refinery services. American Petroleum Institute, Washington, DC, 1989.

8. *NEMA SM-23*, Steam turbines for mechanical drive service. National Electrical Manufacturers Association, Washington, DC.

9. *API Standard 617*, Centrifugal pumps for general refinery services. American Petroleum Institute, Washington, DC.

10. *ASME B73.1M-1991*, Specification for horizontal end suction centrifugal pumps for chemical process. American Society of Mechanical Engineers, New York.

11. Dufour, John W. and Nelson, W., Eds, *Centrifugal Pump Sourcebook*. New York: McGraw-Hill, 1993.

Appendix A
The coin toss case

The coin toss case follows a "binomial" probability distribution. This distribution can be described by the general function,

$$f(x) = C_x^n p^x q^{n-x}$$

where $n =$ number of trials (i.e., six tosses of a coin in our case),
$x =$ number of successes per trial (i.e., 0. . . 6),
$p =$ the probability of success per trial (i.e., 50% in our example),
$q =$ the probability of failure (i.e., 50% or $q = 1 - q$),

$$C_x^n = \frac{n!}{x!(n-x!)}$$

Example. What is the probability of getting 2 heads in 6 throws of a coin?
Solution

$n = 6$

$x = 2$

$p = 0.5$

$q = 0.5$

$$C_x^n = \frac{6!}{2!(6-2)!} = \frac{6!}{2!4!} = \frac{1 \times 2 \times 3 \times 4 \times 5 \times 6}{1 \times 2 \times 1 \times 2 \times 3 \times 4} = \frac{720}{2 \times 24} = 15$$

$$P_{\text{head}} = 15 \times 0.5^2 \times 0.5^4 = 15 \times 0.25 \times 0.0625 = 0.23 = \frac{15}{64} \quad \text{or} \quad 23\%$$

The binomial function can have many useful applications. One of the most common applications is in lot-by-lot acceptance sampling inspections, where the decision is made to accept or reject an entire lot of parts based on the numbers of defectives found in a sample of these parts.

Appendix B

Safety design checklist for equipment reliability professionals

The purpose of this checklist is to provide "memory joggers" for engineers and technicians involved in machinery reliability assessment. Because this list is incomplete, it is the user's responsibility to seek additional input from process and/or project personnel before accepting an installation as complete from safety points of view.

It should also be noted that many equipment safety-related issues must be addressed before even the basic process design is finalized. Accordingly, the various project engineering groups would do well to maintain good and frequent communications with resident maintenance and reliability staff.

Equipment Location and Spacing

1. Have the appropriate site or specification requirements been followed or exceeded for:

   (a) equipment, building, tankage, perimeter fence, etc.
   (b) electrical substations
   (c) electrical equipment
   (d) control houses
   (e) flares

2. Have provisions for containment of spillage been made for:

   (a) pressure storage vessels
   (b) atmospheric storage tanks
   (c) refrigerated storage vessels

513

3. Have any special risks requiring more stringent spacing standards been considered?
4. Has blast proofing of the buildings and control house in accordance with local regulations been considered?
5. Does the design avoid having hydrocarbon lines pass through another unit that will be turned around separately?
6. Has location of existing safety equipment such as safety showers or fire equipment been considered and, if necessary, supplemented with new facilities?
7. Has the influence of this project on future projects been included?
8. Has the prevailing wind direction been considered in the location of equipment likely to release toxic gases or liquids? Has cooling tower location been observed?
9. Is all equipment located outside the dikes area of tankage?
10. Do piping layout and overhead clearances comply with specifications? Accessibility, maintainability, surveillability?
11. Have above-ground wiring systems been routed at least 25 ft from "high fire" risk equipment? If not, has exception been granted and fireproofing provided?
12. Has provision been made for collecting and handling spillage from process equipment and tanks in accordance with good and approved practices?

Equipment Pressure Relieving Facilities

1. Does the design of pressure relieving facilities comply with existing regulations or guidelines?
   Have the following causes of over or under pressure been considered:
   (a) entry of water into hot liquids above 212 °F (100 °C)
   (b) failure of downstream check valve to prevent backflow of fluid into equipment being protected
   (c) tendency of the materials being handled to plug SV inlets or PV vents
   (d) where two or more vessels are protected by one SV, the possibility of one being isolated from the SV by the intermediate block valves or plugged lines
   (e) failure of a level control valve in the open position that could cause over-pressure of the downstream equipment?

2. Where safety valves discharge into an existing system, has the following been considered:

(a) the capability of the existing system to handle the additional load (flare exit velocities, noise, radiation, etc.)

(b) the effect of the increased pressure in the blowdown system on other SV back pressures?

(c) the effect of temperature shock (high temperature or autorefrigeration) on the existing system:

- brittle fracture
- adequate thermal expansion or contraction

3. Will the gas velocity at all points in the SV discharge system remain below 50% of sonic, except in individual discharge headers where velocity must be below 75% of sonic?

4. If the SV is a bellows type, is the back pressure below maximum allowable?

5. Have all safety valve installations been checked for a maximum pressure drop in the inlet system of 3% of set pressure, for subsonic velocities in the outlet system and for heat release in the case of ignition of an atmospheric discharge stream?

6. (a) Has the need for accommodating outlet piping of thermal expansion or contraction been specified?

(b) Does minimum design temperature reflect autorefrigeration in discharge piping to aid in the selection of material of construction?

7. Where SVs and/or rupture discs discharge to atmosphere, have the following points been checked or specified?

(a) acceptability of discharge from environmental pollution considerations

(b) acceptability of noise level

(c) proximity to other equipment and sources of ignition

8. Are bellows-type pressure relief valves clearly indicated and provision made for bellows vent disposal to a safe location?

9. Has compensation been made for the hydraulic head of SVs in liquid service that are located higher than the equipment which they protect?

10. Have SVs that could blow liquid to the atmosphere should a vessel overfill been checked for all applicable regulations and safety guidelines?

11. Does the back pressure acting on the SV comply with good practices?

12. Has the radiant heat level for hydrocarbon vapors discharging via SV vents to atmosphere been checked for radiant heat level at grade and on adjacent equipment?

13. If vapor discharged to atmosphere is combustible, has snuffing steam been provided?
14. Has consideration been given to the need for protection against overpressure resulting from exchanger tube failures?
15. Are safety valve set pressures consistent with equipment design pressures?
16. Have all normal venting and balancing systems been checked for adequacy in all emergencies including external fire?
17. Have safety valves been protected by suitable purge systems against coke formation, wax deposition, and congealing of high pour fluids?
18. Can vacuum conditions result in vessels when there is a loss of heat input or cooling?
19. Does specification require block valves in the relief system to be installed with their stems horizontal or downward?
20. If a rupture disc is being specified, have the following been considered and/or documented:

    (a) basis for selecting a rupture disc as a relieving device
    (b) basis for sizing and operating conditions

21. Is a rupture disc being installed in conjunction with a PR valve? If so, has a PI and vent been provided between the rupture disc and PR? Does the system meet the requirements of the ASTM load, Section VIII, Div. 1?
22. Should a sealed–body type valve be specified? Is a rupture disc required to protect the valve?
23. Is discharge piping of SVs designed to avoid liquid traps? If discharge piping must have liquid traps, are they designed to be safe?
24. Do SVs discharging flammable vapors discharge at least 10 ft above equipment within a radius of 50 ft?
25. Does design make provisions to prevent the accumulation of liquid in piping from discharges of safety valves to the headers of closed systems?
26. Do piping drawings require block valves in the relief system be installed with their stems horizontal or downward?
27. Do drawing notes call for gusseting and seal welding for small piping on SV piping?

Equipment Emergency Isolation Components and Systems

1. Have isolation valves been installed on vessels, pumps, and other equipment where required?
2. Are fire-resistant isolation valves and motors specified?

3. Are check valves installed in the discharges of centrifugal pumps? In other pumps capable of reverse flow?

4. Are tank nozzles and connections below the maximum fill height filled with block valves (except those for atmospheric vents)?

5. Have liquid pull-down lines been specified in accordance with safe practices?

6. Can liquid inventory be disposed of within one hour?

7. Can vessels be depressurized to 50% of design pressure within 15 min?

8. Are all isolation valves of 10 in. and larger provided with motor actuators?

9. Has the minimum closing time of isolation valves been considered from the point of view of water hammer?

10. Have manual block valves been considered for the nozzles of vessels containing cryogenic, high pressure, toxic, and corrosive materials?

11. When two or more pumps are manifolded, are the suction block valves of the proper design pressure?

12. Have drawings been checked to insure block valves are not installed inside skirts of towers and drums?

13. Are battery limit block valves accessible and clearly labeled?

14. Are instrument control lines to RBVs fire proofed within 40 ft of equipment being isolated?

15. Are control lines routed out of range of potential fire sources, or fireproofed?

16. Are all block valves, located adjacent to refrigerated storage vessels in the main process lines, equipped with fireproof motor operators?

17. Has thermal overload removal been specified from motors of emergency block valves?

Hazardous Leak Sources from Equipment

1. Have water drawoff points on pressure storage spheres been specified in accordance with applicable standards?

2. For pressure storage vessels, if circulating facilities are provided, the sample connection should be outside the dike.

3. Has consideration been given to winterizing piping where freezing may occur (particularly important when low or no flow conditions may exist)?

4. For flashing liquid service, has attention been called to protect against vibration? Flash % given?

5. Is the possibility of autorefrigeration considered, e.g. in:

(a) discharge piping of safety valves

(b) low pressure side of heat exchangers

(c) accidental loss of pressure

6. Is some kind of detector used to indicate seal failure on pumps and compressors where failure could cause escape of flammable vapors?

7. Have provisions been made to permit the proper disposal of drainage of process equipment? For example:

(a) light ends and low flash materials to closed drain header

(b) material at temperature below flash point to oil drain 50 ft. from ignition source

8. Are double drain valves specified where appropriate?

9. Have water disengaging drums been specified for water drawoffs from vessels containing light ends and hydrocarbons above their flash points?

10. Has the possibility of water hammer been considered in selecting closing speeds of automatic valves?

11. Are connections to close drain headers provided with check valves to prevent back flow? (Note that springless ball check valves are not usually permitted.)

12. Have thermal expansion relief valves been provided where long sections of liquid lines can be blocked in?

13. Are all utilities connections specified in accordance with applicable standards?

14. Are thermal relief valves in liquid lines set at $1.33 \times DP$?

15. Are double valves used for sample points and drains above 1000°F and 600 psi?

16. Are gas detectors and alarms provided?

17. Have all piping specification breaks been shown, e.g. change from 300 lb to 150 lb flanges?

18. Have provisions been made for the safe isolation of equipment to facilitate onstream maintenance where required?

19. Are check valves provided on injection nozzles for acids, etc.?

20. Have online testing facilities been specified for emergency alarms and shutdown devices?

21. Have fire detection facilities been considered?

22. Has use of gauge glasses been restricted in accordance with risk assessment?

23. Do product streams that go through exchangers en route to tankage have high-temperature alarms to warn of loss of cooling if this could lead to overpressure of storage facilities?

24. Are liquid drawoff lines to the atmosphere double-valved with a quick action valve nearest to the shell (for liquids that can autorefrigerate)?

25. Has consideration been given to upgrading all small piping connections in hydrocarbon service? Schedule 160? Gusseting?
26. Have restrictions on the use of wafer-type valves, butterfly valves, and check valves been applied? Anti-slam devices externally adjustable?
27. Are bleeders provided between block valves and control valves?
28. Has the project been reviewed for the risk of piping failures caused by high flow induced vibrations?

- line size 16 in. or greater: 200,000 lb/h (max.)
- 8–14 in.: 50% sonic, etc.

29. Has piping flexibility been analyzed to prevent overstressing of system components?
30. Have drawings been checked to ensure no bellows-type piping expansion joints have been used in hydrocarbon service in process or high risk areas? Use expansion loops instead.
31. Have all vents, drains, and cleaning connections been designed according to standards?
32. Have lines between tank and dike wall at least one elbow for expansion?
33. Has winterizing been provided on dead legs at the bottom of vessels and other areas in which water can accumulate and freeze?
34. Have vent valves been installed between double blocks?
35. Do water drawoff lines terminate not less than 15 ft from periphery of vessel?
36. Have single layer corrugated metal gaskets been prohibited for bonnet flanges of gate valves?
37. Do plug valves have a position indicator on the stems?
38. Has the use of teflon tape been prohibited?
39. Do sight glasses have a block valve near the vessel?
40. If sight glass connection is more than 15 in. long, is it braced?
41. Has minimum clearance between insulation and other equipment/structures been considered?
42. Have pipe unions been prohibited in hydrocarbon services?

Equipment Blowdown/Flare Facilities

1. Has the possibility of shock chilling or temperatures below $-20\,°F$ occurring in new or existing blowdown lines been considered?
2. Do safety valve and other blowdown lines slope continuously toward blowdown drum (i.e., no liquid traps)?
3. Have drains been provided to drain any trapped sections where these are unavoidable?

4. Have all block valves in the flare/blowdown header that could impede safety valve discharge been identified as CSO (car sealed open)?
5. Have winterizing provisions been made for new connections to blowdown system?
6. Are flare headers that can experience low temperatures of materials of construction resistant to brittle fracture?
7. Are the new flare system connection design pressures according to standards and compatible with the existing systems?
8. Are H_2S-containing materials handled in a specially designed flare system?
9. Are blowdown lines routed outside high risk areas?
10. Are drums and pipe supports in the blowdown/flare system fire-proofed in accordance with standards?
11. If a condensible blowdown is being provided, does it have a backup supply of fire water?
12. Is outlet water seal of condensable blowdown drum, flare seals, and water disengaging drums at least 10 ft or 175% of design pressure?
13. Have high and low level alarms on blowdown drums been provided?
14. Does non-condensable blowdown drum have a high level cut out to close off liquid pull down and closed drain headers?
15. Does flare location meet spacing and height criteria to prevent excessive radiant heat, pollution, etc.?
16. Has effective flashback protection (water seal) been provided in the flare line?
17. Has the problem of taking a flare out of service for maintenance been solved?
18. Is the flare seal drum provided with a low temperature cut-in to provide steam heat to the drum, etc.?

Compressors and Pumping Equipment

1. Are suction drums and interstage KO drums provided with HLCOS and compressors instrumented in accordance with applicable API and supplementary plant standards?
2. Are emergency isolation valves provided on compressors and inter-stage KO drums in accordance with plant standards and safety philosophy?
3. Are emergency isolation valves fireproofed?
4. Can compressor be isolated by stop buttons at least 40 ft from the machine?
5. Has a check valve been provided in the discharge of the compressor?
6. Does the suction line require steam tracing between KO drum and compressor to avoid condensation?

7. Are drains from gas KO drums routed to the flare?
8. Are alarm and shutdown facilities in accordance with DP?
9. Has provision been made to install spectacle blinds in all suction and discharge lines of compressors and turbines?
10. Is the casing protected against overpressure?
11. Have pumps and compressors been checked for minimum allowable flow? Is kickback necessary?
12. Have remote block valves been considered for pumps to block off the flow of hydrocarbons feeding a fire?
13. Is compressor interstage equipment good for or adequately protected against overpressure when the compressor discharge pressure reaches the discharge safety valve set pressure?
14. Are strainers in the lines properly noted and/or specified? (Pump suction strainers must be removed within *days* after startup!)
15. Is compressor interstage equipment adequately protected against overpressure at settling-out pressure conditions?

Furnaces and Boiler Equipment

1. Is furnace instrumentation in accordance with standards?
2. Have bypasses around FGCO valves been identified for CSC?
3. Are snuffing steam valves provided at least 50 ft from furnace?
4. Are drain holes specified for snuffing steam lines?
5. Can fuel be shut off remotely or by block valve 50 ft from furnace? Is valve CSO?
6. Does FG KO drum drain go to flare with a check valve in the line to prevent back flow?
7. Are furnace supports fireproofed and pipe rack supports within 20 ft?
8. Are toe walls and proper drainage specified for furnaces handling liquid hydrocarbons or fuels?
9. Are there isolation valves or check valves in the process outlet of the furnaces to prevent back flow on tube failure?
10. Has provision been made for possible condensation in fuel line downstream of fuel gas KO drum?
11. Is flame cut protection either by a pilot flame from an independent gas supply or flame scanner provided?
12. Are furnace coils protected against low flow?
13. Do all shutdown signals have their own process taps?
14. Is the fire box protected against over-pressure (flames blowing out of inspection holes, etc.)?
15. Are proper blinds provided at all inlets and outlets of process piping to furnaces?

16. Has a note been included specifying additional expansion requirement for furnaces with air/steam decoking?
17. Does pilot gas manifold have PLCO with PLA/PHA to cut-off main fuels on low pressure in pilot manifold if gas supply is not independent of main burner supply?
18. Do valves activated by shutdown circuits have lockout and manual reset provisions?
19. Does low air flow on FD & ID furnaces cut-off main and pilot gas supply?
20. Do fans remain running after burner flame failure?
21. Is there in the control room a PLA/PHA or fuel gas pressure recorder for main fuel pressure downstream of all control valves?
22. With liquid fuels, is there a cut out for the fuel on loss of atomizing steam pressure?
23. Is there an LP alarm downstream of atomizing steam control valve for liquid fuels?
24. Is an emergency switch provided in the control room to cut-off all fuel and pilot gas?
25. Is there an FL (CO) with FLA to shut-off all fuels except pilot burners on low process feed flow?
26. Are all cut-off valves single seat and dedicated solely for shut-off service and not regulatory control?
27. Is the top of the stack at least 10 ft above any platform within 40 ft horizontally of the stack?
28. Has instrumentation been provided to establish visual proof of continuous pilot flame ignition?
29. Have locally mounted manual seal stations been specified for all fuel safety shut-off valves?
30. Is access to the furnace provided in accordance with legislated guidelines (OSHA):
   - stairway access to burners?
   - ladders to sootblowers, etc.?
31. Are there position-indicating handles or pointers on all plug valves and petcocks?
32. Are supporting members of furnace fireproofed between fire box flow and grade?
33. Is the area under the furnace paved and provided with a toe wall 10 ft beyond furnace?

Sewer System Layout Considerations

1. Are acidic wastes prevented from entering industrial and sanitary sewers?

2. Is drainage of diked areas provided as specified by the standards?
3. Have spent caustic drains been routed to the onsite caustic sump to minimize possibility of H_2S generation in sewer system?
4. Has the use of open drains and pits been restricted to an absolute minimum?
5. There should be no catch basins under furnaces.
6. Are catch basins in tankage areas and, are they used for water drawoff, valved and provided with an 8-in. toe wall to prevent overloading the oily water sewer with rain water?
7. Are manhole vents farther than 100 ft horizontally from source of ignition and, if raised 50 ft above grade, 50 ft horizontally from sources of ignition?
8. Do vents discharge away from equipment and overhead pipe racks?
9. Has capacity of sewer been designed according to standards? Is the capacity sufficient for fire water and process water?
10. Is minimum size of sub-laterals 6 in.?
11. Are seals (to prevent flash back in the event of fire) provided at tie-in with offsite main sewer?
12. Are all drains to catch basins and drain headers sealed?
13. Is top of outlet line from manholes lower than top of lowest inlet pipe to prevent trapping?
14. Are floor drains prohibited in control houses and electrical substations?
15. Are layout and sizing criteria for sewer system according to local standards, if applicable?
16. Are sewer pipes buried at least 2½ ft below grade or below the frost line?
17. Has provision been made to trap solid material as close to its source as possible?
18. Do catch basins have a 12-in. debris pot below the seal?
19. Have proper covers been used on catch basins in non-paved areas?

Equipment Fire Fighting Facilities

1. Is there adequate access for fire fighting? At least 50 ft of space on all sides of risk area?
2. Is maximum hydrant spacing less than 300 ft in onsite areas?
3. Are sections of firemain carrying combinations of more than two monitors, hydrants, etc. connected to two separate sections of the fire main?
4. Have other spacing standards for hydrants been met?
5. If the distance from the hydrant to its offtake from the main is more than 50 ft, has a block valve been installed at the main?

6. Are hydrants located within 250 ft of all points that could require water?
7. Are monitors located at least 50 ft from equipment being protected?
8. Are block valves provided so that not more than 1000 ft of pipe containing hydrants shall be lost due to line failure?
9. Are connections to the fire main, other than for fire water, limited to intermittent use as a utility supplement?
10. Is the location of the main operating valve for sprinkler systems at least 50 ft from the area it protects?
11. Do deluge systems on pressurized and refrigerated storage comply with standards?
12. Is half the design firewater capacity to each area maintained with sections taken out of service using hoses not more than 400 ft long?
13. Is pumping capacity equal to maximum demand?
14. Do water rates for offsites and tankage areas comply with conservative design for the media stored?
15. Do water application rates comply with 0.1/gpm sq. ft. or 0.15 to 0.20 for stacked and high density equipment? (Excluding fixed sprays.)
16. Has sewer been checked for capacity to handle fire water load?
17. Is cast iron pipe centrifugally cast and suitable for a working pressure of 150 psig?
18. Are monitors equipped with straight stream fog nozzles?
19. Have fixed sprinkler nozzles at least ½-in. openings with main water supply valve at least 25 ft from equipment?
20. Are foam and cooling facilities for tankage conservatively designed for the fluids stored?
21. Have hydrant guards been provided for hydrants located within 3 ft of roadways?
22. Have hose reels locations been reviewed with fire chief or other agencies?
23. Can hose reels be operated without fully unwinding the hose from the reels?

Exchanger Equipment

1. Have consequences of autorefrigeration that result from tube failure been considered?
2. Have the consequences of hydrocarbon leakage into the steam side of steam heaters been considered?
3. Does baffle design meet TEMA standard to avoid vibrations that may lead to tube failure?
4. Has appropriate pressure relief protection been specified?

5. Have provisions been made to prevent the cold side of the exchanger from being blocked in?
6. Have Charpy test requirements been specified in accordance with low temperature materials guidelines?
7. Have fixed tube exchangers without shell expansion joints been checked per industry guidelines?
8. Is there a potential for high vibration against baffles resulting in tube failure?
9. Has provision been made in the foundations to allow for thermal expansion of the vessel itself and connecting pipework?
10. Have the exchanger and its supports been fireproofed in accordance with local and fluid-related guidelines?

Piping, Instrumentation and Analyzers

1. Have "spec breaks" in line rating been checked to be sure they are correct?
2. Have all valves been shown on the flow plan?
3. Has the possibility of slugs of water reaching pools of hot oil been critically examined?
4. Has equipment and piping been checked for potential upset conditions where significant temperature changes could occur and cause overheating or brittle fracture?
5. Have manual and automatic control valves been checked for wide open conditions?
6. Has the fail-safe position of control valves been analyzed carefully?
7. Are piping and associated equipment suitable for autorefrigeration if it can occur?
8. Have blinds been provided for all atmospheric vents used only during startup?
9. Have drain connections been provided for discharge systems such as water and caustic containing hydrocarbons?
10. Does piping for intermittent venting or blowdown of steam condensate discharge to a safe location?
11. Have tees to allow rodding been installed for lines in fouling service?
12. Have gauge glasses been prohibited on pressure storage tanks and vessels?
13. Do process analyzers meet applicable division of electrical classification standards unless in non-hazardous areas?
14. Has a safe sample disposal system been included for samples not returned to process?
15. Have explosivity alarms been installed in analyzer houses?

16. Has forced ventilation been provided for analyzer houses in hazardous locations?
17. Has a gate valve been installed as close as possible to the vessel in each instrument connection?
18. Have instrument connections been winterized?
19. Have high flow check valves been installed in the connections to gauge glasses operating above 500 psig?
20. Has freezing of the gauge glass been considered?

Appendix C
Machinery system completeness and uptime appraisal forms

Project Phase: P&ID and Design Specification Review
Machine Category: Centrifugal Compressors

Designation: _____

Location: _____

Service: _____

1. Is there an adequate liquid removal system for compressor suction to each stage?

_____
(Yes or No)

The suction drums must be sized to prevent liquid carry-over that could damage the machine. Worst case operating conditions should be considered. For example: highest suction pressure, highest flow, or two machines operating in parallel using the same suction drum should be considered in drum sizing. Use of crinkled wire mesh in drum outlet (appropriately designed, of course) and tangentially entering inlet nozzle to drum should be considered to provide the most effective liquid separation. A tangentially entering inlet nozzle will effectively increase drum flow capacity at identical inlet conditions by about 40%. This may provide a simple upgrade to increase capacity of existing drums equipped with an inlet nozzle that is perpendicular to drum walls.

Review suction piping layout to be certain that no liquid traps exist between suction drum and compressor. Suction lines must be sloped to drain either to compressor or to suction drum and not be flat. Liquid slugging and compressor damage can occur as a result of liquid trapped in these low points.

2. Is there a gauge glass, LHA alarm, and LHA trip on this drum?

_____
(Yes or No)

This instrumentation is extremely critical; catastrophic compressor failure and possible personnel injury can occur if the machine ingests significant amounts of liquid. Failure typically initiates in a thrust bearing and can cause internal rubbing of rotor wheels and diaphragms. Sufficient separation between alarm and trip should be provided so that operators have time to respond to an increase in level before a trip occurs.

3. Is the suction line such a length and size that it can be thoroughly cleaned by water washing?

_____
(Yes or No)

If not, what method do you propose?

Construction debris (hats, bolts, welding rods) can be inadvertently left in suction piping. Additionally, long-term storage of piping outdoors can result in dirt, sand, and rust in suction piping interior. Such material will likely be ingested by the machine during operation causing imbalance or damage to components. Close clearance parts are most vulnerable (seals, etc.). As a result suction piping should be thoroughly cleaned and inspected prior to startup. Visual inspection and removal of larger debris is a mandatory first step for any system cleaning procedure. Piping should be designed to facilitate cleaning.

One relatively inexpensive method of cleaning small debris such as rust, scale, and dirt is to hydroblast with high-pressure water. There are practical limits to lengths of line that can be hydroblasted and limits in piping configurations (hydroblasting typically cannot go through a side outlet "T," for example). Adequate drain piping to remove water after hydroblasting must also be provided. Note that some piping materials may be sensitive to chlorides and other contaminants in water, thus requiring selection of an acceptable source of water.

Another suitable method of cleaning is chemical cleaning. This method is particularly effective for corrosion products and in removing oil-based preservatives. This method, however, is not as effective as hydroblasting in removing larger objects because it often utilizes low-velocity circulation of cleaning solutions.

Access for internal inspection and cleaning by the method selected will likely require additional connections that are for cleaning only (blind flanges at ends of long runs of piping, for example).

4. Does a heat loss calculation indicate that the suction line needs to be traced? If so, is it traced? (allowable condensate is 2% by weight)

_____
(Yes or No)

In general, gases that have no potential for formation of liquids over the range of atmospheric temperatures do not require heat tracing. The ingestion of liquids in excess of 2% by weight of total flow, however, can cause deterioration of thrust bearings and labyrinth seals.

5. If the flow control is by suction butterfly valve, have minimum stops been provided?

_____
(Yes or No)

A centrifugal compressor will surge if the suction butterfly valve is closed. This can damage the compressor. A mechanical stop should be provided. This stop prevents the valve from closing beyond a point that would reduce flow so as to cause surge.

6. Has a discharge check valve been provided?

_____
(Yes or No)

The discharge check valve prevents process gas from flowing backwards through a machine if the machine trips or is shut down. This reverse flow can cause a machine to overspeed, damaging compressor and driver. Bearing damage is also likely. Typically, bearings are designed for one direction of rotation. If a shaft-driven oil pump is used, reverse rotation may occur without lube oil being supplied, leading to more severe damage to lubricated components.

7. Does the antisurge protection come from upstream of a discharge check valve?

_____
(Yes or No)

Compressors may not develop sufficient differential pressure during startup or process upsets to open discharge check valves. Antisurge protection should be upstream of check valve so that flow can be maintained through a machine, via an antisurge loop, when it cannot discharge into normal system. Lower than design molecular weight gas, e.g., could prevent a machine from developing enough differential pressure to open the discharge check valve.

8. Is the antisurge system sized so that it is large enough to prevent compressor surging?

_____
(Yes or No)

Compressor surge systems must be sized so as to prevent compressor surging even if normal process flow is blocked. Vendor-predicted surge flows may not be high enough to use in system sizing. It is suggested that vendor surge flows be increased by at least 50% for surge system sizing purposes. Variable speed machines should have surge systems sized for the highest speed surge flow plus a minimum of 50%.

9. If the antisurge system returns to the suction of the compressor, is it cooled?

_____
(Yes or No)

Does it return upstream of the suction knockout facilities?

_____
(Yes or No)

Compressor operation with gas being continuously recirculated may result in high gas temperatures. Driver horsepower input will essentially be converted into heat energy added to the circulating gas. Higher than design operating temperatures for the compressor can occur in a relatively short period of time when all gas is recirculated. These high temperatures can damage labyrinth seals, and cause rubbing (due to rotor thermal growth relative to the casing).

Liquid that is formed at compressor discharge conditions will be returned directly to the compressor suction and could cause damage. Recirculated gas must first be routed through suction knockout facilities to prevent this problem.

10. Does the system need a discharge safety valve?

_____
(Yes or No)

Consideration should be given to including discharge safety valves on compressors if they can develop discharge pressures in excess of design pressures of downstream equipment. High compressor suction pressure, operation of compressor near surge or at maximum speed, and higher than normal molecular weights are all examples of conditions that will cause higher discharge pressures that may necessitate the use of a relief valve.

11. If the compressor trips out, how high can the suction pressure go even if the discharge check valve does not close tight? Has this pressure been specified?

_____
(Yes or No)

Assume the suction valves are automatic close, isolation-type.

A relief valve or other system design changes may be required to protect upstream equipment if the predicted "settling out" pressure exceeds the pressure rating of piping, vessels, or valves. Also, the compressor suction nozzle rating may be exceeded if suction pressures rise too high. System designers must be aware of compressor casing limitations when selecting pressure relief and pressure control systems.

12. Is there a flow measuring device for the main flow and all other flows?

_____
(Yes or No)

Performance checking after initial startup and performance monitoring will not be possible without these flow loops. Troubleshooting of machines will be difficult without knowledge of operating point. Surge prevention systems must have data from these flow measurement devices in order to function and to provide necessary protection.

13. Is there a PI and a TI on each stage suction and discharge?

_____
(Yes or No)

This instrumentation is essential when troubleshooting or monitoring machine performance.

14. Has the full range of expected conditions been specified? A centrifugal compressor typically has a head rise of only about 6%. Therefore, with a drop in molecular weight of 6% the capacity drops to about 65% of design if system pressures are as specified.

_____
(Yes or No)

Maximum discharge pressure, minimum suction pressure, maximum suction temperature, and minimum molecular weight are conditions that must be considered when determining the head rise which a compressor must develop at design flow rate. These alternative operating conditions increase head required to develop required discharge pressure. Refrigeration compressors and other compressors that raise gas pressure to condensing pressure are affected by heat sink temperatures (i.e., cooling water, air, or other cooling media). An increase in heat sink temperature will increase compressor head requirements at design flow.

15. If the system has a low dynamic head loss, a high head rise to surge is necessary for stability. This often limits off-design capacity. Has the head rise to surge been specified?

_____

(Yes or No)

Compressor control system will not function in a stable manner unless sufficient head rise to surge is provided. Additionally, slight changes in suction and discharge pressure could cause the compressor to surge if insufficient head rise to surge is provided.

16. How close to surge do you wish to operate?

_____

(Yes or No)

If less than 15% above surge flow, will the control system prevent surge? When operating at only 15% above the surge line, it becomes even more difficult to prevent surge if the machine operates at variable speed.

It is generally difficult to operate closer than 15% to the surge line without special control systems. There often exist significant economic incentives, due to reduced power costs, to permit operation at minimum flow. This is generally true if the compressor is expected to operate at lower flows for any significant portion of time. Computer control schemes that correct predicted surge flow for suction temperature, speed, and molecular weight can be used to operate near surge. This system must also include a surge detection device that is reliable so that surge events do not go unnoticed. This surge detection device should alarm and automatically actuate an antisurge or recycle valve to take the machine out of surge if it should occur.

17. Have block valves been provided to isolate the machine for maintenance?

_____

(Yes or No)

Provision for blinding machines should also be made inside isolation block valves. Blinds must be rated at the same pressure as piping system in which they will be installed. Watch settling out pressure.

18. Has the shaft sealing system been specified?

_____

(Yes or No)

(a) *Labyrinth seals*: Use this type whenever possible. These are simple, proven devices that are inexpensive and require no elaborate seal oil system. These types of seals do leak, however, so that product losses or environmental regulations may preclude their use. Labyrinth seals are commonly seen in services where the compressed gas is not of great value or where the sealing pressure is relatively low. Examples are air, nitrogen, steam, and ammonia.

(b) *Mechanical face seals*: These seals are similar to pump mechanical seals. The seal runs in an oil-pressurized cavity. These seals are susceptible to dirt and similar contaminants, so buffer gas should be used in dirty gas service. This type of seal is also more vulnerable to mechanical damage than other types because it sometimes incorporates relatively brittle components.

This seal does have some advantages over other types. One advantage is the ability to contain process gas in the event of a power failure (the compressor must be down, however). Seal oil pressure is not required to contain gas; seal faces close and leakage is minimal. Another advantage is that mechanical face seals do not need seal oil differential pressure controls as precise as liquid film seals. Mechanical face seals are also less likely to lead to rotor dynamics problems that can occur with liquid film seals. Liquid film seals will sometimes act as bearings.

(c) *Dual or redundant seals or liquid film seals*: Typically used for low molecular weight or toxic hydrocarbons or when no atmospheric leakage can be tolerated. These are the most commonly used seal types for petrochemical and hydrocarbon applications. Moreover, these seals are rugged but require an elaborate seal oil support system and precise control of seal oil differential pressure. Dual seals do not leak gas to the atmosphere. Single seals do leak to the atmosphere but flow is dramatically less than leakage of labyrinth seals.

(d) *Gas seals*: This is a relatively new development that is a modification of mechanical face seal technology as applied in pumps.

An elaborate seal oil support system is not required. Due to the recent development and application of these seals, vendor experience should be investigated to verify applicability.

19. If the gas being handled contains H_2S above 10 ppm and water, or can ever do so, specify 90,000 psi yield maximum and state H_2S content on data sheet. Has this been done?

    _____
    (Yes or No)

    Stress corrosion cracking could occur in high stress areas if not in compliance with these limits.

20. If the gas contains over 100 ppm of H_2S, then provision to dispose of contaminated oil from sour oil pots must be made. Has provision of a vacuum dehydrator been considered to recover oil?

    _____
    (Yes or No)

    Internal seal oil leakage will be exposed to process gas, rendering the oil unusable due to high H_2S content. If this oil is reused, it will cause corrosion of some metal components. This oil should be collected in sour oil pots and must either be disposed of or recovered via a vacuum dehydrator. Recovered oil may be reused and could thus provide sufficient economic and environmental incentive to justify a vacuum dehydrator.

21. If the process system cannot tolerate trace quantities of lube oil and seal oil, has provision been made for the trap vents to go to a destination other than the machine suction?

    _____
    (Yes or No)

    Traps are sometimes called sour seal oil pots. They collect oil migrating toward compressor internals from mechanical face seals or liquid film seals. Typically, leakage to traps is about 1–10 gallons/ day. Vents from traps will allow small amounts of oil to enter the compressor suction under normal operation and could allow large amounts of oil to enter if level control instrumentation should malfunction. Vents should be routed to another system that has pressure maintained at or below compressor suction pressure if contamination by trace amounts of oil is a concern. Maintaining a separate disposal system at compressor suction pressure is necessary for drainage of oil to trap.

22. If the compressor can operate under a vacuum has provision been made to maintain the seal system above atmospheric pressure by buffering or some other means?

_____

(Yes or No)

If the compressor is allowed to operate under vacuum there is a possibility of air being drawn into the compressor through seals and suction piping leaks unless seal oil is maintained above atmospheric pressure. This condition can be detrimental to certain processes and potentially dangerous if flammable gases are being compressed. Preventing the inward leakage of air may only require an appropriate adjustment of seal oil pressure.

23. If the gas handled contains more than 50 ppm of H_2S or other constituents which will degrade the seal oil, is a clean gas available which could be used to buffer the seals?
If so, has buffer gas injection from a clean source been specified?

_____

(Yes or No)

Degradation of the oil additive package and viscosity can be caused by certain contaminants in process gases. This is more of a concern if the system is designed to reuse oil that is collected in seal traps. Contamination can also occur by diffusion even when sour oil leakage is collected in traps. Internal oil leakage is exposed to process gas and may absorb gas components that can cause degradation. Buffer gas can be injected into the cavity between the seal and the process gas to prevent contact of seal oil with process gas. Buffer gas must be compatible with the process because it will enter into the process stream. As a result, the selected buffer gas may be the major constituent in the process gas. Of course, the buffer gas comes from a source that has the potential contaminants removed.

24. Have flange finish and facings been specified?

_____

(Yes or No)

Flange facing should generally be consistent with the plant or attached process piping. Raised face flanges, e.g., should be used for compressor flanges if the piping will utilize raised face flanges. Some prefer special finishes on flange sealing surfaces to enhance gasket sealing. Concentric grooves similar in size and appearance to phonograph grooving are used in some instances to enhance sealing. The compressor flange finish should also be consistent with this piping.

25. If a gear is involved in the compressor drive system, has a torsional analysis been specified?

_____
(Yes or No)

Torsional resonances in compressor trains (including all drivers, compressors, and couplings) may be excited by forcing functions generated by a gear. A torsional analysis will indicate if potential problems exist and will allow necessary adjustments in torsional stiffness and damping to be made. Generally, necessary adjustments can be made by changes in couplings. Considerable time may be lost during equipment startup if torsional problems develop (i.e., time required to analyze system and obtain and install parts required to correct the problem). Torsional analysis of the system during the design phase will minimize the possibility of startup delays.

A torsional analysis can be made relatively easily using computer methods if all torsional information is available for the various train components. Determination of optimal corrective action and acceptable differences between significant forcing frequencies and natural torsional frequencies are judgments best made by individuals with experience in this area. A number of qualified consultants are available to assist in verifying vendor calculations if needed.

26. Can the lube/seal oil system for this machine be combined with others? Have dry running gas seals been considered? Can the system be isolated for maintenance? Would the entire installation have to be out of service for seal maintenance?

_____
(Yes or No)

Considerable savings can result if lube/seal oil systems are combined. It is generally not a good idea to combine systems if process gas contains contaminants that could degrade the oil (if bad oil goes undetected it could damage two machines). Use of combined systems is not a good idea if H_2S in process gas exceeds 100 ppm.

Gas seals, if available, and if vendor experience can be verified, might further reduce the complexity and cost of the overall equipment package. The slight amount of atmospheric leakage of process gas, however, could be environmentally unacceptable.

27. Do you want the vendor to supply the lube/seal oil console? If so, has this been specified?

_____
(Yes or No)

It may be necessary to purchase this system from the compressor manufacturer to protect guarantees. (Generally, it is a good idea to purchase this system from the compressor vendor.)

28. Must a separate seal oil console be provided? If so what is the economic justification for this?

_____

(Yes or No)

In some instances it is possible to use lube oil pumps as a source of seal oil; this occurs when compressor suction pressures are relatively low (typically <200 psig).

In these instances the discharge pressure of oil pumps can be increased to the pressure necessary to supply seal oil, thus eliminating the need for separate seal oil pumps. Alternatively, separate seal oil pumps can be provided that take suction from the main lube oil pumps. Economic justification for a separate seal oil console might include the additional energy required for main oil pumps to boost pressure of all oil to seal oil pressure. The main (lube) oil pumps would have to provide only a pressure of typically 40 psig (3 bar) if a separate seal oil console or separate seal oil pumps are provided.

29. Will shop tests specified be witnessed?

(a) Shop inspection _____*
(b) Hydrostatic _____*
(c) Mechanical run _____*
(d) Performance (air) _____
(e) Performance (gas) _____
(f) Auxiliary equipment _____*
(g) Driver tests _____

* Typically witnessed in vendor's shop.

Usually, centrifugal compressors comprise custom-designed, high-value, long-delivery machinery. The equipment user's exposure to potential problems is thus higher and of more serious consequences than for less expensive standard products. It is prudent to monitor certain key production achievements in the vendor's shop during manufacturing and testing to avoid surprises in the field. This is also a good opportunity to become familiar with the details of construction and idiosyncrasies of the machine. All inspection requirements and "hold points" for the customer's inspector should be defined soon after placement of the purchase order. Remember it is much easier and less painful to resolve problems in a vendor's shop than in an operating plant.

30. Are diffusers vaneless or are they of the vane type maximum efficiency design?

    _____

    (Yes or No)

    Vaneless diffusers are less complex and less costly than vaned diffusers. The latter are, however, more efficient and may be justified in non-fouling services.

31. Has the control system been fully described? Will it function correctly under all operating conditions?

    _____

    (Yes or No)

    The mechanical engineer and chemical (process) engineer counterpart must communicate so that the mechanical engineer fully understands all anticipated operating modes of the compressor. This communication should uncover operating conditions that require special controls, alarm, and shutdown devices. The compressor vendor will generally not contribute a great deal to the definition of control systems. The owner's mechanical or chemical engineers who recognize potential problem areas must take the lead in ensuring that control systems are adequately defined. Situations that are likely to lead to problems if not properly addressed in control systems include, but are not limited to, the following:

    - *Startup and shutdown*: Off-design conditions such as low flow or low molecular weight often occur during these periods. Surge control system should be capable of controlling flow to prevent the compressor from surging. This will require correcting variables as needed to determine how close to surge the machine is operating. A surge detection system should also be considered.
    - *Low flow*: Due to reduced unit throughput. Suction temperature control may be required if the antisurge valve recirculates a high percentage of compressor total flow.
    - *Regeneration service*: May require separate instrumentation for compressor antisurge and flow control. Operation in regeneration service is sometimes quite different from normal process operation. Different gases may have to be handled at different operating pressures and temperatures.
    - *Variable speed*: The surge point increases with speed. Also, the compressor performance changes with speed. The control system must compensate for these changes as required to meet process needs. It may be desirable to vary the compressor speed automatically to meet these changing requirements.

32. Unless a pressure-regulated seal has been specified, have all probable suction pressures been specified?

(Yes or No)

Abnormally low suction pressure will affect seal cooling and can affect the seal oil control system. Suction pressure at or above seal oil pressure will allow the process gas to escape if the seal oil pressure is not pressure compensated.

33. If in a single casing the gas is to be removed from the machine, passed through coolers, and/or other processes, and returned, is it essential that the internal bypassing be within certain limits?
    If so, has this limit been specified? Or, should you have specified a back-to-back impeller at the interstage?

(Yes or No)

Cross-contamination of gases with different compositions or at different temperatures will occur due to leakage across interstage labyrinth seal. In some processes this allows clean gas to leak towards the dirty (contaminated) gas stream instead of vice versa. There is a small penalty due to somewhat higher leakage.

34. Are any provisions needed for future requirements?
    If so, have these been specified?

(Yes or No)

Future debottlenecks may be anticipated during this phase of project development. At relatively low incremental cost, key equipment can be sized for or easily modified to a debottlenecked condition. Examples are larger suction knockout drum, larger piping, larger intercoolers, larger driver or a foundation easily modified to accept a larger driver, gear that is oversized or provided with a high service factor. Compressors can also be purchased destaged or with less than maximum diameter impellers. If on-line washing with liquid to clean the compressor is anticipated, then tie-ins can be provided and compressor designed to accept on-line liquid washing.

35. Has the required nozzle position been specified? Would a barrel type compressor have any advantages?

(Yes or No)

Many users prefer bottom-connected compressors to avoid removal of the major process piping when performing maintenance on horizontally split compressors. This minimizes the possibility of introducing pipe strain when major piping is reinstalled. It also reduces time to perform maintenance. Bottom-connected nozzles, however, require relatively costly mezzanine mounting structures and introduce the possibility that the structural resonances may cause vibration problems. Bottom-connected compressors often lead to a piping arrangement that creates a liquid trap between the suction knockout drum and the compressor inlet. If the process gas is near dewpoint, liquid may accumulate and eventually slug the compressor. Additionally, it is not uncommon for tools or parts to be dropped unnoticed into the compressor suction or discharge piping when the compressor top casing is removed; this of course could lead to a compressor failure.

One novel approach that is sometimes attractive is to use a barrel-type casing, even though pressure levels do not dictate this design. The barrel casing arrangement allows removal of all rotating and wearing compressor components without disturbing major process piping even if top connected process piping is provided. Assembly of many critical components can then be performed in a better controlled shop environment instead of in the field. The barrel-type casing virtually eliminates split line leaks sometimes experienced with horizontally split compressors. Use of a barrel-type casing has the advantages mentioned but is somewhat higher in initial cost.

36. If there are any special uprate or standardization considerations with regard to nozzle sizes, types, etc., have these been specified?

_____
(Yes or No)

It may be advantageous to provide flanges on compressor casings that are the same type or the same size as process piping. Special consideration may be given to the possibility of future throughput increases, or standardization requirements of a plant.

37. If this is a critical class of machinery, has a non-standard lube oil system been specified?

_____
(Yes or No)

All vendors will offer their standard lube oil system if equipment specifications allow for competitive reasons. Although the standard system will meet API 614 requirements and may be acceptable in some applications, it will not meet the minimum requirements of certain other applications. This is particularly true for larger unspared compressor trains. Typical potential constraints found in standard systems include:

- self-contained control valves instead of pneumatic control valves and controllers;
- inadequate oil capacity in overhead seal oil reference drums;
- two instead of three lube oil pumps;
- marginally sized oil pumps and/or drivers;
- seal oil and lube oil systems that will not allow a compressor coastdown without damaging bearings or gas leakage from seals in the event of a power failure;
- no facilities to coalesce oil mist emanating from oil reservoir vents;
- instrumentation and control devices that are not normally used in the plant;
- equipment hardware that is not normally used in the plant such as pipe fittings, valves, pumps, flexible couplings, filters, etc.;
- oil coolers that have little or no allowance for cooling water fouling and/or have low water velocities. Additionally, coolers tend to be vendor standard designs and not manufactured in accordance with TEMA (Tubular Exchangers Manufacturers Association).

38. Have all available data on the possible corrosiveness of the gas been specified?

_____
(Yes or No)

Obviously, selection of compressor materials of construction must consider corrosives in process gas. Typical offenders include H_2S, chlorides, and acids. Contaminants during abnormal operation such as startup or regeneration must also be considered. The compressor will be more susceptible to corrosion than stationary equipment due to higher stresses, cyclic stresses, and higher gas velocities; hence, potential corrosive constituents must be identified prior to material selection.

Special coatings may be applied in some applications to standard compressor materials to achieve acceptable corrosion resistance at a reasonable cost. Operating procedures or conditions can be modified in some cases to minimize or eliminate compressor exposure to corrosives (e.g., increasing operating temperature to provide a comfortable margin from the dewpoint of an acid known to be present).

39. Has an API data sheet been prepared?

———————
(Yes or No)

Preparation of the API data sheet will help highlight process and hardware areas that need definition or investigation.

40. Has an API data sheet been prepared for driver and gear (if required)?

———————
(Yes or No)

These data sheets are an excellent means to communicate plant preferences to the vendors. Additionally, these data sheets provide a reasonably comprehensive means to define construction and design details.

41. If the partial pressure of H_2 in the gas is 215 psia or more, has a vertically split case (barrel) been considered?

———————
(Yes or No)

Split line process gas leakage will be increasingly troublesome as the partial pressure of hydrogen increases above 215 psia when horizontally split casings are utilized. Barrel-type casing designs can virtually eliminate these potential leakage problems.

42. If flushing is required, has it been specified?

———————
(Yes or No)

Liquid injection into the compressor (flushing) may reduce wheel and diaphragm fouling in dirty services. The compressor vendor can provide special liquid injection nozzles located to maximize their effectiveness. If required these should be specified to capitalize on vendor expertise.

43. If a maximum casing temperature is required, has this been specified?

———————
(Yes or No)

Emergency, startup, or other unusual operating conditions may expose compressor to abnormally high operation temperatures. The type of casing construction selected and the materials of construction may be affected by the maximum temperature anticipated. Vendor should be informed of such unusual conditions.

44. Has instrumentation gauge panel location and scope of supply been specified?

_____
(Yes or No)

Vendor's scope of supply for all instrumentation should be agreed upon early. Decisions concerning local versus control house instruments should be reviewed with operations and maintenance personnel. Design and fabrication of a control panel is an area of considerable vulnerability; compressor vendor expertise in this area is not always at the same level as the user's. Some users design and provide all protective and control instrumentation. The compressor vendor thus provides appropriately located connections for instrumentation. The compressor vendor would of course be consulted to ensure the control and protective instrumentation is adequate.

45. Has a suction strainer been included in compressor piping? (*Note*: Recommend removing as soon as possible after startup.)

_____
(Yes or No)

A startup strainer should be included to help to prevent construction or other debris from damaging compressor. The strainer mechanical design should be adequate to prevent collapse under maximum anticipated differential pressure. Piping should be arranged to facilitate strainer removal and cleaning. If possible the strainer should be installed in a manner that permits its removal and cleaning without disturbing major process piping; this minimizes the possibility of introducing pipe stress on the compressor casing during strainer maintenance.

46. Does the cost estimate and budget include major spare parts for the compressor?

_____
(Yes or No)

Cost engineers often overlook major compressor spare parts such as rotors, gears, and couplings. These items should be purchased at time of compressor order to obtain best pricing and to ensure critical spare parts will be available for startup. Justification of these essential items later in the project will be difficult and time-consuming, especially if budget problems are being experienced.

47. Have failure modes for instrumentation and control devices been considered in the event that instrument air and control power is lost?

_____
(Yes or No)

The safest failure mode for all compressor instrumentation and control devices should be considered. The failure modes selected should be consistent with failure modes of the related process. For example, should a steam turbine-driven compressor shut down or continue operating in the event of loss of instrument air? This decision is dependent upon process impacts and must incorporate compressor protection impact (i.e., protective devices for compressor may be compromised if the compressor is allowed to operate).

Project Phase: Preorder Review with Vendor
Machine Category: Centrifugal Compressors

Designation: _____

Location: _____

Service: _____

1. Does the vendor have adequate experience?

_____
(Yes or No)

(a) Location of similar machines:

_____

_____

It is important that other users of similar machines be contacted to discuss their experience with vendors of interest. Vendors should be able to provide location and contacts. Information obtained from other users can be very useful in determining performance, maintenance, and vendor support histories. Vendor strengths and weaknesses may be revealed from discussions with other users. If possible a plant visit to discuss (with the user) the history of machines similar to that proposed is an excellent way to learn more about the equipment proposed.

(b) How closely do the "similar" machines conform to the proposal?

	Yours	*Others*
Suction pressure	_____	_____
Suction temperature	_____	_____
Discharge pressure	_____	_____
Discharge temperature	_____	_____
Molecular weight	_____	_____
Head	_____	_____
Tip speed	_____	_____
rpm	_____	_____
Number of stages	_____	_____
Seal types	_____	_____
Settling out pressure	_____	_____
Mach number	_____	_____
Separation of lube and seal oil	_____	_____

It is unlikely that all process conditions for the new compressor will match compressors in service. Comparisons will increase the user's confidence of vendor's capability if the machine proposed has operating conditions that fall between those of machines operating successfully. Operating conditions outside proven operating windows for a proposed compressor must be carefully considered. Use of proven impellers at similar pressure levels and similar molecular weight gases in the proposed compressor reduces the probaility of performance deficiencies.

(c) Have any similar machines experienced difficulty either during testing or in the field?

_____
(Yes or No)

If so, what were the characteristics of the problem and what steps will be taken to avoid a repetition?

All vendors will have some problems. Significant unresolved deficiencies reported would of course render a vendor unacceptable. Unproven fixes for problems reported are suspect. A great deal can be learned about the vendor's overall capability from his past history in identifying problems and resolving them in a timely manner.

(d) Comments of users on these machines:

How long have machines been in service?_____

Operating experience that extends through one compressor overhaul is most valuable. Information concerning premature wear of components, case of maintenance, parts availability, price, and quality will likely be available after the first overhaul. One year of operating experience is the suggested minimum credible experience. It is possible that a vendor may propose a design that has overwhelming advantages but is not totally field-proven. Such advantages might be higher efficiency, lower initial cost, or mechanical simplicity. Alternatives may merit careful consideration if economic incentives are large.

What startup difficulties were experienced?_____

Start-up difficulties are the result of problems in one of several areas. These are:

- basic design problems (such as rotor dynamics);
- quality control;
- instrumentation and controls (associated primarily with lube and seal oil system);
- vendor field support (quality of serviceman and vendor response to questions).

Once problem areas are identified, steps can be taken during procurement, engineering, or manufacturing to reduce vulnerabilities in any one of these areas. For example, independent verification of vendor rotor dynamics calculations is appropriate for a vendor with reported rotor dynamics problems.

If so, was the Service Department's response satisfactory?

_____

_____

Were problems resolved in a timely manner? Were solutions implemented after an adequate engineering analysis was made? Trial-and-error problem-solving techniques are unacceptable. Were appropriate vendor personnel made available for problem-solving? Did the vendor bear a fair portion of the cost to correct vendor-caused problems? Were modified parts furnished quickly, if needed?

Did the machines meet the process specification?

_____

_____

The most reliable source of compressor performance information is the vendor performance test. Field testing is possible but difficult due to dependence on the accuracy of numerous instruments and possible variations in molecular weight. Vendors that are found to overstate efficiency consistently at the proposal stage should be appropriately penalized during evaluation.

What has been the maintenance experience?_____

_____

Are maintenance intervals reasonably spaced or too frequent? How does ease of maintenance compare with machines of similar size and type? Is the use of special tools reasonable? Can routine maintenance be performed by shops that the owner normally uses or must repairs be performed by the original equipment manufacturer? Will owner's craftsmen be able to attain a level of expertise that allows making in-house repairs? Are assembly tolerances reasonable or difficult to attain?

Name and position of contact:_____

_____

The rotating equipment engineer involved in the maintenance and troubleshooting of the compressor of interest is usually a reliable source of information. First line operating management is also a good source of information.

Name and position of contact:_____

_____

(e) Were the machines built in the shop where your machines are to be built?

_____
(Yes or No)

If a different shop is to be used to manufacture your machine, then one should question how the technology has been transferred to a new shop. Have some of the key personnel from the original shop been relocated to the new shop? Have other machines similar to the machine you are purchasing been fabricated in the new shop? Have any manufacturing techniques been altered in adapting to the new shop? If so, how and why, and are the new methods satisfactory? Problems are more likely if your machine is one of the first to be manufactured in a different facility.

(f) If your machine is to be built in a shop other than the vendor's main shop:

- How is the parent company or licensor going to ensure that the machines will be built correctly?

Will he have an inspector (or engineer) in the shop on a full- or part-time basis?

(Yes or No)

How qualified will this person be?

The parent company should provide a qualified inspector and preferably an engineer to follow manufacture of the compressor in the outside shop. Inspection points the original manufacturer intends to enforce should be discussed and agreed upon early by all three parties involved (i.e., the purchaser, vendor, and manufacturing shop). If the particular type of machine is routinely subcontracted or manufactured by a licensee then the potential for problems is minimal.

- How many similar machines have been built in that shop?

Were any problems experienced with the machines built? Were schedules met? The purchaser should be reluctant to allow use of a shop that has little or no experience manufacturing the machine to be purchased.

- How similar were they?_____

Machines of the same frame size, similar molecular weight gas, number of stages, inlet conditions, and speed are ideally similar. Successful installations of such similar machines should increase user confidence in the shop's capability. Machines different by one frame size, especially larger, should be considered similar if all other conditions are approximately the same.

* Who will do the engineering and drafting for the machines?

    The means by which technology will be shared should be discussed with the outside shop and original manufacturer. This should not be a major concern if the shop has previous experience with the machine being purchased; sharing of technology could then be handled as in the past. A review of engineering and drafting packages by the original manufacturer would be one means to minimize potential problems:
* Will the machines be an exact duplicate of the licensor's design?

(Yes or No)

If not, what modifications will be incorporated and why?

    The purchaser should understand all deviations. Some deviations will be obviously acceptable. Other deviations should be field proven. The purchaser should be assured that the latest improvements in equipment design by the parent company are incorporated by the outside shop. Design modifications not approved by the parent company should be rejected.

2. If machines are not identical, how much extrapolation has the bidder done on the established experience limits?

   (a) Are these extrapolations based on experience with other machines?

(Yes or No)

   (b) Do these extrapolations appear sound?

(Yes or No)

    Interpolation of experience (e.g., a vendor offering a 5-stage machine in a frame size that has been in operation in 4 and 6 stages) is preferred to extrapolation. Extrapolation of experience indicates that the offering is outside either the upper or lower limits of proven performance. Extrapolation must be made using sound engineering principles. Extrapolation by more than one standard frame size should be avoided, if possible. Machines that have near maximum number of impellers in a casing should be avoided if successful experience is not demonstrated. Rotor dynamics and aerodynamic performance are the areas of most vulnerability when exceeding the envelope of proven performance.

(c) Is the vendor offering within the limits of any previous experience?

_____
(Yes or No)

(d) Is this experience applicable to the application being considered?

_____
(Yes or No)

Vendor experience in certain specific areas may be of particular value even if frame size or performance is not similar. These areas include such technology as low- and high-temperature applications, low molecular weight gases, dirty services, identical gases, and high-pressure compressors.

Specifications: _____

(a) Does the offering completely meet your specification in all aspects?

_____
(Yes or No)

(b) If not, the vendor's responsibility is to list all deviations. Are there any deviations not listed in the proposal?

_____
(Yes or No)

Vendors attempt to standardize as much as possible and draw from their own operating experience. As a result they often propose deviations from user specifications. Deviations found during proposal review (including discussions with vendor's representatives) but not itemized in proposal indicate that the vendor is not fully aware of specified requirements. Deviations found that are not itemized indicate that others exist. Prior to order commitment the user should be certain that all major vendor exceptions (itemized or not) are understood and resolution reached.

Note that vendor execution in some instances may be acceptable or even superior to specified requirements. Rarely are machines purchased without accepting some vendor deviations from the inquiry specification. The final mechanical design of a machine is almost always negotiated so that it incorporates both the vendor standard execution and the inquiry specified requirements. Do not discount the experience of vendors without due consideration; they are in the best position to offer improvements in design. Vendors are also best able to offer design alternatives that may reduce cost.

3. What design and shop practice modifications have occurred since the similar machines were built?

   - Rotor, including shrink specifications   _____
   - Casing   _____
   - Bearings   _____
   - Seals   _____
   - Clearances   _____
   - Wheel design and manufacture   _____
   - Diffuser design   _____
   - Materials   _____
   - Balancing   _____
   - Tip speeds   _____
   - Overspeed testing   _____

   Which of these changes are definite design improvements and which are cost reduction items? _____

   Any items that have had changes made for cost reduction or design improvement reasons should be tested to see that these changes do not lead to potential problems. Changes in several areas increase the possibility that unexpected problems may arise. It may be necessary to insist that previous practices be adhered to in some instances to increase the probability that any problems experienced will be minimal.

   What types of seals are proposed? (Please refer to P&ID Specification Review, Item 18)_____

   Does the vendor have experience with these seals at a higher pressure?

   _____
   (Yes or No)

Has the vendor operated on gas with similar contaminants?

_____
(Yes or No)

If so, what steps did he take to avoid lube oil contamination?

_____

_____

Examples of steps that might be taken include buffer gas injection, double mechanical seals, and clean-up systems to recover contaminated oil.

How many pressure reducing steps are there?_____

What is the maximum pressure per reducing stage and does prior experience exist?

_____

Are you inside this tolerance?

_____
(Yes or No)

What steps will the vendor take to prevent the bushings acting as bearings?

_____

_____

Seal bushings must be grooved or squeeze film dampers installed unless prior experience can be totally verified.

Has the vendor operated on gas with similar contaminants?

_____
(Yes or No)

If so, what steps did he take to avoid seal oil contamination?

_____

How many pressure reducing stages are there?   _____

What is the design maximum pressure per reducing stage?_____

Are you inside this tolerance?

_____
(Yes or No)

What steps will the vendor take to prevent the bushings acting as bearings? _____

_____

If none, what action will he take if this occurs?‗‗‗‗‗‗‗

‗‗‗‗‗‗‗‗‗‗‗‗‗‗‗‗‗‗‗‗‗‗‗‗‗‗‗‗

Why cannot he do this before problem arises?‗‗‗‗‗‗‗

‗‗‗‗‗‗‗‗‗‗‗‗‗‗‗‗‗‗‗‗‗‗‗‗‗‗‗‗

If the seals are bushing type, will overhead tanks be provided?

‗‗‗‗‗‗‗‗‗

(Yes or No)

If the differential is controlled by overhead tanks, a bladder must be inserted between the reference gas and the main seal system if gas in the seal zone will contaminate the seal oil.

4. How are machine expansions handled to minimize thrusts due to coupling slip forces and expansion (gear couplings)?

(a) Where is casing anchored axially? ‗‗‗‗‗‗‗
    Where is the thrust bearing? ‗‗‗‗‗‗‗
    Are these on opposite ends of casing? ‗‗‗‗‗‗‗

If not, there will be a large shaft movement into the coupling. Is the coupling adequate to take this movement? (Question wisdom of this approach.)

(b) In which direction will coupling slip force act relative to internal thrust forces? ‗‗‗‗‗‗‗

(c) Assuming the coupling slip force to be

$$F = 0.3\frac{T}{d} = \frac{18{,}900 \times P}{\text{rpm} \times d}$$

where T = torque in lb-in.,
$\quad\quad\ P$ = horsepower,
$\quad\quad\ d$ = shaft diameter at coupling in inches,

are all thrust bearings designed for normal thrust forces plus or minus the coupling slip force?

‗‗‗‗‗‗‗‗‗

(Yes or No)

(d) Can the coupling slip force completely cancel the normal thrust and position the shaft on the "inactive" side of the thrust bearing?

‗‗‗‗‗‗‗‗‗

(Yes or No)

(e) Would a non-lubricated coupling be more appropriate?

_____
(Yes or No)

If no, check again with designated machinery specialist.

5. How will the job be handled from proposal through design, drawings, fabrication, inspection, testing, and shipping?

(a) Has a wheel layout already been established?

_____
(Yes or No)

(b) If so, who did it and how?_____

_____

(c) Will the wheel layout be rechecked by others before drafting?
Who?_____

How?_____

(d) If drafting selects the parts from a range of standard parts, who checks that these parts will meet the requirements?_____

_____

(e) What part does Design Engineering play in the machine selection?_____

_____

(f) If wheels must be cut who specifies the final diameter? _____

_____

_____

Will this final diameter be rechecked?

_____
(Yes or No)

If so, who will do it?_____

_____

(g) If sidestreams are involved, who designs the sidestream inlets?

_____

Are these rechecked for Δp?

_____
(Yes or No)

If so, who rechecks?_____

(h) Is there a contract engineer who has full responsibility for following the unit through all phases of design, manufacture, inspection, and test?

_____
(Yes or No)

How much experience has the individual selected for our machines had?_____

How many other machines is this individual responsible for at this time?_____

(i) Who is responsible for inspection?_____

What function does inspection report to?_____

(j) What function is responsible for testing?_____

Will the vendor's shop loading affect delivery?

_____
(Yes or No)

Ask to see the schedule on the compressor superimposed on a loading diagram for: Engineering, Drafting, Purchasing, Machining, Assembly, and Testing.

Are all of these functions without overload to the degree that they may delay delivery?

_____
(Yes or No)

Potential problem areas.

What analysis work will be done to ensure no problems due to:

- Rotor flexibility and stability_____
- Torsional vibrations (if more than two elements tied together)

_____

- Lateral criticals_____
- Thrust loadings_____

Do these approaches seem satisfactory?_____

6. If more than 10 ppm of H_2S has been specified in the process gas, what steps will be taken to ensure the 90,000 psi max. yield or $22R_c$ hardness are not exceeded? _____

How will hardness in heat-affected zone of welded impellers be checked?_____

_____

7. What overspeed tests will be conducted on the wheels? _____

_____

Will the wheels be spun to a given % above mechanical design or a given % above max. continuous for the machine?

_____
(Yes or No)

What will be the percentage overspeed?_____

Will the assembled rotor be tested at overspeed?

_____
(Yes or No)

Project Phase: Contractor's Drawing Review
Machine Category: Centrifugal Compressors

Designation: _____

Location: _____

Service: _____

P&IDs

1. Is there a KO drum on the suction and any cooled interstages?

_____
(Yes or No)

2. Do the recycle lines re-enter upstream of the KO drums?

_____
(Yes or No)

3. Are the KO drums equipped with a gauge glass, LHA, and shutdown switches?

_____
(Yes or No)

4. Have all the alarm and shutdown switches specified been provided?

Low lube oil pressure alarm and auxiliary pump start-up actuation   _____

Low lube oil pressure alarm and trip   _____

Low seal oil level (or pressure) alarm and auxiliary pump start-up actuation   _____

Low seal oil level (or pressure) alarm and trip   _____

High seal oil level alarm   _____

High discharge temperature alarm   _____

Others (detail)   _____

---

5. Are TIs and PIs specified for suction and discharge of each stage?

_____
(Yes or No)

6. (a) Is there a check valve in the discharge downstream of the anti-surge recycle?

_____
(Yes or No)

   (b) If there are two machines in parallel, is there a check valve on the discharge of each?

_____
(Yes or No)

7. Is there a flow meter on each feed and discharge stream from the machine?

_____
(Yes or No)

The mass flow on each stage is to be measured. More than one meter is only required if the mass flow changes from section to section.

8. Have pressure taps been provided up and downstream of the temporary suction strainers?

_____
(Yes or No)

9. Will all instruments be changeable on the run and are shutdown circuits testable on the run?

_____
(Yes or No)

10. Are all lines to remote pressure gauges valved at the tie-in to the main line?

_____
(Yes or No)

11. Is there an isolatable closed circuit including a cooler at the machine so that the pre-start-up run-in can be performed?

_____
(Yes or No)

12. If the machine is not in closed circuit and is not an atmospheric air machine, is there a suction flare release to dump the suction gas in the event of shutdown?

_____
(Yes or No)

13. If the process is "flow controlled," are the metering elements outside the recycle loop?

_____
(Yes or No)

14. Is the antisurge metering element inside the recycle loop?

_____
(Yes or No)

15. On refrigeration machines the TICs must close on driver trip. Do they?

_____
(Yes or No)

16. On motor driven refrigeration machines the casing pressure must be reduced to about 40 psig before starting to prevent driver overload. Liquid in the suction drums impedes the pressure reduction. Can the liquid from the drums be pumped into the accumulator?

_____
(Yes or No)

17. Is there a safety valve on the discharge of the machine if the downstream equipment cannot stand the machine's discharge pressure under the combined conditions of:

Trip speed_____

High mol. wt._____

High suction pressure_____

Low temperature_____

Layouts

1. Are the main and interstage suction lines from the KO drums cleanable by the method proposed? _____
   (Yes or No)

2. By the time of this review the moment of inertia of compressor and driver should be available. Is the hydraulic energy inside the check valves less than 1.3 of the kinetic energy of the shafts?
   _____
   (Yes or No)

   If not, the check valves will either have to be relocated or additional ones installed.

3. Are there drain valves at the low points of the suction and discharge lines?
   _____
   (Yes or No)

4. Are all the check valves horizontal?
   _____
   (Yes or No)

5. Are all the check valves damped or equivalent?
   _____
   (Yes or No)

6. Will the recycle valves pass the compressor design flow?
   _____
   (Yes or No)

7. Are the control valve actuators adequate for contemplated operating conditions?

_____

(Yes or No)

(Users have had trouble on both straight control valves and butter-flies.)

8. If the machine is an atmospheric air compressor, does the antisurge vent have a silencer and is the intake of a sound attenuating type?

_____

(Yes or No)

9. Is the foundation separate from that of reciprocating machines?

_____

(Yes or No)

10. Can the temporary strainers be removed without disconnecting any piping?

_____

(Yes or No)

11. Are the suction, discharge, and compressor drains connected either to a blowdown system or vented to a safe place?

_____

(Yes or No)

12. Unless the compressor vendor has given dispensation, is there a straight section of at least three pipe diameters on the suction flanges?

_____

(Yes or No)

13. Are the pipe stresses and moments within the levels allowed by the vendor?

_____

(Yes or No)

14. Are all the piping supports and anchors as described in the piping stress calculation?

_____

(Yes or No)

15. Has sufficient allowance been made in the stress calculation for friction of supports?

_____
(Yes or No)

16. On refrigeration machines, are the liquid injection points at a sufficient distance from the drums to ensure vaporization of all the liquid?

_____
(Yes or No)

17. On refrigeration machines, do the liquid level control valves have blocks and bypasses?

_____
(Yes or No)

18. On refrigeration machines, it is necessary to adjust the TIC controllers during start-up. On motor-driven machines, is there a single switch to commission the TICs immediately after start-up? Is it readily accessible from the platform?

_____
(Yes or No)

On turbine-driven machines are the TICs readily accessible from the platform? (Adjustments to the set points are necessary during run-up.)

_____
(Yes or No)

19. Have ROVs been provided on all lines which can feed hydrocarbons to a fire at the machine and do not have block valves at least 25 ft horizontally from the machine?

_____
(Yes or No)

Have the electrical conduit and valve operator to such ROVs been fireproofed sufficiently to permit operation of the valves after a 10-min fire?

_____
(Yes or No)

20. *Consider operability*

(a) Are all instruments clearly visible?

_____
(Yes or No)

(b) Has the operator safe and easy access to all bearings?

_____

(Yes or No)

(c) Has the operator safe and easy access to the handwheels on the ROVs, the flow control devices, and the recycle valves?

_____

(Yes or No)

(d) Run through a start-up sequence. Can all the operations required be done by one person?

_____

(Yes or No)

(e) Is there safe access to the suction and discharge line and all casing drains?

_____

(Yes or No)

(f) Can the oil drain sight glasses be readily seen?

_____

(Yes or No)

(g) If there are overhead seal tanks, has the operator a clear view of the level gauge?

_____

(Yes or No)

If the oil level control has to be put on hand control using the bypass, will the operator be able to see the level gauge from his position at the valve?

_____

(Yes or No)

(h) Can either seal trap be taken out of service for repairs with the other trap draining both seals?   _____

(Yes or No)

21. *Consider maintenance*

(a) Is there a suitable location where the casing top half can be put without interfering with maintenance?

_____

(Yes or No)

Note: If the machine has multiple casings or if it has a turbine driver, all top halves may be off at the same time.

(b) If the casing is a barrel type, is there room to pull the barrel internals *in situ*?

—————————

(Yes or No)

(c) Is the crane big enough to carry the largest maintenance weight? Usually the top half of the largest casing.

—————————

(Yes or No)

(d) Can the top halves be moved to the storage area without passing over operating machinery?

—————————

(Yes or No)

(e) If the machine is motor-driven, is there access to that end so that the motor rotor can be pulled, if necessary, using portable equipment?

—————————

(Yes or No)

(f) Are the motor cooling ducts so positioned that they do not unnecessarily interfere with the crane movement?

—————————

(Yes or No)

(g) If the compressor is at grade with overhead piping, can the piping spools be readily removed and swung out of the way leaving vertical lift for casing?

—————————

(Yes or No)

Have lifting provisions been made to facilitate this?

—————————

(Yes or No)

(h) Can the rotors be removed to the maintenance shop without passing over operating machinery?

—————————

(Yes or No)

(i) Piping should not run unnecessarily over parts which must be removed for maintenance. Has this been complied with?

—————————

(Yes or No)

22. *Consider instrumentation*

Are the TIs installed in such a way that they will measure the correct temperature? If the lines are two phase, will they see the correct phase?

_____
(Yes or No)

Project Phase: Mechanical Run Tests
Machine Category: Centrifugal Compressors

Designation: _____

Location: _____

Service: _____

The mechanical run test for centrifugal compressors is basically a balance check. In some cases data on the vibration characteristics of a machine will also be disclosed.

The basic procedure should be to run up to 110% of max. continuous for turbine-driven machines, and run for a minimum of 15 min. Then drop back to max. continuous speed and make the overall test 4 h. For motor-driven machines, max. continuous speed is design speed.

	Conditions	
	Design	*Test*
(a) Speed rpm	_____	_____
(b) L.O. inlet pressure	_____	_____
(c) L.O. inlet temperature	_____	_____
(d) Max. vibration	_____	_____
(e) Calc. critical	_____	_____
(f) Max. noise level	_____	_____

1. Do the first four test conditions match the design conditions to your satisfaction?

_____
(Yes or No)

2. Is the actual critical speed within calculated value?

_____
(Yes or No)

3. Is the temperature rise across each bearing less than 60 °F?

$$\overline{\qquad\qquad}$$
(Yes or No)

4. *Vibration*

    (a) Frequency survey when running at max. continuous speed. Note vibration at running speed and other frequencies.

Probe location	*Magnitude (mil)*	*Frequency (cpm)*
_____	_____	_____
_____	_____	_____
_____	_____	_____
_____	_____	_____
_____	_____	_____

    (b) Is there undue vibration at critical frequency, also at frequencies between 35 and 50% of running speed?

$$\overline{\qquad\qquad}$$
(Yes or No)

    (c) Shaft and bearing vibration attenuation:

	A, shaft	*B, Housing*	*A/B, attenuation*
I.B. bearing	_____	_____	_____
O.B. bearing	_____	_____	_____

Is the attenuation less than 4? (If no, mention in report.)

$$\overline{\qquad\qquad}$$
(Yes or No)

    (d) Vibration readings (mil):

	110% design speed	*Max. continuous*	*Difference*
I.B. bearing	_____	_____	_____
O.B. bearing	_____	_____	_____

Is the difference in vibration levels at these speeds less than 20%?

$$\overline{\qquad\qquad}$$
(Yes or No)

5. Is a thorough check being made for oil leaks?

_____

(Yes or No)

6. *Seals*

   (a) Is the oil collected from each seal drain less than 5 gal/day?

_____

(Yes or No)

     (This is significant only on carbon seals or bushing seals with normal differential pressure.)

   (b) Is the seal oil outlet temperature less than 180°F?

_____

(Yes or No)

     If not, insist that it be lowered. However, operation of normal running seals at low pressures may make the outer bushing run hot. An assurance that this is the cause should be accepted.

7. Bearing inspection. Are all bearing surfaces showing normal running pattern?

_____

(Yes or No)

     Demand replacement otherwise.

8. Seals to be inspected only if it is the vendor's standard practice.

9. Internal inspection to be carried out if a spare rotor is to be fitted and run (normally specified).

     Is there an absence of rubbing?_____

     If rubbed, demand a clearance check.

10. Check the internal alignment and clearance data from final assembly.

   (a) Is alignment good, clearances within tolerances? _____

   (b) Likewise for a spare rotor if it has been fitted. _____

11. Were copies of vendor's test log sheet and final internal clearance diagrams obtained?

_____

(Yes or No)

12. When witnessing a test, always try to find out if any difficulties occurred in preparing for it. Such problems could be repeaters.

13. Gas Pressure Test following mechanical run.

(a) Is the test being carried out with visible soap bubbles?

_____
(Yes or No)

(b) Is the shaft being rotated to check for freedom of seals?

_____
(Yes or No)

Project Phase: P&ID and Design Specification Review
Machine Category: Reciprocating Compressors

Designation: _____

Location: _____

Service: _____

P&IDs

1. Based on project philosophy (availability required) are two machines called for?

_____
(Yes or No)

2. Have KO facilities been specified on all suction and interstages? Are you aware of King-type coalescer-separators?

_____
(Yes or No)

3. Do recycle lines re-enter upstream of the KO facilities?

_____
(Yes or No)

4. Are KO facilities equipped with a gauge glass, LHA, and shutdown switches?

_____
(Yes or No)

5. Are the suction lines traced between the KO drums and the machine flanges, including pulsation bottles?

_____
(Yes or No)

6. Do KO facilities have automatic drains and bypasses to allow checking?

_____
(Yes or No)

7. Is there a safety valve on each compression stage and are there both pressure and thermal relief valves on the coolant header?

_____
(Yes or No)

8. Has a remote shutdown switch been provided?

_____
(Yes or No)

9. Have all the alarm and shutdown devices been specified?

    (a) Low lube pressure alarm _____

    (b) Low lube pressure trip _____

    (c) High-temperature alarm on each cylinder discharge _____

    (d) Low cylinder lube flow

_____
(Yes or No)

Others (detail):_____

_____

10. Are TIs and PIs specified for suction and discharge of each stage and TIs on discharge of each cylinder and on the coolant outlets from each cylinder?

_____
(Yes or No)

11. On machines which are specified to be reaccelerated, have controls been provided to unload the cylinders during reacceleration?

_____
(Yes or No)

12. If these controls are of the bypass type, is there a check valve on the compressor discharge?

    <u>   </u>
    (Yes or No)

13. Have automatic unloading facilities been specified for start-up?

    <u>   </u>
    (Yes or No)

14. Are there coolant block valves on each cylinder?

    <u>   </u>
    (Yes or No)

15. Does each machine have double block valves and a vent to avoid blinding for valve repairs?

    <u>   </u>
    (Yes or No)

16. Have pressure taps been provided around the spool which will contain the temporary suction screen?

    <u>   </u>
    (Yes or No)

17. Do the oil coolers have provision for back-flushing the water side?

    <u>   </u>
    (Yes or No)

18. Will all instruments be changeable on the run?

    <u>   </u>
    (Yes or No)

19. Are all lines to remote pressure gauges valved at the tie into the main lines?

    <u>   </u>
    (Yes or No)

20. Are individual packing vents used and is each packing vent discharging to a safe location?_____

    _____

    If to atmosphere, will this cause a pollution or safety problem?

    _____

21. Is there an isolatable closed circuit at the machine so that the pre-start-up run-in can be performed?

_____
(Yes or No)

22. (a) Is the suction drum adequate?

_____
(Yes or No)

   (b) Has a variable or high density crinkled wire mesh been specified?

_____
(Yes or No)

   (c) Has a coalescing section been specified?

_____
(Yes or No)

23. How is the suction pressure controlled?

   (a) On underload (less gas than design)_____

   (b) On overload (more gas than design)_____

24. (a) If the compressor is handling a flammable gas and is not suction pressure controlled or if so controlled and the control fails, will the suction remain above atmospheric?

_____
(Yes or No)

   (b) With variations in the suction pressure will rod failure be avoided?

_____
(Yes or No)

   (c) With variations in the suction pressure will excessive compression ratios and temperatures be avoided?

_____
(Yes or No)

   (d) Based on (a), (b), and (c) above should the machine have a low suction pressure shutdown?

_____
(Yes or No)

25. Has the possible need for superior filtration been thoroughly considered?

    _____
    (Yes or No)

    If the KO drum is a long distance from the machine it may be cheaper to install a filter than clean a long suction line.

26. Can the suction line be thoroughly cleaned by the method proposed? Should it be coated?

    _____
    (Yes or No)

    Describe the method proposed_____

    _____

27. (a) Has an API data sheet been prepared for the compressor?

    _____
    (Yes or No)

    (b) Have all applicable specifications been indicated? Normally there should be several in-house specifications.

    _____
    (Yes or No)

    (c) Have all required accessories been specified?

    _____
    (Yes or No)

    Always specify pulsation bottles. Specify interstage piping on two-stage machines whenever this pressure level is not used for process. Specify frame intercoolers and moisture separators where practical for multistage compression. Consider use of frame aftercoolers on air machines. Specify cooling water manifolding. Specify instrument panel. Specify sight flow indicators on all systems.

    (d) Have capacity control requirements been specified?

    _____
    (Yes or No)

    Do not use suction valve lifters. Suction valve unloaders must be air-operated if control is manual. If this is what you want, indicate suction valve unloading; indicate manual and note in remarks. Unloaders to be air-operated.

Note in remarks if reacceleration is required.

(e) Has distance piece requirement been specified?

_____
(Yes or No)

Normally standard with solid cover. Abnormal conditions would be over 0.1% H_2S or some other contaminant which could degrade the lube oil. Extra long distance pieces are required for non-lube machines.

(f) Have site data been completed?

_____
(Yes or No)

In remarks, indicate cooling water temperature range, especially on open systems.

(g) Has a helium test been called up if partial pressure of H_2 at the discharge is 100 psia?

_____
(Yes or No)

(h) Has the gas analysis data sheet been completed?

_____
(Yes or No)

Does this fully describe the full range of foreseeable operation?

_____
(Yes or No)

Do the sheets of operating conditions also fully describe this range?

_____
(Yes or No)

(i) Has the driver data sheet been prepared?

_____
(Yes or No)

(j) If TEMA inter- and/or aftercoolers are required, has this been noted?

_____
(Yes or No)

Project Phase: Preorder Review with Vendor
Machine Category: Reciprocating Compressors

Designation: _____

Location: _____

Service: _____

1. *Vendor experience*

   (a) Location of similar machines _____

   _____

   (b) How closely do the machines resemble yours?

	Yours	*Others*
Suction (psia)	_____	_____
Discharge (psia)	_____	_____
Suction temperature (°F)	_____	_____
$k(C_P/C_V)$	_____	_____
Bore and stroke	_____	_____
rpm (max. 600)	_____	_____
Discharge temperature (°F)	_____	_____
Piston speed (ft/min) (max. lubricated, 850; unlubricated, with piston rings, 650)	_____	_____
Valve velocity (ft/min)	_____	_____
Piston rod loading (pounds at design conditions)	_____	_____

   The last five are significant. If data for the other machine are lower than yours, it is not a suitable machine for comparison.

   (c) Have any similar machines experienced difficulty in the field?

   _____
   (Yes or No)

   If so, what were the characteristics of the problem and what steps were taken to avoid a repetition?_____

   _____

(d) Comments of users on their machines.

How long have machines been in service (min. 1 year)? _____

What start-up difficulties were experienced? _____

If so, was the Service Department's response satisfactory?

_____
(Yes or No)

Did the machines meet the process specifications?

_____
(Yes or No)

What has been the maintenance experience? _____

_____

Name and position of contact _____

(e) Were the similar machines built in the same shop where your machines are to be built?

_____
(Yes or No)

(f) If the machine is to be built in a shop other than the vendor's main shop:

- How is the parent company or licensor going to ensure that the machines will be built correctly?_____

_____

Will he have an inspector (or engineer) in the shop on a full- or part-time basis?

_____
(Yes or No)

How qualified is this person?_____

_____

- How many similar machines have been built in that shop?

_____

_____

- How similar are they?_____

_____

- Who will do engineering and drafting for the machines? _____

- Will the machines be an exact duplicate of the licensor's design?

(Yes or No)

If not, what modifications will be incorporated and why? _____

2. If the comparison machines are not identical, how much extrapolation has the bidder done on the established experience limits?_____

  (a) Are these extrapolations based on experience with other machines?

(Yes or No)

  (b) Do these extrapolations appear sound?

(Yes or No)

  (c) Are any extrapolations within any previous limits or experience?

(Yes or No)

  (d) Do these extrapolations appear sound?

(Yes or No)

  In all cases of extrapolation, consult with your responsible Machinery Specialist.

3. What design and shop practice modifications have occurred since the comparison machines were built?

  (a) Bearings_____

  (b) Valves_____

  (c) Piston to rod attachment_____

  (d) Rod to crosshead attachment_____

(e) Liner location device_____

Which of these are definite design improvements and which are cost reduction items?_____

_____

Consult with your responsible Machinery Specialist on all cost reduction items.

4. Will the vendor conduct a torsional analysis?

_____
(Yes or No)

If not, what evidence is he providing to comply with applicable API specification?_____

_____

5. Do the cylinder materials meet experience requirements and conservative industry guidelines, i.e.

Relief valve setting	*Cylinder material*
0–1000 psig	CI, cast, or forged steel
1000–2500 psig	Cast or forged steel
2500 psig	Forged steel

6. Does the vendor agree to run rod load reversal checks at all expected loadings?

_____
(Yes or No)

7. Will the motor be adequate to run the machine at all expected loading conditions?

_____
(Yes or No)

8. If the machine is for vacuum service, is the motor big enough for the drawdown peak hp?

_____
(Yes or No)

If not, how will the machine be started and will unloading be acceptable to process?_____

_____

Also, is it possible to trip one unit without causing an overload trip on the other?

_____
(Yes or No)

9. Does the vendor take any exceptions to the specifications? _____

_____

If so, consult with your responsible Machinery Specialist on all exceptions. Check data sheet if specifying against no negative tolerance.

10. Does the expected performance meet with that specified? _____

_____

Note: Many vendors quote flow as 3% higher than specified to get around API's no negative tolerance.

11. What is rod stress at area of rod under thread root?

$$\text{Stress} = \frac{0.785[\text{cylinder bore}^2(P_d - P_s) + \text{rod diameter}^2 P_s]}{\text{area at root (in square in.)}} \quad \underline{}$$

Does this comply with your specification, i.e. maximum stress?

_____
(Yes or No)

> 8000 psi for rolled threads
> 7500 psi for ground threads
> 7000 psi for cut threads

On multistage machines, check rod loadings at all unloading conditions to ensure that the above values are not exceeded.

12. The requirements of 11 are for the worst design conditions. Will the rod loadings at SV setting be within the vendor's rod loading limits?

_____
(Yes or No)

13. If the piston has a tail rod, is it retained by a substantial steel cover?

_____

(Yes or No)

(A broken rod outboard of the piston can cause the tail rod to be driven out like a projectile.)

14. Are the main bearings either babbitted or aluminum?

_____

(Yes or No)

15. Are the cross-head rubbing surfaces either babbitted or aluminum?

_____

(Yes or No)

Are these surfaces replaceable?

_____

(Yes or No)

16. Will the cylinder lube connections in the water jacket run through solid metal?

_____

(Yes or No)

(Piping through the water jacket is not permitted.)

17. Are there external connections for the coolant between the cylinder jackets and heads?

_____

(Yes or No)

(Internal cooling connections are not permitted except on air compressors.)

18. Does the machine have vented packings?

_____

(Yes or No)

If the gas will degrade the lube oil (H_2S or chlorine, etc.), is a double distance piece required?

_____

Accept vendor's recommendation for this.

19. Will the valve gaskets be proven execution?

_____

(Yes or No)

20. Is there a full flow lube oil filter?

_____
(Yes or No)

On engines, is the filter twinned?

_____
(Yes or No)

Is the filter of the type which can be cleaned on the run?

_____
(Yes or No)

If not, how long is it expected to last? (6 months minimum) ____

Will there be pressure gauges on both inlet and outlet of the filter?

_____
(Yes or No)

21. If needed, where will piping pulsation study be performed?_____

_____

22. Will the cylinders have outboard supports not attached to the heads?

_____
(Yes or No)

23. Will the suction and discharge valves be non-interchangeable?

_____
(Yes or No)

24. Will the machines be suitable for rail mounting?

_____
(Yes or No)

25. Will the compression cylinders be equipped with dry type liners?

_____
(Yes or No)

26. Will hollow pistons be easily ventable for disassembly or are safer, continuously vented types furnished?

_____
(Yes or No)

27. If the specification calls for reacceleration on power failure, will the starting unloaders automatically operate?

_____

(Yes or No)

28. If strainers are required on packing coolers, will they be twinned?

_____

(Yes or No)

29. If intercoolers and/or aftercoolers are to be supplied for flammable or toxic gas, will they be to TEMA R?

_____

(Yes or No)

Gas engines
This section is only to be completed if an integral or separate gas engine is involved.

30. If the engine is to burn refinery gas (i.e., fuel gas, hydrogen, propane, etc.), will the engine compression ratio be less than 7:1?

_____

(Yes or No)

31. Will the ignition system be the solid state type (i.e., no magneto)?

_____

(Yes or No)

32. On integral engines will there be anti-blowback type explosion doors on the crankcase?

_____

(Yes or No)

33. On integral engines compressing hydrocarbons will the distance pieces be of the two-compartment type?

_____

(Yes or No)

34. Will the engine have a proven electronic governor? State make.

_____

(Yes or No)

35. Will an electrical tachometer be provided?

_____
(Yes or No)

Project Phase: Contractor's Drawing Review
Machine Category: Reciprocating Compressors

Designation: _____

Location: _____

Service: _____

P&IDs

1. Have KO facilities been installed on all suction and interstages?

_____
(Yes or No)

2. Do recycle lines re-enter upstream of the KO facilities?

_____
(Yes or No)

3. Are KO facilities equipped with a gauge glass, LHA and shutdown switches?

_____
(Yes or No)

4. Are the suction lines traced between the KO drums and the machine flanges, including pulsation bottles? Alternatively, is cylinder cooling medium (preferably water/glycol mixture) warmer than incoming gas?

_____
(Yes or No)

5. Do KO facilities have automatic drains?

_____
(Yes or No)

6. Is there a safety valve on each compression stage and on the coolant header?

_____
(Yes or No)

7. Has a remote shutdown switch been provided?

_____
(Yes or No)

8. Have all the alarm and shutdown devices specified been provided?

_____
(Yes or No)

(a) Low lube pressure alarm   _____

(b) Low lube pressure trip   _____

(c) High-temperature alarm on each cylinder discharge_____

(d) Low cylinder lube flow alarm   _____

(e) Others (detail)   _____

9. Are TIs and PIs specified for suction and discharge of each stage and TIs on discharge and coolant outlet of each cylinder?

_____
(Yes or No)

10. On machines which are specified to be reaccelerated, have controls been provided to unload the cylinders during reacceleration?

_____
(Yes or No)

11. If these controls are of the bypass type, is there a check valve on the compressor discharge?

_____
(Yes or No)

12. Have unloading facilities been provided for start-up?

_____
(Yes or No)

13. Is there a coolant block valve on each cylinder with a low point drain and a high point vent?

_____
(Yes or No)

14. Does each machine have double block valves and a vent to avoid blinding for valve repairs?

_____

(Yes or No)

15. Have pressure taps been provided around the spool which will contain the temporary suction screen?

_____

(Yes or No)

16. Do the oil coolers have provision for back-flushing the water side?

_____

(Yes or No)

17. Will all instruments be changeable on the run?

_____

(Yes or No)

18. Are all lines to remote pressure gauges valved at the tie-in to the main line?

_____

(Yes or No)

19. Is there an isolatable closed circuit at the machine so that the pre-start-up run-in can be performed?

_____

(Yes or No)

20. Are there purge connections which allow gas-freeing the compressor in preparation for maintenance?

_____

(Yes or No)

Layout, etc.

1. Are the main and interstage suction lines from the KO drums or filters cleanable by the method proposed?

_____

(Yes or No)

If the lines are becoming quite long it may be cheaper to put filters adjacent to the machine to reduce cleaning cost.

2. Are all the cylinder, snubber, and gas cooler supports from the machine foundation?

_____

(Yes or No)

Note: This requirement does not apply to remote coolers.

3. Are all the piping supports either on the machine foundation or on separate footings going down below the frost line?

_____

(Yes or No)

4. The pulsation study will indicate which gas lines have high shaking forces and an estimated magnitude of these forces. Are these lines suitably supported and clamped?

_____

(Yes or No)

5. Is there sufficient clearance to pull all pistons and cooler bundles?

_____

(Yes or No)

6. Is all contractor's piping such that it does not interfere with access to any valve, distance piece, crosshead, or crankcase cover?

_____

(Yes or No)

7. Can all piping to each cylinder be removed to permit cylinder removal?

_____

(Yes or No)

8. Where the contractor is doing some of the oil piping, is the piping between the filters and the machine stainless steel?

_____

(Yes or No)

9. Are packing vents run separately to a safe location?

_____

(Yes or No)

If these vents run into a disposal header under pressure, is there a block valve and check valve at the connection to the header?

_____

(Yes or No)

Has each packing vent line a means of individually monitoring for leakage?

_____

(Yes or No)

10. Have distance piece vents and drains been provided?

_____

(Yes or No)

11. Have all small connections been gusseted?

_____

(Yes or No)

12. Has piping been modified as required per acoustic study results?

_____

(Yes or No)

Process piping layout: suction piping

13. Are the main and interstage suctions steam traced and insulated?

_____

(Yes or No)

14. Are the machine laterals taken off the top of the header?

_____

(Yes or No)

15. Is there a manual drain on the header?

_____

(Yes or No)

16. If the compressor is handling a flammable gas, are the isolation valves 25 ft from the machine?

_____

(Yes or No)

Is the spool piece for the temporary strainer readily accessible for both inspection and removal?

_____

(Yes or No)

17. Normally, it should be mounted immediately adjacent to the suction bottle. A strainer is required at each stage unless frame type intercoolers are used.

18. Are blind flanges available at each end of the suction headers to facilitate cleaning, inspection, etc.?

_____
(Yes or No)

19. If the contractor's piping ties directly onto CI cylinders, are the flanges flat-faced?

_____
(Yes or No)

20. Are all valves supported?

_____
(Yes or No)

21. Vertical unbraced lines can result in excessive vibration. Has unnecessary flexibility been avoided wherever possible?

_____
(Yes or No)

22. Is the suction KO drum within 50 ft of the machine, or if not, has a proven separator been provided within this distance?

_____
(Yes or No)

Discharge piping system

23. Are all fittings such as oil separators adequately supported?

_____
(Yes or No)

24. Is the piping system flexible enough to keep thermal load stresses on cylinders within acceptable limits?

_____
(Yes or No)

25. The liquid condensate on intercoolers and aftercoolers will not run uphill into KO facilities. Assume that it will mist. Therefore, on intercoolers and on aftercoolers where condensate removal is

desired, have liquid separation facilities been provided at the low point in the line between the cooler and the KO drum?

_____

(Yes or No)

26. On machines mounted above grade with mezzanine floor, has grating been provided for access to all machines valves, distance pieces, and block and bypass valves?

_____

(Yes or No)

27. Cylinder ends sometimes blow out. Is all equipment requiring operator attention such as instrument panels, instruments, and block and bypass valves out of direct line with the cylinder ends?

_____

(Yes or No)

28. Consider maintenance on the machines. Is there a way of removing all sections of the machines for maintenance? Suitable laydown available for cylinders? If on mezzanine is deck strong enough?

_____

(Yes or No)

29. *Consider operability*

    (a) Are all instruments clearly visible?

_____

(Yes or No)

    (b) An operator's primary sense is touch. Can he feel all valve covers?

_____

(Yes or No)

    (c) Run through the starting sequence. Can all the operations required be done by one person?

_____

(Yes or No)

30. For Engine Compressors consider regrouting. During the life of the machine it will likely have to be regrouted. Is the building design such that heavy lifting equipment can be brought in to lift and move the whole unit?

_____

(Yes or No)

Project Phase: Mechanical Run Test in Shop
Machine Category: Reciprocating Compressors

Designation: _____

Location: _____

Service: _____

The purpose of an in-shop running test on reciprocating compressors is to determine and correct any flaws or errors in the machinery manufacture which would delay commissioning of the machine on site. Test should consist of a flush check of frame lubrication cleanliness before run, with a 100-mesh or smaller screen before filter; a run of approximately 8 h; and a physical examination of machine internals after the run to ascertain that all working parts will operate satisfactorily in the field. Usually, because of horsepower limits of shop driver, the test will be run with the machine unloaded. Screens should be inserted in suction and discharge, but valves should be left in. The surge bottles need not be mounted. The manufacturer must have taken all necessary steps to satisfy himself that the machine is ready to run before the witness test is begun. Special note: The review engineer may wish to include certain of the following checklist items in his pre-purchase review with vendors.

1. Is lube screen free of foundry sand and weld slag? If not, crankcase and oil cooler should be reopened to find where it came from.

_____
(Yes or No)

2. Does the lube screen show sufficiently clean that the machine may be safely run? Minor amounts of matter on the screen that will be removed by frame filter are acceptable.

_____
(Yes or No)

3. Is auxiliary lube pump capable of supplying pressure?

_____
(Yes or No)

4. Is the pressure drop across frame oil filter normal?

_____
(Yes or No)

5. During the run is there sufficient oil on rod to indicate satisfactory operation of the lubricator?

_____

(Yes or No)

6. Are oil drops showing in all lubricator pumps and is lubricator developing full pressure?

_____

(Yes or No)

7. Are all valve covers cool and do all valves appear to be operating properly?

_____

(Yes or No)

8. Is the machine free of knocks or undue vibrations? If not, machine must be stopped immediately and problems corrected.

_____

(Yes or No)

On completion of run request a contact thermometer, dial indicators, suitable micrometers, and feeler gauges. Have the valve covers, valves, cylinder heads, and crankcase covers been removed and have all packing cases been pulled out of bores? Record measurements.

9. Have all measurements been recorded?

_____

(Yes or No)

10. Are rod clearances in head and partition bores large enough to prevent scraping of rod after normal piston and crosshead wear?

_____

(Yes or No)

11. Are all gasket seats free of paint, casting defects, or tool chatter marks? (Use a flashlight oriented on its side and look for radial shadows.)

_____

(Yes or No)

12. Are valve gaskets solid metal type?

———————————
(Yes or No)

13. Is the bore for stuffing box in frame end head free of casting defects?

———————————
(Yes or No)

14. Is the rod securely fastened to piston and crosshead such that it cannot back off?

———————————
(Yes or No)

15. Is there sufficient space in distance piece to readily change packing in cylinder, intermediate diaphragm, and oil scraper rings?

———————————
(Yes or No)

16. Are separate covers supplied for each space in distance pieces for packing access?

———————————
(Yes or No)

17. Are vent and drain connections from packing and distance pieces securely piped and labeled?

———————————
(Yes or No)

18. Is the cylinder head gasket face free of any openings to cored water passages in either cylinder or head?

———————————
(Yes or No)

19. Are the gas passages cored such that there are no cavities or depressions for a liquid trap?

———————————
(Yes or No)

20. Are the cylinders firmly supported on the distance piece or cylinder body and not on the head?

———————————
(Yes or No)

21. Is the frame oil circulation connected such that the filter is last in the stream before injection to bearings?

(Yes or No)

22. Is there an easily removable plug in the piston?

(Yes or No)

23. Is there any indication of penetration of stud drill or tap in the drilling for cover studs?

(Yes or No)

24. Are the cylinder analyzer holes drilled and located properly? They should be open to bore through liners and not covered by rings when piston at end of stroke.

(Yes or No)

25. Are the suction and discharge valves truly non-reversible?

(Yes or No)

26. Are the supports for the lubricator and piping, and frame oil cooler and piping firm enough to prevent vibration in operation and sturdy enough to withstand shipping without damage?

(Yes or No)

27. Are the liners, pistons, rods, and crossheads free of any score marks deep enough to catch a finger nail? If not, item must be washed clean and score must be smoothed off.

(Yes or No)

28. Is the crankcase nameplate securely fastened and does it indicate machine description and serial number and will it be sufficiently viewable when on-site?

(Yes or No)

29. Does each cylinder carry a nameplate securely fastened and does it indicate cylinder description and serial number and piston end clearances?

—————————
(Yes or No)

30. Are the rails or sole plates for the machine precoated with epoxy paint for proper adherence to epoxy grout?

—————————
(Yes or No)

31. Are laminated shim packs of stainless steel available for shipment with machine?

—————————
(Yes or No)

32. Is rust protection being applied to machine before shipment and is it equal to that specified?

—————————
(Yes or No)

33. Is the shipping crate sturdy enough to protect the machine, especially the external piping and gauges? All integral pipes and lines must remain installed and connected for shipment.

—————————
(Yes or No)

34. Question the supplier if the shipper has adequate experience and if shipping method is completely satisfactory to prevent any damage to the machinery in transit.

—————————
(Yes or No)

Measurements taken immediately after run test (dimensions in thousandths of an inch)

Name of Inspector:_____

Date:_____ Manufacturer:_____

Compressor No.:_____ Model:_____

Serial No.:_____ Size:_____

Cylinder numbers

Temperatures	*1*	*2*	*3*	*4*	*5*	*6*	*7*	*8*
Oil at pump <180°F								
Main bearing lube end <225°F								
Main bearing behind throw to cylinder number <225°F								
Conrod shell to cylinder number <225°F								
If temperature excessive, rerun until satisfactory and then examine part for wipes								
Rod drop to cylinder (piston high (*H*) low (*L*) with dial indicator) <5 mils								
Rod centerline runs below head bore centerline. Micrometer in stuffing box bore top and bottom of rod								
Rod clearance on bottom-cylinder head bore. Feeler gauge reading								
Clearance at crosshead shoe-top. Long feeler gauge. Ample clearance required, but must be uniform across the whole shoe								
Piston clearance in liner with feeler gauge. (If rider rings supplied must be ample to allow for wear)								

Frame oil pressure before filter/after filter (<5 psi across filter)_____

Standby lube pump pressure_____

Cylinder lubricator pressure if available_____

Project Phase: P&ID and Design Specification Review
Machine Category: Special-purpose Steam Turbines

Designation: _____

Location: _____

Service: _____

Note: The following sections are in two parts. The questions raised under "General" apply to all turbines. Those under "Condensing Turbines" are additional for those machines.

P&IDs

General

1. Is there a warm-up vent (at least 1 1/2 in.) on the inlet line?

_____
(Yes or No)

2. Does the inlet block have a 1-in. bypass for line warm up?

_____
(Yes or No)

3. Does the exhaust valve have a 1-in. bypass for warm up? (Back pressure turbines only.)

_____
(Yes or No)

4. Is there a trap and bypass upstream of the trip and throttle valves?

_____
(Yes or No)

5. Is there a trap and bypass on the steam chest of single valve turbines?

_____
(Yes or No)

6. Is there a trap and bypass on the low point of the exhaust casing?

_____
(Yes or No)

7. Is there a low pressure seal vent line on both seals?

_____
(Yes or No)

8. (a) What devices cause a trip of the turbine other than the built-in ones?

_____

  (b) Have these been specified? Have you considered 2-out-of-3 trip logic?

_____

(Yes or No)

  (c) All special-purpose turbines have a separate trip and throttle valve which is a shutdown device and can be actuated by an electrical signal. Do the process safety shutdowns utilize this device?

_____

(Yes or No)

  (d) Is this provision called out on the turbine data sheet?

_____

(Yes or No)

9. What does the overspeed trip pressure switch actuate?_____

_____

  Have these been specified?

_____

(Yes or No)

10. (a) Has an exhaust line safety valve been provided between the turbine and the exhaust block valve?

_____

(Yes or No)

  (b) Is the safety valve setting at or below the maximum design exhaust system pressure?

_____

(Yes or No)

11. All special-purpose turbines should have either an approved/redundant electronic or a hydromechanical/electric governor. If automatic control of the turbine speed is desired, has provision of an air head been called out on the data sheet?

_____

(Yes or No)

If an increase in signal is not to increase speed, has this requirement been specified?

_____

(Yes or No)

12. Has the following instrumentation been specified?

(a) Inlet and exhaust TIs _____

(b) Inlet and exhaust PIs _____

(c) Steam chest PI (only on single valve units) _____

(d) 1st stage pressure _____

(e) Other stage pressures _____

(f) Steam flow

_____

(Yes or No)

13. Have the requirements for turbine washing been specified?

_____

(Yes or No)

Based on experience, multistage turbines may often require washing during first year of operation. Consider full wash facilities. For some locations only tie-ins may have to be considered. All large (over 2000-hp) turbines will probably be multistage on 600 psi to 125 psi steam.

14. (a) Has the API data sheet been prepared?

_____

(Yes or No)

(b) Has location been specified?

_____

(Yes or No)

(c) Have applicable specifications been detailed?

_____

(Yes or No)

(d) Have the instruments required on the local panel been specified?

_____

(Yes or No)

(e) Has the lube oil specification sheet been completed if this is not to be from the driven equipment?

_____

(Yes or No)

Condensing turbines

1. Has a suitable shaft seal system been provided?

_____

(Yes or No)

Usually, the sealing steam will be taken from a pressure tap on the casing but for start-up a pressure-controlled live steam supply must be provided. Adequate traps must be installed to keep water out of the seals.

2. Is there an exhaust pressure safety device to relieve pressure in the event of cooling water failure?

_____

(Yes or No)

3. Is this safety device water sealed?

_____

(Yes or No)

4. Is there a minimum flow recycle arrangement on the condensate pumps?

_____

(Yes or No)

5. Is the condenser level control and bypass (if valved) arranged for max. pumpout on air failure?

_____

(Yes or No)

6. Are the pump glands sealed by discharge pressure?

_____

(Yes or No)

7. Is the pump suction chamber vented back to the condenser steam space?

_____

(Yes or No)

Project Phase: Preorder Review with Vendor
Machine Category: Special-purpose Steam Turbines

Designation: _____

Location: _____

Service: _____

1. Are there any deviations to your specification other than those previously disclosed?_____

   _____

   If yes, list and discuss with your Machinery Specialist.

2. Does the vendor have adequate experience?

   _____
   (Yes or No)

   (a) Location of similar machine_____

   (b) How closely does it conform to the proposal?_____

   _____

	Yours	*Others*
Blading	_____	_____
Inlet pressure	_____	_____
Inlet temperature	_____	_____
Exhaust pressure	_____	_____
Exhaust temperature	_____	_____
rpm	_____	_____
Tip speed	_____	_____
Blade passing frequency	_____	_____
Number of stages	_____	_____
Type of seals	_____	_____
Governor type	_____	_____
Reviewed by Machinery Specialist	_____	_____

(c) Have any similar machines experienced difficulty either on test or in the field?

_____
(Yes or No)

If so, what were the characteristics of the problem and what steps were taken to correct it and to avoid repetition?_____

_____

(d) Comments on users and their machines.

How long have machines been in service (min. 1 year)_____

What start-up difficulties were experienced?_____

_____

Is so, was the Service Department's response satisfactory?

_____
(Yes or No)

Did the machines meet the specified duty and efficiency?

_____
(Yes or No)

What has been the maintenance experience?_____

_____

Name and position of contact:_____

(e) Were the comparison machines built in the shop where your machines are to be built?

_____
(Yes or No)

(f) If your machine is to be built in a shop other than the vendor's main shop:

- How is the parent company or licensor going to ensure that the machines will be built correctly?_____

_____

- How many similar machines have been built in that shop?____

_____

- How similar are they?_____

  _____

- Who will do the engineering and drafting for the machines?

  _____

- Will the machines be an exact duplicate of the licensor's design?

  _____

  (Yes or No)

- If not, what modifications will be incorporated and why?

  _____

  _____

(g) *Tip speed*

- Is the tip speed above 825 ft/sec at maximum design speed?

  _____

  (Yes or No)

- If so, API-612 says an integrally forged shaft and wheel arrangement is preferred. Is this what you are getting?

  _____

  (Yes or No)

If not, consider following up on the vendor's experience in this area. Look especially at the rpm of "similar" machines because centrifugal forces increase as the square of rpm while tip speed is a linear function. Also check on steam temperatures.

3. If the comparison machines are not identical, how much extrapolation has the bidder done on the established experience limits?

   (a) Are the extrapolations based on experience with other machines?

   _____

   (b) Do these extrapolations appear sound?

   _____

   (Yes or No)

   (c) Describe extrapolations beyond previous limits of experience___

   _____

(d) Do these extrapolations appear sound?

_____

(Yes or No)

4. What design and shop practice modifications have occurred since the similar machines were built?

(a) Rotor including shrink specifications_____

(b) Casing_____

(c) Bearings_____

(d) Shaft seals_____

(e) Running clearances_____

(f) Wheel design and manufacture_____

(g) Diaphragms and fixed nozzles_____

(h) Materials_____

(i) Balancing_____

(j) Tip speeds_____

(k) Overspeed trip design_____

(l) Overspeed testing of rotor_____

Which of these changes are definite design improvements and which are cost reduction items?_____

_____

Consult with your designated Machinery Specialist on all items which are cost reduction items.

5. *Shaft sealing*

(a) What type of external shaft seals are proposed?_____

_____

(b) If carbon ring:

• Is the rubbing speed below a conservative maximum of 160 ft/sec?

_____

(Yes or No)

- The shaft should be hard chrome or ceramic coated in the seal zone. Is it?

  _____
  (Yes or No)

- How many rings are there?_____

  Maximum pressure per ring is 35 psi and minimum number of rings is 4.
- Is there a satisfactory vent to atmosphere part way down the seal to prevent lube oil contamination if the seal partially fails?

  _____
  (Yes or No)

(c) If labyrinth packing:

- Has a vacuum vent system been provided?

  _____
  (Yes or No)

- Is there an inter-packing vent which could be connected to the 15-psig steam system?

  _____
  (Yes or No)

  (Manufacturers often underestimate seal leakage and vent condenser is therefore undersized.)
  Is the labyrinth compatible with the shaft bearing surface?

  _____
  (Yes or No)

6. How are the machine expansions handled to minimize thrusts due to coupling slip forces and expansion?

   (a) Where is the casing anchored axially?_____

   Where is the thrust bearing?_____

   If these are not at opposite ends of the casing, there will be a large shaft movement into the coupling. Is the coupling adequate to take this movement? (Question the acceptability of this approach.)
   (b) In which direction will the coupling slip force act relative to the internal thrust forces?_____

(c) Assuming the coupling slip force to be

$$F = 0.3\frac{T}{d} = \frac{18,900 \times P}{\text{rpm} \times d}\ \text{lb}$$

where T = torque (lb-in.),
   P = horsepower,
   d = shaft diameter at the COUPLING (in.).

Are all thrust bearings designed for normal thrust forces plus and minus the coupling slip force?

_____
(Yes or No)

If not, why not?_____

(d) What residual thrust capacity is available in the bearings to overcome the thrust due to blade deposits?_____

_____

Is this capacity sufficient to cover fouling which would reduce the hp output by 20%?

_____
(Yes or No)

Are the wheel discs provided with balance holes to reduce thrust increase due to fouling?

_____
(Yes or No)

7. *Machine integrity*

(a) As mounted on the baseplate, will the machine retain its internal alignment during shipping?

_____
(Yes or No)

(b) Will the machine have to be opened up on site for final adjustments before starting?

_____
(Yes or No)

(This is a contractor's problem, but allowance must be made in the schedule for it. Also, the turbine service persons will be required.)

8. *Governing*

   (a) Is the governor to be a hydromechanical/electric type or an electronic model?

   _____
   (Yes or No)

   (b) How stable will it be at minimum specified temperature?____

   _____

   (c) If the governor oil system is separate, has an oil heater been provided?

   _____
   (Yes or No)

   (d) How is the maximum speed stop override operated in testing the overspeed trip? Does this seem a controlled operation?_____

   _____

   (e) Is a minimum speed stop provided?

   _____
   (Yes or No)

9. *Critical speeds*

   (a) Have the critical speeds been determined and do these meet your requirement of a 20% speed margin from all running speeds?

   _____
   (Yes or No)

   (b) How were the criticals calculated?

   Rigid supports_____

   Flexible supports_____

   Only flexible support calculations are acceptable.
   (c) What bearing stiffness was used in the critical calculation?_____

   _____

   This stiffness should be between 1 and 10×10^6 lb/in.
   (d) How closely have actual criticals established on test agreed with calculations?

   _____

If there is any doubt about the turbine vendor's ability to calculate criticals, insist that the vendor with train responsibility check the calculations.

10. *Blade vibrations*

   (a) Has the vendor submitted Campbell diagrams for all turbine blades over 5-in. long?

   _____
   (Yes or No)

   These diagrams must show all three fundamentals (tangential, axial, and torsional) and their harmonics up to the maximum frequency of excitation (usually blade passing frequency). Margins of 10% must be maintained. Excitation due to harmonics above passing frequency can be tolerated because of low energy levels.

   (b) How does the vendor propose to meet nozzle and blade experience criteria?

   By test_____

   By demonstration_____

   Be very cautious about demonstration. Are there the same number of nozzles and blades? Is the speed range the same (5% may be critical).

11. *Journal bearings*

   (a) What is the journal speed?_____ ft/min
   (b) What is the bearing loading?_____ psi
   (c) If over, what type of bearing does the vendor propose to use?__

   _____

   (d) Can he demonstrate that he has used exactly the same bearing in a similar condition of load and speed and with a critical speed as low or lower in percent of running speed than our application?

   _____

   This bearing design problem is critical. We would like to have tilting pad bearings but many turbine manufacturers have no experience with them or will not use them. If the investigation leaves one in doubt about the bearing experience, try to insist on tilting pad journal bearings.

12. *Combined T&T valve or separate T&T valve*

    (a) Manufacturer's name and model number_____

    (b) Does the valve have a built-in 5-mesh monel strainer?

_____

(Yes or No)

    Note: An additional separate external strainer would be advantageous.

    (c) Does the valve have a pilot arrangement to assist opening?

_____

(Yes or No)

    If not, you should either get another type of valve or put a small bypass on the isolation valve for run-up.

    (d) Does the valve have the partial stroke feature as specified?

_____

(Yes or No)

13. *Starting*

    If the turbine is to operate at an exhaust pressure of 30 psi or above:

    (a) Will heating the casing from the exhaust header with the turbine stationary be acceptable?

_____

(Yes or No)

    (b) If not, how must the turbine be heated to prevent shaft distortion?

_____

    Dependent on the answers to 13(a) and (b), check the flow plan to ensure that provision has been made to run up in the correct manner.

    (c) Has the entire train been checked out to see that turbine warm-up procedure will not cause lubrication problems in any bearing in the train?

_____

(Yes or No)

    (d) What procedure will be followed in design to ensure that the vibration when passing through the critical will not damage the labyrinths?_____

_____

Most manufacturers have programs which will predict shaft movement but may not use them unless you ask.

14. *Shaft thermal stability*

 (a) What is the chance of the shaft developing a thermal bow if the machine is tripped with the exhaust left open?_____

 _____

 (b) Is there some time limit to get the machine rolling again?

 _____
 (Yes or No)

 (c) If there is a strong possibility of a thermal bow, should a turning gear be provided?

 _____
 (Yes or No)

 If the vendor thinks a thermal bow is likely, then turning gear should be provided.

 (d) If turning gear is provided, will the speed be sufficient to provide hydrodynamic lubrication to all the bearings in the train?

 _____
 (Yes or No)

 If not, what is proposed to ensure adequate lubrication?_____

 _____

15. *Lube viscosity*

 (a) Is there any incompatibility between the lube viscosity requirement of the turbine and of the driven equipment?

 _____
 (Yes or No)

 There should be no problem with this but the question should be asked.

16. *Allowable piping forces*

 (a) Will the allowable piping forces be to NEMA SM-21?

 _____
 (Yes or No)

(b) What evidence can the vendor supply to indicate that these forces will not cause excessive casing strains?_____

There has been feedback from a number of sources which indicates that many turbines will not tolerate the forces allowable under SM-21 as calculated by piping stress analysis.

(c) Does the vendor have sufficient knowledge of piping arrangements that he could analyze the contractor's proposed arrangement and suggest necessary modifications?

_____

(Yes or No)

(d) Would the contractor accept this arrangement?

_____

(Yes or No)

17. *Tachometer*

(a) Whose tachometer will be provided?_____

API-612 says the tachometer will be the pulse counter type or equal. The vibrating reed unit is not equal.

(b) Will the tachometer be mounted so that it can be seen by an operator at the T&T valve?

_____

(Yes or No)

18. *Shaft access for vibration*

(a) How does the vendor propose to provide access to the shaft at both bearings to permit shaft vibration readings with a hand-held pickup?_____

19. *Nozzles*

(a) Are all bladed nozzles replaceable in the field? They must be.

_____

(Yes or No)

(b) Will all the nozzle block bolts be wired to prevent unscrewing?

_____

(Yes or No)

If not, how will they be fixed?_____

_____

If in doubt, request wiring.

(c) Are any of the trailing edges in the blades less than 0.02-in. thick?

_____

(Yes or No)

If they are, insist they be beefed up to 0.02 in.

20. *Washing*

Will turbine washing produce any problems with:

(a) Thermal expansion?_____

(b) If the water were to fail during the wash when the steam inlet was at saturation, would this result in insufficient differential expansion to cause failure?_____

(c) Thrust bearing load?_____

(d) What load could be expected from the turbine with the inlet saturated?_____

Project Phase: Contractor's Drawing Review
Machine Category: Special-purpose Steam Turbines

Designation: _____

Location: _____

Service: _____

Note: The following sections are in two parts. The questions raised under "General" apply to all turbines. Those under "Condensing Turbines" are additional for these machines.

P&IDs

General

1. Is there a warm-up vent (at least $1 \frac{1}{2}$ in.) on the inlet?

_____

(Yes or No)

2. Does the inlet block have a 1-in. bypass for line warm up?

_____
(Yes or No)

3. Does the exhaust valve have a 1-in. bypass for warm up? (Back pressure turbines only.)

_____
(Yes or No)

4. Is there a trap and bypass upstream of the trip and throttle valve?

_____
(Yes or No)

5. Is there a trap and bypass on the steam chest of single valve turbines?

_____
(Yes or No)

6. Is there a trap and bypass on the low point of the exhaust casing?

_____
(Yes or No)

7. Is there a low pressure seal vent line on both seals?

_____
(Yes or No)

8. If the vendor has specified a pressure for this vent, has satisfactory control been provided?

_____
(Yes or No)

9. What devices cause a trip of the turbine other than the built-in ones?

_____

Is this as specified?

_____
(Yes or No)

10. What does the overspeed trip pressure switch actuate?_____

_____

Is this as specified?

_____
(Yes or No)

11. If there is no built-in strainer in the trip and throttle valve, is there a Y-type strainer in the inlet line? (Prefer to have both!)

_____
(Yes or No)

12. Has an exhaust line safety valve been provided between the turbine and the exhaust block valve if the exhaust pressure is above 75 psig?

_____
(Yes or No)

Is the safety valve setting at or below the max. design exhaust system pressure (including exhaust casing)?

_____
(Yes or No)

13. Back pressure turbines with labyrinth seals must have an eductor and condenser. Are these shown?

_____
(Yes or No)

14. Has the following instrumentation been provided?

Inlet and exhaust TIs_____

Inlet and exhaust PIs_____

Steam chest PI (only on single valve units)_____

1st stage pressure_____

Steam flow_____

15. Have turbine washing facilities been provided?

_____
(Yes or No)

Condensing turbines

1. Has a suitable shaft seal system been provided?

_____
(Yes or No)

Usually the sealing steam will be taken from a pressure tap on the casing, but for start-up a pressure-controlled live steam supply must be provided to keep water out of the seals.

2. Is there an exhaust pressure safety device to relieve pressure in the event of cooling water failure?

_____
(Yes or No)

Is this safety device water sealed?

_____
(Yes or No)

3. Is there a minimum flow recycle arrangement on the condensate pumps?

_____
(Yes or No)

4. Is the condenser level control and bypass (if valved) arranged for max. pumpout in case of air failure?

_____
(Yes or No)

5. Are the pump glands sealed by discharge pressure?

_____
(Yes or No)

6. Is the pump suction chamber vented back to the condenser steam space?

_____
(Yes or No)

7. Has the following additional instrumentation been provided?

   (a) Vacuum gauge on inlet and interstage of ejector_____

   (b) Seal steam pressure gauge_____

Piping layouts

1. Is the inlet steam taken off the top of the main header and is there a trapped dead leg on the header downstream of the turbine take-off?

_____
(Yes or No)

2. Does the inlet slope continuously between the washing de-superheater connection and the machine flange with no pockets of any kind to trap water?

_____
(Yes or No)

3. Can the inlet pipe be readily diverted outside for initial blowout?

_____
(Yes or No)

The piping not to be blown must be capable of being thoroughly inspected internally.

4. Can the trip and throttle valve be manipulated easily from the main platform?

_____
(Yes or No)

This valve is used to start up the turbine and control its speed when out of governor range.

5. The turbine exhaust safety valve must be removable for testing with the machine in service. Is it located so that if it is dropped it will not cause damage to other equipment?

_____
(Yes or No)

6. On air blower drivers, are all steam vents on both inlet and exhaust lines away from and above the air intake hood?

_____
(Yes or No)

7. Has the inlet and exhaust pipe been provided with sufficient direction anchors so that all piping growth will be away from the turbine?

_____
(Yes or No)

8. Are the inlet and exhaust pipe supports, guides, anchors, etc. as described in the piping stress calculation?

_____
(Yes or No)

Has sufficient allowance for friction been made in the stress calculation?

_____
(Yes or No)

9. Before start-up the operator must blow all steam line and casing drains. Are all these valves accessible for him to do this?

_____
(Yes or No)

These drains are normally taken to a funnel. Is the location of the funnel such that the steam venting from it will not interfere with other equipment nearby or cause a hazard?

_____
(Yes or No)

10. Piping must not run unnecessarily over parts which must be removed for maintenance, i.e. bearing covers, top half casing, governor, and trip and throttle valves. Has this problem been avoided and have crane capacity and lifting height been checked?

_____
(Yes or No)

11. Is there a place where the casing top half can be set down during maintenance without interfering with maintenance?

_____
(Yes or No)

Can it be moved to this location without passing over other equipment which might be running?

_____
(Yes or No)

Can the rotor be removed to the maintenance shop without passing over running equipment?

_____
(Yes or No)

12. Is the exhaust steam trap adequate to dispose of all the water required to make the inlet steam 1% wet?

_____
(Yes or No)

Is it located on the low point of the exhaust?

_____

(Yes or No)

(If the casing will readily drain into the line the trap should be on the line. Otherwise, it should be on the casing.)

13. Can the operator safely open the inlet and exhaust block valves?

_____

(Yes or No)

14. Will the operators be able to manipulate the turbine washing system safely and in a controlled manner?

_____

(Yes or No)

Consider operability

1. Are all instruments clearly visible?

_____

(Yes or No)

2. Does the operator have safe and easy access to all the bearings?

_____

(Yes or No)

3. Has he clear access to the manual trip?

_____

(Yes or No)

4. Can he see clearly the tachometer from the governor overspeed device?

_____

(Yes or No)

5. Can he see clearly the tachometer from the trip and throttle valve?

_____

(Yes or No)

6. Run through a starting sequence. Can all the operations required be done by one man?

_____

(Yes or No)

Project Phase: Mechanical Run Tests
Machine Category: Special-purpose Steam Turbines

Designation: _____

Location: _____

Service: _____

The mechanical run test for steam turbines is basically a means of checking rotor balance, controls, safety trips, and checking for leaks.

The basic procedure should be to run up to speed with steam conditions as close to design as possible. When conditions stabilize, including bearing and lube oil temperatures, the turbine should be operated for a period of 1 h, with no further rise in bearing and lube oil temperatures.

(On condensing turbines, steam inlet temperature or test period may be reduced to prevent excessive temperature in the turbine casing.)

	Conditions	
	Design	*Test*
1. Steam inlet pressure	_____	_____
2. Steam inlet temperature	_____	_____
3. Steam exhaust pressure	_____	_____
4. Steam exhaust temperature	_____	_____
5. Lube oil pressure	_____	_____
6. Lube oil inlet temperature	_____	_____
7. Speed		
(a) Max. continuous	_____	_____
(b) Normal operating	_____	_____
(c) Trip	_____	_____
(d) Calculated critical	_____	_____
8. Max. vibration	_____	_____

    1. Do conditions 1, 2, 3, 5, 6, 7, and 8 on test match the design conditions to your satisfaction?

_____
(Yes or No)

2. Is the turbine half coupling (with adaptor, if necessary) fitted for the test?

_____

(Yes or No)

3. Is there a steam strainer on the inlet?

_____

(Yes or No)

4. Is the actual critical speed within 5% of the calculated value?

_____

(Yes or No)

5. Bearing temperature: Is the temperature rise across each bearing less than 60 °F?

_____

(Yes or No)

6. *Vibration*

   (a) Frequency survey when running at max. continuous speed. Note vibrations at running speed and other frequencies.

Probe location	*Magnitude (mil)*	*Frequency (cpm)*
_____	_____	_____
_____	_____	_____
_____	_____	_____
_____	_____	_____
_____	_____	_____

   Is there an absence of vibration at critical frequency, also at frequencies between 35% and 50% of running speed?

_____

(Yes or No)

   (b) Shaft and bearing housing vibration attenuation:

	A, shaft	*B, housing*	*A/B, attenuation*
I.B. bearing	_____	_____	_____
O.B. bearing	_____	_____	_____

Is the attenuation less than 4? (If not, state actual figures in the report.)

_____

(Yes or No)

(c) Vibration readings (mil):

	Just below trip speed	*Max. continuous speed*	*Difference*
I.B. bearing	_____	_____	_____
O.B. bearing	_____	_____	_____

Is the difference in vibration levels at these speeds less than 20%?

_____

(Yes or No)

7. *Overspeed trip*

(a) The overspeed trip must be actuated at least three times. Is the difference between the highest and the lowest trip speeds less than 0.5% of the highest?

_____

(Yes or No)

(Note any problems with adjustment of trip and operation of the trip valves.)

Note the actual final setting of the trip.   _____ rpm

Is this approximately 110% of max. continuous speed?

_____

(Yes or No)

8. Are all auxiliary trips being tested – low lube oil pressure, etc.?

_____

(Yes or No)

9. Is a thorough check being made for leaks, steam, oil, air to governor, etc.? Finally satisfactory?

_____

(Yes or No)

10. Is the machine stable at all speeds? (Hunting within 0.5%.)

_____
(Yes or No)

11. Is the speed control operating satisfactorily?

_____
(Yes or No)

12. (a) Are the data being taken, including the linearity of speed versus control signal?

_____
(Yes or No)

Record	Signal												
	Speed												

(b) On loss of control air, is the result as specified?

_____
(Yes or No)

13. Bearing inspection: Do bearing surfaces show normal wear patterns?

_____
(Yes or No)

14. If there is a spare rotor to be run, request an internal inspection (normally specified).

(a) Is there an absence of rubbing?

_____
(Yes or No)

If rubbed, demand a clearance check.

15. Check the internal alignment and clearance data from final assembly drawing (if available).

(a) Is alignment good; clearances within tolerances?

_____
(Yes or No)

(b) Likewise for a spare rotor if it has been fitted.

_____
(Yes or No)

16. Have copies of vendor's log sheets, and final internal clearance diagrams been obtained?

_____

(Yes or No)

17. When witnessing a test, always try to find out if any difficulties occurred in preparing for it. Such problems could be repeaters.

Project Phase: Mechanical Run Tests
Machine Category: Lube and Seal Oil Consoles

Designation: _____

Location: _____

Service: _____

Lube and seal oil consoles are inspected during fabrication and erection, and the flushing operation carried out in the manufacturer's shop.

On consoles and seal oil drainer packages, the following procedure should be followed by the inspector:

1. Is the oil piping at the pumps rigidly supported?

_____

(Yes or No)

2. With pump piping flanges unbolted, is the piping alignment satisfactory?

_____

(Yes or No)

3. Are all valves and strainers accessible?

_____

(Yes or No)

4. Have all piping and valves been inspected internally, probing with a magnet? Finally, no machining chips, welding spatter, machining burrs, weld dross, burn through, flux, and other contaminants?

_____

(Yes or No)

5. Has all lacquer been removed from bends and fittings?

_____
(Yes or No)

6. Have cooler bundles been pulled and checked for cleanliness?

_____
(Yes or No)

7. (a) Reservoir interior clean?

_____
(Yes or No)

(b) Paint work satisfactory?

_____
(Yes or No)

8. *Pumps*

(a) Alignment to drivers satisfactory?

_____
(Yes or No)

(b) With dial indicators on pumps, is alignment satisfactory while jumping on console base?

_____
(Yes or No)

(If not, check whether console is to be grouted, or is supported as level in shop as expected in the field.)

9. Whole system successfully hydrostatically tested?

_____
(Yes or No)

10. *Flushing*

(a) Are all console discharges connected to the tank by temporary bypasses?

_____
(Yes or No)

(b) Is the flushing oil temperature at 180°F?

_____
(Yes or No)

(c) Are vibrators being used to shake pipework?

_____
(Yes or No)

(d) Are all control valve bypasses, four-way valves on filters and coolers, being swung periodically?

_____
(Yes or No)

(e) Is this initial flushing carried out for 8 h uninterrupted?

_____
(Yes or No)

(f) Control function tested successfully?

_____
(Yes or No)

(g) After initial flush, system checked by installing felt pads, backed up with SS mesh in the temporary bypasses and filter outlets. System re-flushed for 2 h, oil at 180 °F, flow as high as possible, pipework vibrated?

_____
(Yes or No)

(h) Bypass pads clean?

_____
(Yes or No)

If no, flushing must continue.
(i) Filter outlet pads clean?

_____
(Yes or No)

If no, filters to be overhauled to determine the cause of the filter leakage.

11. Will the console be shipped with the temporary bypasses installed?

_____
(Yes or No)

(Wanted so that flushing can start in the field as soon as the console is set on its foundation.)

12. *Drainer packages*

(a) Are all valves accessible?

_____
(Yes or No)

(b) Have all volume chambers, sight glasses, traps, pipework, and valves been inspected internally, probing with a magnet?

_____
(Yes or No)

(c) Check made for no evidence of lacquer?

_____
(Yes or No)

**Project Phase: Field Handling, Storage, and Installation
Machine Category: All**

Designation: _____

Location: _____

Service: _____

1. Are machinists available to assist in checking for damage, etc.?

_____
(Yes or No)

2. Has machine/unit been checked for transit damage?

_____
(Yes or No)

3. Are blinds on flanged openings still tight? (If not, retighten or renew.)

_____
(Yes or No)

4. Are all other openings plugged or blinded?

_____
(Yes or No)

5. Is the paint covering on machine/unit still good? No signs of rust? (Rust should be removed and area repainted.)

_____
(Yes or No)

6. Check all items against packing list.

(a) Anything short?

_____
(Yes or No)

(b) Anything damaged?

_____

(Yes or No)

(c) Has this been reported?

_____

(Yes or No)

7. Have all loose items been restored in closed boxes?

_____

(Yes or No)

(a) Have these been stored in a limited access area?

_____

(Yes or No)

(b) Has a record been made of where these items are stored?

_____

(Yes or No)

8. How much time is expected between receipt and start of installa-tion? Give strong consideration to using oil mist as a "preservative blanket" for all machine internals!

9. Have specific instructions regarding rotation of rotors, crankshafts, etc. been included in the vendor's service manual?

_____

(Yes or No)

10. Have oil reservoirs been checked for presence of water and drained if necessary?

_____

(Yes or No)

11. Have oil reservoirs been topped up with lube oil or rust preventative?

_____

(Yes or No)

12. Following (11) above, has the rotor or crankshaft been turned two complete revolutions? (This includes small pumps as part of a pack-age.)

_____

(Yes or No)

(a) On reciprocating compressors, operate the hand pump if available, and crank the cylinder lubricator if the machine has the cylinders installed.

13. Has a program been established for regular rotation of shafts – two turns at 2-week intervals? Include draining water from oil.

   _____
   (Yes or No)

14. If time in (8) above is over 1 month, have blinds been removed and machine internals inspected to determine condition of protective coatings? Renew if necessary. This should be repeated at 2-month intervals.

   _____
   (Yes or No)

15. Has plastic sheeting been placed over casings?

   _____
   (Yes or No)

   Has a breathing space been left open?

   _____
   (Yes or No)

16. Are all exposed machined surfaces coated with rust preventative?

   _____
   (Yes or No)

17. Have reciprocating compressor valves been stored in a container of light oil?

   _____
   (Yes or No)

18. Has it been arranged that lube and seal units will be installed as soon as possible in order to put them into operation?

   _____
   (Yes or No)

   (These can be flushed by discharging directly back to the tank while waiting for hook up to machinery piping.)

19. Has all major equipment not stored within warehouse, etc. been stored in a place where damage from construction activities and

traffic is least likely? Again, has oil mist preservation been considered?

_____

(Yes or No)

20. Gear units – preserved in vendor's shop for extended storage – should be stored such that unit will not be turned. No oil to be added until finally installed. Consider use of appropriate diester or polyalphaolefin synthetic lubricant.

Construction Phase: Installation Completeness Reviews

Machine Category: As Noted

Installation completeness checklists are intended to be used by reliability-conscious maintenance staff reviewing the adequacy of field installations. These checklists reflect good practices. They are not to be confused with mandatory requirements.

Of course, even the most comprehensive review checklist will be of limited usefulness if its desirable features have not previously been part of the owner's equipment purchase and installation specifications. Note that our checklists will enable you to critically examine if your existing specifications reflect the procurement and implementation guidelines applied by the reliability-minded competition.

Horizontal Baseplates

All baseplates shall:

1. Be of solid construction and design.

2. Have all metal shims removed from under base.

3. Have all anchor bolts tight.

4. Be level and *grouted with no voids*.

5. Have driver pedestals lower than driven pedestals for proper equipment alignment.

6. Have stainless steel alignment positioning screws at the corners of each driver pedestal for drivers (including gears) 100 hp and

greater. Eight positioning screws are required for four directional adjustment. Lugs shall be attached to baseplate.

7. Have provisions to collect and drain packing or seal leakage if driven equipment is in corrosive service.

Baseplates for Vertical Column Pumps

1. The steel mounting plate shall have a $3/4$-in. pipe connection to vent the space between the outer barrel and the foundation.

2. The plate shall be rectangular and completely enclose the outside diameter of the can or barrel.

3. The plate shall be attached to the can or barrel with a continuous weld.

4. The mounting plate shall be separate from the main body flange, and located sufficiently below the flange to permit use of through-bolting on the body flange.

Centrifugal Pumps

Case

1. The pump shall be reasonably easy to remove for maintenance.

2. The pump casing shall be furnished with a drilled and tapped drain opening.

3. Connection for seal flush lines shall be $3/8$ in. NPS or larger.
   Exception: Drilled or tapped openings in high velocity areas are not permitted if the specified corrosion allowance of the pump casing is more than $1/8$ in. or if the pump is in *acid or erosive service*. Suction and discharge nozzles and stuffing boxes are not considered high velocity areas.

Lubrication

1. Non-pressurized oil-lubricated bearings shall have adjustable constant level oilers with transparent containers.

2. Wet sump or "purge mist" lubricated bearings. Typically sleeve bearings only. Verify correct oil level and orifice size.

3. Dry sump lubricated bearings, when specified, do not require constant level oilers. Verify correct orifice size.

4. Grease lubricated bearings, where specified, shall have accessible grease fittings. Verify accessible *drain* plug as well.

Seals and Packing

1. Seal glands shall be labeled to identify cooling, flushing, vent and drain connections.

2. Connections for seal oil and flush lines shall be $1/2$ in. NPS or larger.

3. *Flexible hose* to the quench gland for conventional packing is *not permitted.*

Guards

1. Coupling guards shall be fabricated from 12-gauge, galvanized expanded sheet metal. Alternatively, guards shall be provided with hinged door for coupling inspection.

Piping, General

1. Centrifugal pump suction and discharge piping installation shall be as described below:

   - All handwheels on block valves shall be reasonably convenient for operation.
   - All pressure and temperature indicators shall be reasonably easy to remove without interference.
   - Pipe flange ratings shall be consistent for suction and discharge pressure of pump:
     150# flange—240 psig
     300# flange—700 psig

Suction Piping

1. Minimum sloping in horizontal suction lines shall be 1:50.

2. Suction valves shall be line size.

3. Strainer (temporary or permanent) shall be installed between the suction flange and the pump block valve.

4. A ¾-in. valved connection shall be provided on each side of the suction strainer. The upstream connection may be common to the main pump and its spare.

5. Reducers in horizontal suction lines shall be eccentric and installed with the flat on top.

Discharge Piping

1. A check valve shall be installed in the discharge line between the pump and the block valve.

2. Check valves and discharge valves shall be line size. *Exception*: If discharge line is *two or more sizes larger* than pump discharge nozzle, the valves may be the next standard size smaller.

3. Check valve shall be drilled if pumping temperature exceeds 300 °F or is below −50 °F or if pump is in automatic startup service (regardless of pumping temperature).

4. In lieu of drilled checks, a valved bypass may be used on pumps with discharge nozzles 2 in. NPS diameter and larger, if high differential pressure exist.

5. Warmup or cooldown lines shall be installed if the pumping temperature exceeds 300 °F or is below −50 °F. This is in addition to the drilled hole in the check valve.

6. Warmup lines shall be steam or electric traced.

7. For double block valve installations, valves shall be provided with a body drain or with a drain installed in a spool piece between valves.

8. A $^3/_4$-in. valved connection shall be installed between the check valve and the discharge flange.

Strainers

1. Strainers may be cone- or basket-shaped and shall be installed between the suction flange and the suction block valve.

2. Mesh size of strainers shall be selected to stop all objects too large to pass through the pump main flow passage.

3. Temporary strainers shall be used during flushing and initial operating periods unless permanent strainers are specified.

4. Piping layout shall permit removal of strainers without disturbing pump alignment.

5. The design and location of permanent strainers shall permit cleaning without removing the strainer body.

6. Arrangement of strainers shall permit cleaning without interrupting the pumping service.

7. For installations equipped with a spare pump, a permanent strainer shall be installed in the suction line of each pump.

8. Twin strainers or self-cleaning strainers may be used for pumps without spares.

9. Y-type strainers shall be restricted to 2 in. maximum size.

10. Suction lines for proportioning pumps shall be chemically or mechanically cleaned to permit operation without strainers.

Seal Flush Piping

1. Fully threaded or "all threaded" nipples are not permitted.

2. Pipe bushings are not permitted.

3. Carbon steel pipe shall be schedule 160, $1/2$ in. NPS minimum. Stainless steel pipe shall be schedule 80S, $1/2$ in. NPS minimum.

4. Flanges or threaded connections shall be provided to permit complete removal of the piping system.

5. Flanges are preferred over threaded connections.

6. Pipe nipples at branch connections shall be 4 in. long maximum and 2 in. long minimum.

7. All valves, pressure gauges, gauge glasses, etc. shall be kept a minimum distance from the connection to the line, machine, or vessel.

8. Piping shall be adequately supported. Maximum unsupported span shall be as follows:

   $1/2$-in. lines—2 ft 0 in.
   $3/4$-in. lines—2 ft 6 in.
   1-in. lines—3 ft 0 in.

9. Piping branch connections less than 2 in. NPS shall be gusseted.

10. Excess flow valves and/or check valves shall be installed in flush lines for light ends services and for hydrocarbons above 300 °F. The

valves shall be located adjacent to the gusseted valved connection into the suction or discharge lines.

11. Seal welding is required for all flushing and seal oil lines except for connections directly at the machinery.

12. Socket-welded systems are acceptable.

13. Seal welding shall cover all exposed threads, $1/4$ in. minimum.
    - Flanges shall be installed between the block valves and the seal oil pot.
    - Restriction orifice shall be installed in vent from seal oil pot.

14. Tandem seal piping shall be per sketch supplied by owner's machinery engineer.

Small Bore Piping

1. Refer to small bore piping section.

Pipe Supports

1. Refer to pipe supports section.

General Purpose Steam Turbines

Case

1. Blowdown valves shall be provided for low point drains in the turbine casing.

Piping, General

1. Inlet and exhaust piping installation shall be as per Figures 1, 2, or 3.

2. Inlet and exhaust valve installation shall permit the complete drainage of accumulated condensate before valve opening. Pipeline or body drains may be installed on the pressure side of the valve.

3. Valves installed in horizontal pipe runs shall be positioned with stems extending vertically upward or horizontal, or any intermediate position, providing the stem is above the horizontal center line.

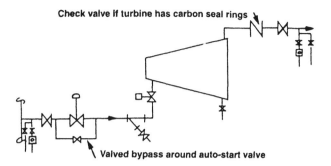

Figure C-1. Auto-start turbines.

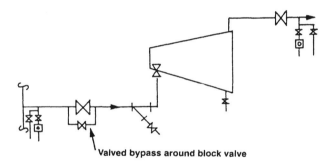

Figure C-2. Manual start turbines with integral trip valve.

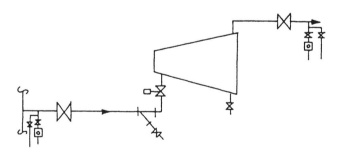

Figure C-3. Manual start turbines with trip and throttle valves.

General Purpose Steam Turbines

1. Inlet pipe of all turbines shall have:

   - Block valve
   - Y-type strainer
   - Low point drains and/or traps.

2. Exhaust pipe of all back-pressure turbines shall have:

   - Block valve
   - Low point drains and/or traps.

Inlet Piping

1. A block valve shall be installed in the inlet piping.

2. Steam traps and drains shall be installed at all low points in inlet piping.

3. A Y-type strainer, with a valved blowdown to a safe location, shall be installed between the inlet block valve and the inlet flange.

4. A control valve shall be provided for auto-start turbines and shall be located near the turbine.

5. A valved bypass shall be installed around the auto-start control valve.

Exhaust Piping

1. A block valve shall be installed in the exhaust piping on all back pressure turbines.

2. Steam traps and drains shall be installed at the low points of the exhaust piping of all back pressure turbines.

3. Auto-start turbines with carbon seal rings shall have a swing-type check valve in the exhaust piping.

4. A 1-in. valved bypass shall be installed around the check valve on auto-start turbines.

5. An exhaust hood shall be installed for turbines exhausting to atmosphere.

6. Exhaust hood shall be located a suitable distance away from the turbine.

7. A sign stating "CAUTION—DO NOT CLOSE THIS VALVE UNLESS INLET BLOCK VALVE IS CLOSED AND TURBINE CASE BLEEDERS ARE OPEN" shall be placed at the exhaust steam block valve.

Small Bore Piping

1. Refer to Small Bore Piping Section.

Lubrication

1. Horizontal turbine bearings shall be pressurized or non-pressurized, oil lubricated.
2. Non-pressurized oil lubricated bearings shall have adjustable constant level oilers with transparent containers.
3. Wet sump or "purge mist" lubricated bearings: refer to Oil Mist Guidelines specifically prepared for a job.
4. Dry sump lubricated bearings, when specified, do not require constant level oilers. Refer to Oil Mist Guidelines specifically prepared for a job.

Guards

1. Same as in Centrifugal Pumps Section.

Baseplates

1. Refer to Baseplate Section.

Pipe Supports

1. Refer to Pipe Support Section.

Special Purpose Centrifugal Fans

Casing

1. Fan housing and inlet box construction shall be as follows:
   - Carbon steel, $3/16$ in. minimum thickness.
   - Housing to be split design to permit removal of the rotor without dismantling the ductwork.
   - Internal bolting to be self-locking.
   - Interior walls to be painted.
2. Suction and discharge connections shall be flanged, with the joints designed for through-bolting and full-face gaskets.

3. Access doors shall be provided as follows:

- Hinged doors to allow insertion of lance for steam or water washing of fan impeller during operation.
- A minimum of one inspection door located in the lower half of the fan housing permitting access to the impeller.
- An inspection door located for access to the inlet guide vanes.

4. A valved drain connection shall be provided at the low point of the fan casing and inlet box.

Lubrication

1. Bearing housing shall be arranged for oil lubrication.

2. A pressurized lubrication system shall be provided for any fan unit supplied with a steam turbine or gear in the main drive system.

3. Non-pressurized oil lubricated bearing shall have constant level-sight feed oilers with transparent containers and protecting wire cages.

4. Wet sump or "purge mist" bearing housings shall have constant level oiler and a mark indicating oil level. Refer to oil mist instructions which will be supplied by owner.

5. Inlet guide vane components requiring periodic lubrication shall be furnished with lubrication fittings that are accessible for maintenance while the fan equipment is in operation.

Bearings

1. Radial or thrust antifriction (rolling contact) bearings for the fan shaft require owner's engineer's written approval.

2. Shaft bearings and seals shall be accessible without dismantling duct work or fan casing.

3. Induction draft fans operating in hot gas service shall be provided with a deflector plate between the shaft seal and the bearing housing to deflect hot gas leakage away from the bearing housing.

Guards

1. Same as in Centrifugal Pump Section.

Screen

1. Forced draft fans shall have inlet screens installed at the suction connection.

2. Screen material shall be galvanized steel unless a more resistant material is specified.

3. Screen mesh size shall be approximately 1 $1/2$ in.

Outlet Duct

1. For parallel fan operation, each fan shall be furnished with an outlet guillotine shutoff gate or louvered damper with a spectacle blind, as specified.

2. If a louvered damper is specified:
   - Each damper leaf shall be supported by and continuously welded to a shaft spindle.
   - Spindles shall be supported externally at both ends by permanently lubricated type bearings.
   - Manual operation of the damper from grade level is required.

Baseplates

1. Sole plates shall be provided for each fan bearing pedestal.

2. Driver and gear combinations shall be mounted on a common baseplate. Refer to Baseplate Section.

Small Bore Piping

1. Refer to Small Bore Piping Section.

Small Bore Pipe and Tubing

Pipe and Tubing, General

1. All piping and tubing shall be rigidly supported to avoid vibration. Maximum unsupported span shall be as follows:

$1/2$-in. lines—2 ft 0 in.
$3/4$-in. lines—2 ft 6 in.
1-in. lines—3 ft 0 in.

Exceptions: Longer unsupported spans for piping runs between machines, lube oil consoles, and instruments panels.

*2. All piping systems shall have valved high point vents, valved low point drains, and valved bleeds on gauge glasses and pressure gauges.

*3. A block valve shall be installed between each instrument and the connection and shall be located close to the line, vessel, or machinery.

4. Stainless steel shall be used for control, lube, and seal oil lines downstream of the filters.

5. All openings shall be plugged or blanked to keep out foreign matter.

Pipe

1. Threaded reducing bushings are not permitted.

2. All-thread nipples are not permitted.

3. Nipples shorter than 2 in. (including thread and/or socket lengths) are not permitted.

4. Hex head plugs are not permitted. Round barstock plugs shall be used.

5. Unions are not permitted. *Exceptions*: In cooling water service and where the union can be isolated without machinery shutdown.

*6. The minimum size of piping connections shall be $1/2$ in. NPS.

*7. Piping branch connections less than 2 in. NPS shall be gusseted. Refer to sketch, to be provided by owner's engineer.

8. Flanged connections shall be provided to permit removal or assembly of equipment and shall be accessible and close to machinery.

9. Cooling water, nitrogen and air piping 3 in. NPS and smaller shall be galvanized.

10. All flanges mating to cast iron flanges shall be flat-faced and full-faced gaskets shall be used.

11. Spiral wound gaskets shall be used downstream of oil filters in lube, seal, and control oil systems.

---

* *Note*: These are higher potential problem areas. Refer any non-compliance to the reliability group.

Tubing

1. Welding of tubing is not permitted.

2. The use of tubing is permitted only for connections to instruments.

3. All tubing connectors shall be type 316 stainless steel.

4. Tubing shall not be used to support valves, instruments, etc.

5. The maximum run length of tubing should be limited to 3 ft and have no more than two bends in each tubing run.

Glossary

Accelerated life testing – Testing to verify design reliability of machinery/equipment much sooner than if operating typically. This is intended especially for new technology, design changes, and ongoing development.

Acceptance test (Qualification test) – A test to determine machinery/equipment conformance to the qualification requirements in its equipment specifications.

Accessibility – The amount of working space available around a component sufficient to diagnose, troubleshoot, and complete maintenance activities safely and effectively. Provision must be made for movement of necessary tools and equipment with consideration for human ergonomic limitations.

Allocation – The process by which a top-level quantitative requirement is assigned to lower hardware items/subsystems in relation to system-level reliability and maintainability goals.

Assets – The physical resources of a business, such as a plant facility, fleets, or their parts or components.

Asset management – The systematic planning and control of a physical resource throughout its economic life.

Availability – The probability that a system or piece of equipment will, when used under specified conditions, operate satisfactorily and effectively. Also, the percentage of time or number of occurrences for which a product will operate properly when called upon.

CBM – See condition-based maintenance.

Changeout – Remove a component or part and replace it with a new or rebuilt one.

CMMS – Computerized maintenance management system.

Component – A constituent part of an asset, usually modular or replaceable, that is serialized and interchangeable.

Concept – Basic idea or generalization.

Condition-based maintenance – Maintenance based on the measured condition of an asset.

Confidence limit – An indication of the degree of confidence one can place in an estimate based on statistical data. Confidence limits are set by confidence coefficients. A confidence coefficient of 0.95, for instance, means that a given statement derived from statistical data will be right 95% of the time on the average.

Configuration – The arrangement and contour of the physical and functional characteristics of systems, equipment, and related items of hardware or software; the shape of a thing at a given time. The specific parts used to construct a machine.

Corrective maintenance – Unscheduled maintenance or repair actions, performed as a result of failures or deficiencies, to restore items to a specific condition. See also Unscheduled maintenance and Repair.

Cost-effectiveness – A measure of system effectiveness versus life-cycle cost.

Critical – Describes items especially important to product performance and more vital to operation than non-critical items.

Defect – A condition that causes deviation from design or expected performance.

Dependability – A measure of the degree to which an item is operable and capable of performing its required function at any (random) time during a specified mission profile given item availability at the start of the mission.

Discounted cash-flow analysis – A method of making investment decisions using the time value of money.

Distributions – See Probability distribution.

Downtime – That portion of calendar time during which an item or piece of equipment is not able to perform its intended function fully.

Durability life (Expected life) – A measure of useful life, defining the number of operating hours (or cycles) until overhaul is expected or required.

EAM – Enterprise Asset Management.

Emergency maintenance – Corrective, unscheduled repairs.

Engineering – The profession in which knowledge of the mathematical and natural sciences is applied with judgment to develop ways to utilize economically the materials and forces of nature.

Environment – The aggregate of all conditions influencing a product or service, or nearby equipment, actions of people, conditions of temperature, humidity, salt spray, acceleration, shock, vibration, radiation, and contaminants in the surrounding area.

Equipment – All items of a durable nature capable of continuing or repetitive utilization by an individual or organization.

ERP – Enterprise Resource Planning (ERP Software).

Exponential distribution – A statistical distribution in logarithmic form that often describes the pattern of events over time.

Failure – Inability to perform the basic function, or to perform it within specified limits; malfunction.

Failure analysis – The logical, systematic examination of an item or its design, to identify and analyze the probability, causes, and consequences of real or potential malfunction.

Failure effect – The consequence of failure.

Failure mode – The manner by which a failure is observed. Generally a failure mode describes the way the failure occurs and its impact on equipment operation.

Failure Mode Effect Analysis (FMEA) – Identification and evaluation of what items are expected to fail and the resulting consequences of failure.

Failure rate – The number of failures per unit measure of life (cycles, time, miles, events, and the like) as applicable for the item.

Fault tree analysis (FTA) – A top–down approach to failure analysis starting with an undesirable event and determining all the ways it can happen.

FMEA – See Failure mode effect analysis.

FMECA – Failure mode, effect, and criticality analysis – a logical progressive method used to understand the causes of failures and their subsequent effects on production, safety, cost, quality, etc.; see also failure mode effect analysis.

Function – A separate and distinct action required to achieve a given objective, to be accomplished by the use of hardware, computer programs, personnel, facilities, procedural data, or a combination thereof; or an operation a system must perform to fulfill its mission or reach its objective.

Hardware – A physical object or physical objects, as distinguished from capability or function. A generic term dealing with physical items of equipment – tools, instruments, components, parts – as opposed to funds, personnel, services, programs, and plans, which are termed "software."

Hazard function – The instantaneous failure rate at time, t.

Infant mortality – Early failures that exist until debugging eliminates faulty components, improper assemblies, and other user and manufacturer learning problems, and until the failure rate lowers.

Item – A generic term used to identify a specific entity under consideration. Items may be parts, components, assemblies, subassemblies, accessories, groups, equipment, or attachments.

Life cycle – The series of phases or events that constitute the total existence of anything. The entire "cradle to grave" scenario of a product from the time concept planning is started until the product is finally discarded.

Life-cycle cost – All costs associated with the system life cycle, including research and development, production, operation, support, and termination.

Life units – A measure of use duration applicable to the item (e.g., operating hours, cycles, distance, lots, coils, pieces, etc.).

Maintainability – The inherent characteristics of a design or installation that determine the ease, economy, safety, and accuracy with which maintenance actions can be performed. Also, the ability to restore a product to service or to perform preventive maintenance within required limits.

Maintainability testing – Maintainability testing is used to demonstrate MTTR (mean time to repair). Once MTTR of a critical component is defined and the appropriate personnel trained in the proper procedure, we can test to investigate if the function can be performed in the stated MTTR. It should be stressed that this is not a test of the person's skills but rather a test of the procedure and design of the equipment.

Maintenance – Work performed to maintain machinery and equipment in its original operating condition to the extent possible; includes scheduled and unscheduled maintenance, but does not include minor construction or change work.

Management – The effective, efficient, economical leadership of people and use of money, materials, time, and space to achieve predetermined objectives. It is a process of establishing and attaining objectives and carrying out responsibilities that include planning, organizing, directing, staffing, controlling, and evaluating.

Material – All items used or needed in any business, industry, or operation as distinguished from personnel.

Mean time between failure (MTBF) – The average time/distance/events a product delivers between breakdowns.

Mean time between maintenance (MTBM) – The average time between both corrective and preventive actions.

Mean time between replacement (MTBR) – Average use of an item between replacements due to malfunction or any other reason.

Mean time to repair (MTTR) – The average time it takes to fix a failed item.

Median – The quantity or value of an item in a series of quantities or values, so positioned in the series that, when arranged in order of numerical quantity or value, there are an equal number of values of greater magnitude and of lesser magnitude.

Mission profile – A time-phased description of the events and environments an item experiences from initiation to completion of a specified mission, to include the criteria of mission success or critical failure.

Model – Simulation of an event, process, or product physically, verbally, or mathematically.

Modification – Change in configuration.

Normal – Statistical distribution commonly described as a "bell curve." Mean, mode, and median are the same in the normal distribution.

MTBR – See mean time between repair.

MTTR – See mean time to repair.

On-condition maintenance – Inspection of characteristics which will warn of pending failure, and performance of preventive maintenance after the warning threshold but before total failure.

Operating time – Time during which equipment is performing in a manner acceptable to the operator.

Overhaul – A comprehensive inspection and restoration of machinery/equipment, or one of its major parts, to an acceptable condition at a durability time or usage limit.

Predictive and preventive (Scheduled, planned) maintenance – All actions performed in an attempt to retain a machine in specified condition by providing systematic inspection, detection, and prevention of incipient failures.

Predictive maintenance – Predictive maintenance is a maintenance method that involves a minimum of intervention. In its simplest form it is based on the old adage "don't touch, just look." In the context of process machinery, predictive maintenance is practiced through machinery health monitoring methods such as vibration and performance analysis.

Preventive maintenance (PM) – Actions performed in an attempt to keep an item in a specified operating condition by means of systematic periodic inspection, detection, and prevention of incipient failure. See also Scheduled Maintenance.

Proactive – A style of initiative that is anticipatory and planned for.

Probability distribution – Whenever there is an event E which may have outcomes $E_1, E_2, \ldots, E_n$, whose probabilities of occurrence are $p_1, p_2, \ldots, p_n$, one speaks of the set of probability numbers as the p.d. (probability density) associated with the various ways in which the event may occur. The word "probability distribution" refers therefore to the way in which the available supply of probability, i.e. unity, is "distributed" over the various things that may happen.

Production – A term used to designate manufacturing or fabrication in an organized enterprise.

Random – Any change whose occurrence is not predictable with respect to time or events.

Re-rating – Alteration of a machine, a system, or a function by redesign or review for change in performance; mostly, but not always, for increased capacity, etc.

RCFA – See Root Cause Failure Analysis.

RCM – See reliability-centered maintenance.

Rebuild/recondition – Total teardown and reassembly of a product, usually to the latest configuration. See also revamp.

Redundance (Redundancy) – Two or more parts, components, or systems joined functionally so that if one fails, some or all of the remaining components are capable of continuing with function accomplishment; fail-safe; backup.

Refurbish – Clean and replace worn parts on a selective basis to make the product usable to a customer. Less involved than rebuild.

Reliability (*R*) – The probability that an item will perform its intended function without failure for a specified time period under specified conditions.

Reliability-centered maintenance – Optimizing maintenance intervention and tactics to meet predetermined reliability goals.

Reliability growth – Machine reliability improvement as a result of identifying and eliminating machinery or equipment failure causes during machine-testing and operation.

Reliability modeling – A model that uses individual component reliabilities to define reliability of a subsystem. Allows for analysis of parallel versus series systems, and defines low reliability components of a subsystem.

Reliability testing – Reliability testing is used to demonstrate MTBF (mean time between failure). Once the MTBF of a critical component is defined, a test can be performed (with a measure of confidence) to demonstrate this MTBF. A measure of confidence is built into statistically designed test plans, guaranteeing that if the MTBF requirement has not been achieved, there is a low probability that the test will be passed.

Repair – The restoration or replacement of components of facilities or equipment as necessitated by wear, tear, damage, or failure. To return the facility or equipment to efficient operating condition.

Repair parts – Individual parts or assemblies required for the maintenance or repair of equipment, systems, or spares. Such repair parts may be repairable or non-repairable assemblies or one-piece items. Consumable supplies used in maintenance, such as wiping rags, solvent, and lubricants, are not considered repair parts.

Repairable item – Durable item determined by application of engineering, economic, and other factors to be restorable to serviceable condition through regular repair procedures.

Replaceable item – Hardware that is functionally interchangeable with another item but differs physically from the original part to the extent that installation of the replacement requires such operations as drilling, reaming, cutting, filing, or shimming in addition to normal attachment or installation operations.

Return On Capital Employed – ROCE.

Return On Net Assets – RONA.

Revamp – Change as to upgrade or modernize.

ROCE – Return On Capital Employed.

RONA – Return On Net Assets.

Root Cause Failure Analysis – A formalized systematic approach effort to determine the underlying cause of a failure. This effort is generally separate from repair activities but should be part of the repair cycle. It usually entails a detailed technical analysis of the failure mode by a team of experts.

Safety – Elimination of hazardous conditions that could cause injury. Protection against failure, breakage, and accident.

Scheduled maintenance – Preplanned actions performed to keep an item in specified operating condition by means of systematic inspection, detection, and prevention of incipient failure. Sometimes called preventive maintenance, but actually a subset of PM.

Scheduled (planned) downtime – The elapsed time that the machine is down for scheduled maintenance or turned off for other reasons.

Spares – Components, assemblies, and equipment that are completely interchangeable with like items and can be used to replace items removed during maintenance.

Specifications – Documents or verbal communication that clearly and accurately describe the essential technical requirements for materials, items, equipment, systems, or services, including the procedures by which it will be determined that the requirements have been met. Documents may include performance, support, preservation, packaging, packing, and marking requirements.

Standards – Established or accepted rules, models, or criteria by which the degree of user satisfaction of a product or an act is determined, or against which comparisons are made.

Standard deviation – A measure of average dispersion or departure from the mean of numbers, computed as the square root of the average of the squares of the differences between the numbers and their arithmetic mean. It is also a measure of uncertainty when applied to probability density distribution.

Standard item – An item for common use described accurately by a standard document or drawing.

Standby – Assets installed or available but not in use.

Surveillability – A qualitative factor influencing reliability. It contains such considerations as accessibility for surveillance and monitoring of a machine or its function(s), etc.

System – Assembly of correlated hardware, software, methods, procedures, and people, or any combination of these, all arranged or ordered toward a common objective.

Time – The universal measure of duration.

Time to repair (TTR) – Total clock time from the occurrence of failure of a component or system to the time when the component or system is restored to service (i.e., capable of producing good parts or performing operations within acceptable limits). Typical elements of repair time are diagnostic time, troubleshooting time, waiting time for spare parts, replacement/fixing of broken parts, testing time, and restoring.

Total downtime – The elapsed time during which a machine is not capable of operating to specifications.

Total downtime = scheduled downtime + unscheduled downtime

Total productive maintenance – Company-wide equipment management program emphasizing operator involvement in equipment maintenance and continuous improvement in equipment effectiveness.

TPM – See total productive maintenance.

Training – The pragmatic approach to supplementing education with particular knowledge and assistance in developing special skills. Helping people to learn to practice an art, science, trade, profession, or related activity. Basically more specialized than education and involves learning what to do rather than why it is done.

Troubleshooting – Locating or isolating and identifying discrepancies or malfunctions of equipment and determining the corrective action required.

Unscheduled (unplanned) downtime – The elapsed time that the machine is incapable of operating to specifications because of unanticipated breakdowns.

Unscheduled maintenance (UM) – Emergency maintenance (EM) or corrective maintenance (CM) to restore a failed item to usable condition. Often referred to as breakdown maintenance.

Useful life – The number of life units from manufacture to when the item has an unrepairable failure or unacceptable failure rate.

Up – In a condition suitable for use.

Uptime – The capacity to produce and provide goods and services.

Utilization factor – Use or availability.

Warranty – Guarantee that an item will perform as specified for at least a specified time.

Wear out – The process that results in an increase of the failure rate or probability of failure with increasing number of life units.

Note: Some of the definitions in this glossary were selected from MIL-STD-721C and the SAE publication *Reliability and Maintenance Guidelines for Manufacturing Machinery and Equipment.*

Index

Printed and bound by CPI Group (UK) Ltd, Croydon, CR0 4YY

14/05/2025

01871125-0001